UNDERSTANDING

INDUSTRIAL

DESIGNED

EXPERIMENTS

2nd Edition

CQG Ltd Printing, Longmont, Colorado

Library of Congress Catalog Card Number: 88-92885

The authors recognize that perfection is unattainable without continuous improvement. Therefore, we solicit comments as to how to improve this text. To relay your comments or to obtain further information, contact:

STEPHEN R. SCHMIDT
14220 Gleneagle Dr
Colorado Springs, CO 80921
(719) 488-3554

Understanding Industrial Designed Experiments

TABLE OF CONTENTS

Chapter 2   Statistical Techniques

Chapter 3   Design Types

Chapter 4 Design Evaluation

Chapter 5   Optimization

Chapter 6   Taguchi Philosophy, Design and Analysis

Chapter 7   Case Studies

Chapter 8   Statistical Tables

Preface

The time is quickly approaching when you will not be considered a competent engineer or technical manager without a working knowledge of Experimental Design. The demands of increased efficiency of processes, lower product cost and shortened development cycles will dictate that we use simple but powerful tools to get the most out of our experiments. No longer do we have the luxury of running one–factor–at–a–time experiments or experiments with excessive sample sizes. In competitive environments, only those groups which **apply** experimental design approaches efficiently and effectively will survive.

Once an engineer decides to start using orthogonal or nearly orthogonal designs, they are quickly faced with a plethora of competing design and analysis strategies. Full factorial, Plackett–Burman, Taguchi, and CCD are just some of the competing design types. What are the strengths and weaknesses of each? What is the niche that each strategy best fills? How about analysis approaches? Should the engineer use "Pick the Winner", plots of marginal means, ANOVA, Normal Probability Plots, or regression analysis? In which situations do the competing analysis strategies best fit? Should the objective of the experiment influence the design type and analysis approach to be used? Some possible objectives for an experiment are Screening, Modeling, and "Robust" Design. Is the Brainstorming session critical? Who should be involved?

Industry needs a handbook which answers the above questions! The following chapters offer what we believe are appropriate answers to the questions raised. Our approach is not to compulsively follow one specific philosophy of experimentation. Rather, the authors attempt to select the best from competing strategies, combining seemingly dichotomous techniques so as to produce a synergism which leads to an improved approach.

CHAPTER 1 : FOUNDATIONS

1.1 Intended Audiences for this Handbook

This handbook was written for managers and engineers. It is intended to bridge the gap produced by texts that are too mathematical and those that attain simplicity by omitting important concepts. Mathematical and statistical theory are not emphasized; rather, the presentation is intended to provide conceptual understanding of designed experiments and related topics in statistical quality control. For years statisticians have inundated us with textbooks written for understanding statistical theory,written only for those well versed in the statistical jargon and mathematics. Those texts are recommended for anyone seeking more depth into these subjects.

The presentation of material in this handbook is arranged to include an introduction (Chapter 1), basic statistical techniques (Chapter 2), types of designs (Chapter 3), analysis of experimental data (Chapter 4), optimization (Chapter 5), Taguchi philosophy and analysis techniques (Chapter 6), case studies (Chapter 7) and statistical tables (Chapter 8).

1.2 Quality and the Role of Statistical Techniques

The American Society for Quality Control has defined quality as "the totality of features and characteristics of a product or service that bear on its ability to satisfy stated or implied needs." A simple interpretation of this definition is that quality is a measure of how well the user's needs are met. The user's interpretation of quality focuses on whether the product functions as advertised with no variation in performance. There are several aspects of quality: the quality of the product design; the quality of the process design; the quality of incoming parts; and the quality of the manufactured product. In the design phase of the product or process, it is important to test which factors affect the quality of the design and determine the settings of these factors that optimize the desired output. It is in this phase where experimental designs and response surface techniques are used to develop a process which will produce a quality product. The quality of incoming parts and the quality of the manufacturing process are achieved through the use of statistical process control techniques described in [3] and [6].

This text will focus on quality of design and present both sides to the growing controversy between Taguchi Methods and traditional techniques. The authors are not committed to either side but advocate an approach which finds the "best" method to solve the problem at hand. In addition, new ideas are presented on how to combine certain aspects of both approaches so as to produce a synergism leading to improved methods of experimental design.

1.3   What is Experimental Design

Experimental design consists of purposeful changes of the inputs to a process in order to observe the corresponding changes to the response (output). The process is defined as some combination of <u>machines</u>, <u>materials</u>, <u>methods</u>, <u>people</u>, <u>environment</u> and <u>measurement</u> which used together form a service, produce a product or complete a task. Thus, experimental design is a scientific approach which allows the researcher to better understand a process and how the inputs affect the response.

An alternative to experimental design is to watch a process and wait for changes in the response. This approach is very slow in providing useful information and frequently leads to effects of each of the inputs becoming so entangled that their true relationship with the response cannot be determined.

1.4   Why Use Experimental Design

The manager should be interested in experimental design to achieve the following: 1) improved performance characteristics; 2) reduced costs; and 3) shortened product development and production time. Improved performance characteristics result from the identification of the critical factor levels that optimize the mean response and minimize response variability. This improved performance also leads to the reduction of scrap and rework, which greatly reduces costs. Understanding which factor levels are critical to improved performance allows for relaxing tolerances on benign factors and selection of less expensive materials. Shortened lead times are accomplished by understanding what the customer really wants and using efficient testing procedures to optimize the design of the product and process. Historical use of one–factor–at–a–time experimentation and/or incomplete full factorial designs has resulted in very inefficient and ineffective attempts to understand and optimize product designs and processes. No longer can these archaic methods be tolerated if a company intends to keep up in a highly competitive market.

For the engineer, testing different design strategies of a new product or trouble shooting problems in an on-line process, experimental designs are used as: 1) efficient methods for gaining an understanding of the relationship between the input factors and the response; 2) a means of determining the settings of the input factors which optimize the response; and 3) method for building a mathematical model relating the response to the input factors. Perhaps the easiest way to convey to the reader the importance of learning experimental design is to review some alternatives and discuss their weaknesses. The reader is presented 3 cases which sequentially illustrate the pitfalls associated with historical data, one-at-a-time designs and full factorials. In addition, Case 4 reveals the need for improvement beyond traditional techniques in the pursuance of robust designs.

Case 1: A plant manager wishes to know if changes in temperature during a production process are related to the number of defects of the product. Since there was an abundance of historical or "happen-stance" data available it was decided to use simple linear regression (see Chapter 2 for a detailed discussion) to correlate temperature and defect rate. The regression results indicated that a high positive correlation existed between these two variables, implying that as you increase the temperature the defect rate would increase. Since the goal was to minimize defects, the engineer accordingly directed that temperature be maintained at some appropriate low level. What was not revealed from the data analysis was undocumented intervention in the production process by the operator. If the defect rate started to increase (for any reason) he would raise the temperature to minimize this increase. After the number of defects started to decline he would reduce the temperature back to the standard operating settings. A low temperature had originally been specified so as to keep costs down; however, no experimentation had taken place to accurately examine the relationship of temperature and defect rate. Analyzing this non-experimental data results in a confounding of temperature and operator manipulation which makes it impossible to separate the two inputs to the process. The apparent positive correlation of temperature and defect rate did not represent the true relationship but rather a relationship induced by the operator. Historical data will typically lead to this type of faulty conclusion because it probably does not provide sufficient manipulation of input factors and it frequently contains factors which are highly interrelated.

Case 2 : The plant manager is now convinced that historical data can lead to erroneous results so he has decided to design an experiment to test temperature and pressure effects on defect rate. The design consists of the following: (1) select two levels of temperature, T_1 and T_2 and two levels of pressure P_1 and P_2. (2) Holding P_1 constant, test the product at both T_1 and T_2 as shown in table 1.1.

table 1.1		
Pressure	Temperature	Response
P_1	T_1	.03
P_1	T_2	.015

Since a more desirable defect rate occurred for T_2, set the temperature at T_2 and test the product at pressure P_1 and P_2. The results are displayed in table 1.2.

table 1.2		
Pressure	Temperature	Response
P_1	T_2	.015
P_2	T_2	.010

At first glance you would say that the ideal settings of temperature and pressure are T_2 and P_2 resulting in a 1 percent defect rate; however, this design, referred to as a "one-at-a-time" design, did not allow for testing the interaction effect of temperature and pressure. Suppose that the untested combination P_2T_1 would produce a defect rate of 0.005. This response for P_2T_1 would produce an interaction as displayed in figure 1.1. The interaction effect of temperature and pressure is interpreted as follows: as temperature changes from T_1 to T_2 the change in the defect rate is different depending on pressure settings of P_1 and P_2, i.e., the lines in the graph are non-parallel.

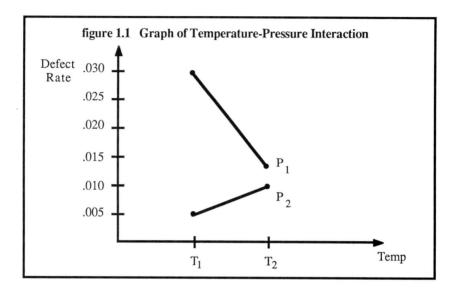

figure 1.1   Graph of Temperature-Pressure Interaction

A more appropriate design would be to test all four combinations of temperature and pressure, referred to as a factorial design (discussed further in Chapter 3).

Case 3:   After understanding the pitfall discovered in case 2 and brainstorming which other factors could potentially affect the defect rate, the manager requests further testing of temperature, pressure and 10 other input factors of interest. If all combinations are to be tested for each factor measured at two levels, there will be $2^{12} = 4096$ experimental conditions in the full factorial design. This presents a new problem because of resource limitations in conducting the experiment. Fortunately, there is a solution to the problem. The material in this manual includes fractional factorials as one efficient alternative to full factorial designs. Obviously, you cannot get something for nothing. You'll see in Chapter 3 what you give up when conducting a fractional factorial of 32 experimental combinations versus the full factorial of 4096.

The idea of resource limitations also brings to mind that different temperature and pressure settings may result in different overhead costs for production. These costs, along with other response variables of interests should be considered when designing experiments.

Case 4: The manager and his engineering staff have given the 12 factors in case 3 some more thought and they agree that 3 of the 12 factors cannot be controlled in the operating environment of the product. What is needed is a design to determine the 9 controllable factor settings such that the defect rate and the variability of defect rate across the 3 uncontrollable factors are minimized. The solution to their problem is found in the study of robust designs which are described somewhat in Chapter 4 and in detail in Chapter 6.

These four cases demonstrate that unless the researcher is well informed on all the statistical tools available and his experiments are well designed, results can be misleading and large amounts of resources consumed unnecessarily. An even more compelling reason for educating American industrial managers and engineers in the use of designed experiments is based on the need to compete with countries who have successfully adopted these techniques. In 1964, American industries were basking in a 6 billion dollar trade surplus, not realizing they were under attack by aggressive overseas companies. By 1984, the U.S. was looking at a 123 billion dollar trade deficit. Over this period, U.S. productivity increased a mere 35% as compared to about 60% for some European countries and an amazing 120% for Japan. The same country that had a reputation for making "junk" in the 50's and 60's was threatening the survival of U.S. industry in the 70's and 80's by producing high quality products at a lower price. How did this happen? It is becoming well known that this remarkable change in the quality of Japanese products is attributed to designing quality into the product through the use of designed experiments and monitoring quality during production by way of statistical quality control. Ironically, these techniques were developed mostly in the U.S. and Europe over 50 years ago, but U.S. companies failed to implement them. American statisticians, having been unsuccessful at convincing industry to adopt these methods, soon directed their efforts toward developing new theory which did not always have an industrial application. Most U.S. industries became less and less interested in statistics and continued to flounder. Japan, however, listened to two American statisticians, W. Edward Deming and Joseph M. Juran, and the Japanese soon became the quality experts and their companies prospered. A few American companies did implement some statistical techniques but not to the extent that the Japanese did. In Japan, the use of statistical methods are recognized by management to be important. Everyone in the company, from management to on-line workers, are taught these techniques in training sessions and they all use them not just to monitor quality in production but to steadily improve quality throughout the production process. This philosophy was initiated through Deming as shown in his 14 points as a guide for management to improve quality and productivity.

The following 14 points are taken from Hogg and Ledolter [3] and can be further studied in Deming's book *Quality, Productivity and Competitive Position*.[2]

1. Create a constancy of purpose toward the improvement of product and service. Consistently aim to improve the design of your products. Innovation, money spent on research and education, and maintenance of equipment will pay off in the long run.

2. Adopt a new philosophy of rejecting defective products, poor workmanship, and inattentive service. Defective items are a terrible drain on a company; the total cost to produce and dispose of a defective item exceeds the cost to produce a good one, and defectives do not generate revenues.

3. Do not depend on mass inspection because it is usually too late, too costly and ineffective. Realize that quality does not come from inspection, but from improvements on the process.

4. Do not award business on price tag alone, but also consider quality. Price is only a meaningful criterion if it is set in relation to a measure of quality. The strategy of awarding work to the lowest bidder has the tendency to drive good vendors and good service out of business. Preference should be given to reliable suppliers that use modern methods of statistical quality control to assess the quality of their production.

5. Constantly improve the system of production and service. Involve workers in this process, but also use statistical experts who can separate

special causes of poor quality from common ones.

6. Institute modern training methods. Instructions to employees must be clear and precise. Workers should be well trained.

7. Institute modern methods of supervision. Supervision should not be viewed as passive "surveillance", but as an active participation aimed at helping the employee make a better product.

8. Drive out fear. Great economic loss is usually associated with fear when workers are afraid to ask a question or to take a position. A secure worker will report equipment out of order, will ask for clarifying instructions, and will point to conditions that impair quality and production.

9. Break down the barriers between functional areas. Teamwork among the different departments is needed.

10. Eliminate numerical goals for your work force. Eliminate targets and slogans. Setting the goals for other people without providing a plan on how to reach these goals is often counterproductive. It is far better to explain what management is doing to improve the system.

11. Eliminate work standards and numerical quotas. Work standards are usually without reference to produced quality. Work standards, piece work, and quotas are manifestations of the inability to understand and provide supervision. Quality must be built in.

12. Remove barriers that discourage the hourly worker from doing his or her job. Management should listen to hourly workers and try to understand their complaints, comments and suggestions. Management should treat their workers as important participants in the production process and not as opponents across a bargaining table.

13. Institute a vigorous program of training and education. Education in simple, but powerful, statistical techniques should be required of all employees. Statistical quality control charts should be made routinely and they should be displayed in a place where everyone can see them. Such charts document the quality of a process over time. Employees who are aware of the current level of quality are more likely to investigate the reasons for poor quality and find ways of improving the process. Ultimately, such investigations result in better products.

14. Create a structure in top management that will vigorously advocate these 13 points.

Other philosophies such as those of Juran, Crosby and Taguchi may vary from Deming but all appear to center on three main inputs to improved quality: teamwork, customer needs and statistical methods.

There are many reasons why U.S. industry failed to implement the techniques and philosophies described above. The authors suggest the following 6 reasons:

(1) Failure of management to recognize the value of these techniques.

(2) Failure of statisticians to teach these techniques in an easy to understand fashion.

(3) Misperception that statistically designed experiments are costly, time-consuming and impractical.

(4) Fear and mistrust of unfamiliar techniques on the part of managers and engineers.

(5) Lack of technical resources in industry.

(6) Lack of competition until recent years which would finally force U.S. industry to investigate ways to improve quality.

1.5 Variation and its Impact on Quality

It should not be a surprise to the reader that even in well controlled production processes, there will be variation in the final product. The variation is either caused by uncontrolled factors or production noise. Too much variation degrades the quality of the product and causes a loss to the company. If large numbers of the product are produced outside specificatons, and if products are not inspected before they are shipped, then (1) complaints will increase, (2) extra resources will be expended to repair items under warranty and (3) eventually customers will become discouraged and seek a more reliable product. Historically, the approach to this problem has been to set up specification limits and perform inspections of finished products to ensure zero defects out the back door. This approach will dichotomize the quality aspect of any product into either acceptable (within spec) or unacceptable (out of spec). Loss to the company is based on being out of spec as shown in figure 1.2.

figure 1.2

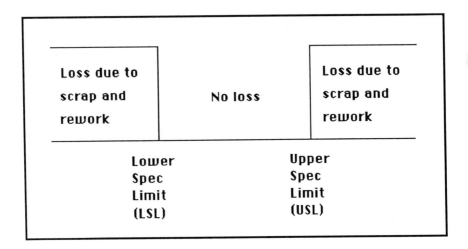

The optimal solution to this problem will not be to increase inspections at the end of the production line. According to Juran, these inspections are only 80% effective. Furthermore, this approach requires more manpower and results in rework and scrap costs that can become prohibitive. In addition, maintaining profits under these conditions will require sales prices to increase, resulting in a continued decrease in consumer satisfaction. It is also unreasonable to assume that a product just inside a specification limit results in no loss to the company. For example, consider building window glass and window frames. If the glass is at the upper limit and the frame at the lower limit (or vice versa) the quality of the product is reduced. To develop a quality product at a reduced cost, quality must be designed into the process and focus must be on the target (nominal) value instead of just being in spec [5] as depicted in figure 1.3.

figure 1.3

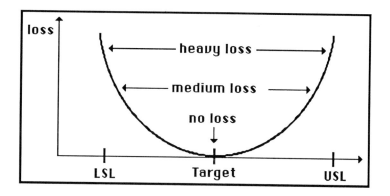

To minimize the loss incurred due to deviation from the target, the variability of the product about the target must also be reduced as shown in figure 1.4. Product B has a much higher proportion of measured values close to the target. Therefore, Product B will result in less loss.

In order to effectively and economically find the optimal settings of input factors such that this variability about the target value is reduced, properly designed experiments are required. Experimental designs test input factors, and determine which factors need tight controls and which ones can be allowed more variation without affecting the output. After testing factors as to their importance, it is reasonable to expend resources on expensive control devices and control charts to closely monitor critical input factor settings. Once you have completed a quality design, monitor the product quality as it goes through the production process and make necessary adjustments to the input factors. If the relationship of the input factors with the output should change, resulting in reduced quality, more experimentation may be conducted to find the new factor settings which optimize the output.

figure 1.4

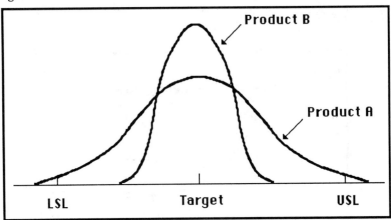

Two measurements used to quantify variability with regard to lower spec limits (LSL), target and upper spec limits (USL) are C_p and C_{pk}. The C_p index was designed to compare variability with only LSL and USL; the formula is

$$C_p = \frac{(USL - LSL)}{6\,\hat{\sigma}} \qquad \text{where} \qquad \hat{\sigma} = \sqrt{\frac{\sum y_i^2 - n\bar{y}^2}{n - 1}} \; ; \quad \bar{y} = \frac{\sum y_i}{n}$$

According to accepted standards in industry, C_p values less than 1.00 are unacceptable, values between 1.0 and 1.3 are marginally satisfactory and values greater than 1.3 are desired. Since the C_p value does not take the target value into consideration, a new measurement C_{pk} was developed. To calculate, use the lesser of $\left(\dfrac{USL - \bar{y}}{3\,\hat{\sigma}}\right)$ or $\left(\dfrac{\bar{y} - LSL}{3\,\hat{\sigma}}\right)$. Interpretation of C_p values is the same as that of C_{pk} values. See table 1.3 for an approximate relationship between C_{pk} and number of defects.

table 1.3	
C_{pk}	Number of defects
.50	133,600 PPM
.67	71,800 PPM
.80	16,400 PPM
.90	6,900 PPM
1.00	2700 PPM
1.33	66 PPM
1.67	<1 PPM
2.00	<1 PPM
3.00	<1 PPM
4.00	<1 PPB

This information was obtained from an R&M 2000 Variability
Reduction Program slide prepared by HQ USAF/LE-RD

1.6 Objective of Experimental Design

When testing the effects of input factors on the output, it is desired to be able to ascertain causality, not just correlational relationships. Just as pointed out in section 1.4, a high correlation does not necessarily relate to the cause of the problem. To ascertain the *causality* of an input factor it is important to ensure that changes in the output occurred because of changes in the input and not some other related factor. To accomplish this task, experiments must be designed such that input factors are uncorrelated and all *nuisance variables* (variables not controlled) have their effects averaged out over all experimental conditions. This procedure is referred to as *randomization* and consists of running experimental conditions in a random sequence. The advantage of randomization is that it averages out the effects of all nuisance variables. This is especially important when the brainstorming process fails to identify factors and/or some factors cannot be measured or controlled in the experiment.

As an example demonstrating the need for randomization, consider the following: A new product is being tested using two different bonding agents.

Assume that there are three batches of raw material being used to make these products and the batch factor was uncontrolled in the design. After testing bond strength, the data appear as shown in table 1.4.

table 1.4				
		Bonding Agent (A, B)		
	A	Batch #	B	Batch #
	2.4	3	2.7	3
	2.5	3	3.3	1
	2.1	3	3.1	1
	3.1	1	3.7	2
	2.7	3	3.5	2
	2.9	3	3.6	1
	3.0	2	2.8	2
Mean	2.670		3.243	
STD Dev	.359		.391	

Initially, it appears that bonding agent B provides a higher mean bond strength and therefore, it is the better product. However, consider the consequences of batch #3 being defective. Since more observations using bonding agent A were conducted with batch #3 than any other, the results are now suspect. Therefore, we cannot state unequivocally that bonding agent differences caused substantial differences in bond strength because the batch effect is confounded with bonding agents. To avoid this problem, we could have randomly assigned materials to bonding agents in an attempt to average out the effects of the defective batch over the two treatments. It is also possible to control the different batches beyond the randomization process by including them as an input factor. As a final note, randomization does provide test validity (controls for systematic bias) but it does not decrease the error variability which is only decreased by including all important factors in the design.

1.7   Organizing the Experiment

The reason for any designed experiment is to provide data that can be used to answer well thought out questions. Failure to allocate sufficient time and thought in formulating these questions often results in wasting resources throughout the remainder of the experimental process. In many situations, up to 50% of the overall effort should go into the planning phase of the design. It is important that all of the key players (management, engineers, experts in the area, and analysts) be involved in the planning phase. The goals should be specific and statements of how the goals will be attained (objectives) must be complete.

One of the most effective approaches to the planning phase is the use of the "brainstorming" technique. This technique consists of an uninhibited method for creating ideas that pertain to the goals, objectives and the variables involved. A list of independent variables (input factors) and dependent variables (output or response measures) must be obtained. In addition, how the variables are measured and how many levels per independent variable must be decided. Through the use of brainstorming and a team effort there is a higher likelihood that the planning phase will be a success. If the planning phase fails, there is little chance of salvaging the experiment and further efforts could be a waste of resources. Barker's text, *Quality by Experimental Design* [1], presents the guidelines for brainstorming given in figure 1.5. While the brainstorming process appears simple, there are a few additional concerns to ensure success. Brainstorming groups should be small, i.e., 4 to 10 people. Someone should record the ideas and everyone should participate. Ideas are not to be challenged until after the brainstorming session is over. When the brainstorming is complete the list can be critiqued and reduced to a workable number of ideas. For a more structured brainstorming session, you can use fishbone diagrams. [4]

GUIDELINES FOR BRAINSTORMING

TEAM MAKEUP

Experts

"Semi" experts

Implementers

Analysts

Technical Staff who will run the experiment

DISCUSSION RULES

Suspend judgement

Strive for quantity

Generate wild ideas

Build on the ideas of others

LEADER'S RULES FOR BRAINSTORMING

Be enthusiastic

Capture all the ideas

Make sure you have a good skills mix

Push for quantity

Strictly enforce the rules

Keep intensity high

Get participation from everybody

Figure 1.5

At the completion of the brainstorming session, a form similar to the one in Appendix 1A should be initiated and continued throughout the remainder of the experiment. This form is an expanded version from that of Barker [1]. The actual form you use may be a modification; however, it is imperative that a similar document be used to facilitate an organized flow of events.

It is important that any experimental effort be aimed at design or production problems where there is the most need. For example, assume a product consists of components A, B, C, D and E. Failures of this product are the "company's" number one concern. Management directs the use of experimental design to solve the problem. You could attempt to make variable adjustments to A, B, C, D and E and build a design to see which component deviations contribute to increased defect rates, or you could use historical data presented in a Pareto diagram as shown in figure 1.6. Thus, historical data can sometimes be used to identify critical problem areas. This is an acceptable use of historical data, whereas using it to make inferences of causality is not. Figure 1.6 reveals that component D contributes to 60% of the defects, component C contributes to 20% of the defects, etc. Since component D contributes the most to product defects, the best approach would be to build an experimental design to (1) determine the cause of defects within component D, and (2) improve the quality of component D thereby improving the product quality. After completing the work on product D, the product quality is again

figure 1.6

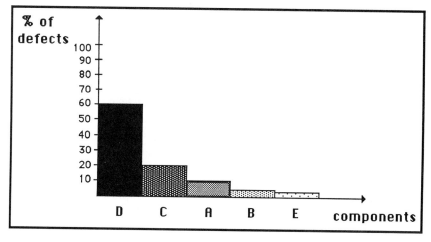

monitored and defects categorized as to component contributions. If necessary, further experimentation can be conducted. This process of continued improvement will lead to a high quality item. In summary, you want to attack that aspect of your product where the most gains can be made.

1.8 Selecting the Response

Selection of the response, often referred to as the quality characteristic, is very important to good experimentation. For example, continuous responses are typically much better than non-continuous or categorical type responses. If the response is ordered categorical (e.g. poor, fair, good, very good, excellent), you can assign the catagories ordered integers (e.g. 1, 2, 3, 4, 5) and conduct the analysis in the same manner described in chapters 2, 4 and 5. However, if binary responses are used (e.g. defective/non-defective) you will need large numbers of replications of each experimental condition in order to avoid ambiguous results. The authors have found that, on occasion, continuous variables related to the binary response can be used in the analysis resulting in large savings of experimental resources. For example, the vibration of a device during processing can be highly related to whether the device will be defective or non-defective. Therefore, minimizing vibration will have the same effect as minimizing defects while decreasing the number of required data points. Some interesting guidelines [7] in the selection of the proper response are as follows:

1. The quality characteristic should be related as closely as possible to the basic engineering mechanism of the technology.

2. Use continuous responses if possible.

3. Use quality characteristics which can be measured precisely, accurately, and with stability. (The authors have found this can be a major stumbling block in effective experimentation with leading-edge technologies.)

4. Ensure the quality characteristics cover the important dimensions of the ideal function of the technology.

Selection of the right response requires solid engineering insight so as to include critical discussions of the proper response. The success of any experimentation requires a liberal blending of technical engineering knowledge, statistical prowess and common sense.

1.9   Conducting the Experiment

The importance of exercising extreme discipline in the actual conduct of the experiment cannot be overstressed. During the brainstorming process, it is recommended that a wide open, creative environment in which the free flow of wild ideas is welcomed. Conducting the experiment, however, requires a rigorous, compulsive approach. The authors find that groups are rarely accustomed to functioning at this level. Factor levels must be set as close as possible to the specific design settings. Samples must be prepared in exactly the same way each time. Data sheets need to be clearly prepared in advance. All parties involved in conduct and analysis of the experiment must be well versed in the intimate details of how the experiment is conducted. You, as the champion of experimental design within your area, must communicate and enforce the need for rigor in the conduct of the experiment without alienating the other key team members. All too often, a lack of communication, leadership, or control can doom an otherwise brilliantly conceived experimental scheme.

1.10   A Quality Improvement Example

Suppose you are responsible for vehicle maintenance in a large organization. As an initial attempt to implement continuous process improvement (CPI), you would probably form a team to assist you in developing a process flow diagram for vehicle repair. The diagram, shown in figure 1.7, can be used to improve quality by identifying non-value added steps to be modified or eliminated. An obvious CPI objective would be to ensure that "the right repairs are done right the first time". Just imagine how many steps could be eliminated from the process flow diagram if you could educate, train and motivate your people to perform in this manner.

After reviewing the process flow diagram, assume you decide to implement a gas mileage improvement strategy. To benchmark your current state of vehicle quality, you would probably direct that gas mileage data be recorded on each vehicle. The best way to display this data is shown in figure 1.8.

figure 1.7

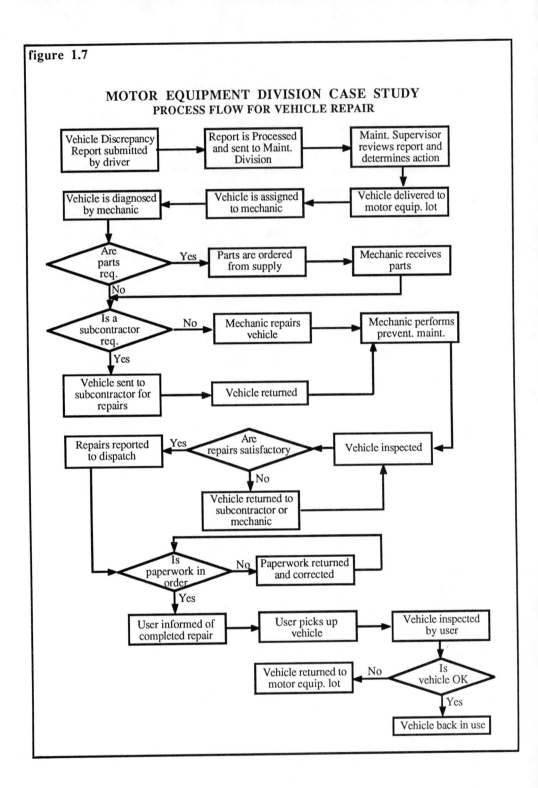

MOTOR EQUIPMENT DIVISION CASE STUDY
PROCESS FLOW FOR VEHICLE REPAIR

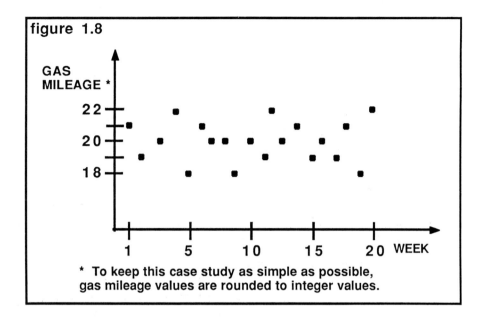

figure 1.8

GAS
MILEAGE *

* To keep this case study as simple as possible,
gas mileage values are rounded to integer values.

As you can see, gas mileage does vary from week to week. Your concern is whether this variability is natural or due to specific causes. The use of statistical process control (SPC) can assist you in answering this question.

To present a simple overview of SPC, we will assume that the 20 weeks of data in figure 1.8 are from a vehicle whose performance is currently known to be under control. Computing the average gas mileage over the 20 weeks will give us a feel for the vehicle's expected gas mileage. This average is found as shown below.

$$\bar{y} = \frac{\sum_{1}^{20} y_i}{20} = \frac{[21 + 19 + 20 + \cdots + 22]}{20} = \frac{400}{20} = 20$$

To obtain a measure of the variability in vehicle gas mileage, we can calculate the variance, s^2, of the 20 mileage values. The equation for s^2 is

$$s^2 = \frac{\sum (y - \bar{y})^2}{n - 1}$$

which can be described as the sum of the squared deviations from the mean, divided by (n – 1). The deviations from the mean and the variance calculations are displayed in figure 1.9.

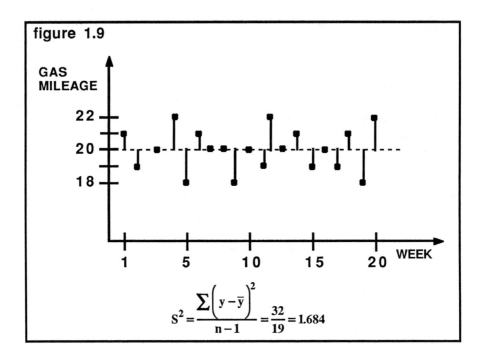

figure 1.9

$$s^2 = \frac{\sum\left(y - \bar{y}\right)^2}{n-1} = \frac{32}{19} = 1.684$$

Other formulas for variance can also be used, such as

$$s^2 = \frac{\sum y^2 - \left(\sum y\right)^2 \big/ n}{n-1} = \frac{\sum\limits_{i>j}(y_i - y_j)^2}{n(n-1)}$$

    A common technique which is a simple, yet powerful, way to get a visual feel for the way data is distributed is to draw a histogram. A histogram is a graph which displays how frequently a given outcome occurs. The histogram for our data is shown in figure 1.10.

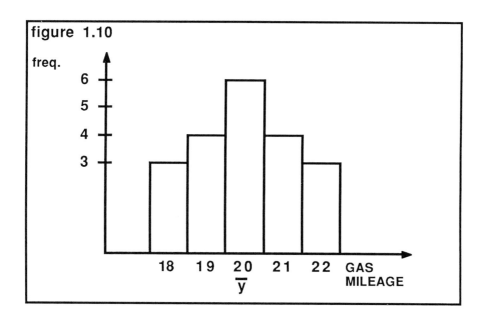

figure 1.10

If a smooth line were superimposed on the histogram it would appear to be symmetric and bell shaped, much the same as a normal curve. Assuming, then, that our gas mileage values are approximately normally distributed allows us to make use of the empirical rule shown in figure 1.11.

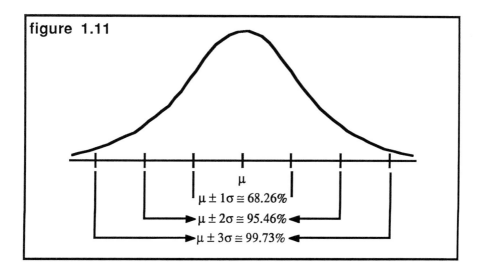

figure 1.11

$\mu \pm 1\sigma \cong 68.26\%$
$\mu \pm 2\sigma \cong 95.46\%$
$\mu \pm 3\sigma \cong 99.73\%$

The normal distribution of gas mileage values supports our assumption that the vehicle's performance is in control. We can now use the empirical rule to assist us in developing control limits. For example, $\bar{y} \pm 3s$ results in a lower control limit (LCL) of 16.1 and an upper control limit (UCL) of 23.9. The empirical rule states that approximately 99.73% of all future gas mileage values for this vehicle should fall within these limits. Therefore, we can use these limits to evaluate the future performance of this specific vehicle. For instance, if week 21 produced a gas mileage of 23, chances are that nothing specific has caused this value; however, a value of 16 is outside our control limits (i.e., outside the natural variability) and indicates that there is a 99.73% chance this value was caused by some specific change in the vehicle. Other SPC rules exist for determining out of control conditions due to trends or increases in variability [6].

Once an out of control condition is determined, the obvious next step is to determine its cause. To obtain this information, a brainstorming session with the appropriate members should be conducted. The results of such a session might resemble the cause and effect diagram displayed in figure 1.12. After completion of the brainstorming, the group should try to reduce the information to those inputs most likely to affect the gas mileage. You now have a set of input factors to the process which can be investigated each time the mileage control chart detects a change.

figure 1.12

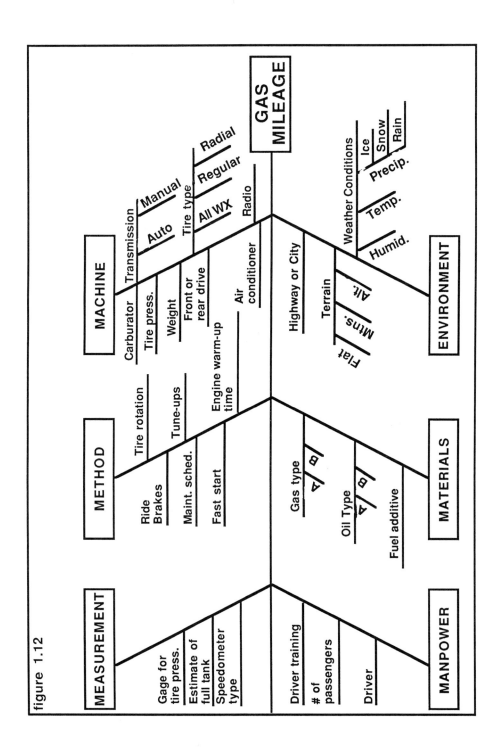

Assume your investigation revealed an extremely low tire pressure which you determine to be the cause of the out of control mileage value. You would obviously correct the low tire pressure problem and possibly begin to chart the tire pressures on a weekly basis. In this way, you are now controlling a critical input to the process instead of waiting for a substantial change in the process output. Thus, you have now entered a problem prevention mode versus detecting poor quality through inspection of output values.

Control charting is an excellent tool; however, engineers are not always sure which inputs are the important ones (vital few) and which ones are unimportant (trivial many). Furthermore, using control charting to gain an in–depth understanding of your process is a long drawn out procedure. What is needed is a faster method for achieving "profound knowledge" of the gas mileage process. The statistical tool which provides this type of in-depth knowledge of a process in a timely manner is experimental design. The basic idea is to characterize the process by determining which inputs have a critical impact on the response. The design of experiments approach can be used effectively after SPC has been implemented; however, it is advantageous to characterize the process before production begins. This strategy allows for quality to be designed into the product from the beginning of the design engineering phase.

For our gas mileage problem, assume the brainstorming resulted in four input factors to be used in the experiment. To investigate all possible combinations of four factors at two levels would require the 16 runs shown in table 1.5. The (–) values indicate a factor set at its low value and the (+) values represent a high factor setting. Notice that the design columns are all balanced vertically; i.e., an equal number of (+) and (–) values in each column. In addition, for any factor at its (–) or (+) value, the other factors have settings that are balanced, i.e., horizontal balancing. These balancing properties result in the design matrix being orthogonal. The importance of orthogonality is that it allows us to estimate the effects of each factor independent of the others.

table 1.5				
	FACTORS			
RUN	A	B	C	D
1	−	−	−	−
2	−	−	−	+
3	−	−	+	−
4	−	−	+	+
5	−	+	−	−
6	−	+	−	+
7	−	+	+	−
8	−	+	+	+
9	+	−	−	−
10	+	−	−	+
11	+	−	+	−
12	+	−	+	+
13	+	+	−	−
14	+	+	−	+
15	+	+	+	−
16	+	+	+	+

The design in table 1.5 is referred to as the full factorial and it consists of all possible factor combinations. The total number of effects which can be estimated are four linear effects, six two-way linear interactions, four three-way interactions and one four-way interaction. Infrequently do experimenters find interactions beyond two-ways to be significant and thus time and resources can be conserved by the use of a fraction of the full factorial design. For the gas mileage experiment, it was decided to use a fractional factorial design with eight runs. The resulting design matrix is shown in table 1.6. Notice that this eight-run design also has the balancing properties required for orthogonality. The column for factor D was formed by using the elements of the ABC interaction column. The result is that factor D is aliased with the ABC interaction. This, in turn, produced other aliasings as shown in table 1.7. More will be said regarding aliasing once we have analyzed the data.

table 1.6

FACTORS

RUN	A	B	C	D
1	−	−	−	−
2	−	−	+	+
3	−	+	−	+
4	−	+	+	−
5	+	−	−	+
6	+	−	+	−
7	+	+	−	−
8	+	+	+	+

table 1.7

ALIAS PATTERN

effect	alias
A	BCD
B	ACD
C	ABD
D	ABC
AB	CD
AC	BD
AD	BC
BC	AD
BD	AC
CD	AB
ABC	D
ABD	C
ACD	B
BCD	A

The four factors to be used in the experiment and the low (−) and high (+) levels of each are shown in table 1.8.

table 1.8

FACTOR	LEVELS
A: TIRE PRESSURE	28, 35
B: TIMING SETTING	LOW, HIGH
C: TYPE OF OIL	1, 2
D: TYPE OF GAS	1, 2

The completed experimental matrix with the response values (taken over three weeks for each run) are displayed in table 1.9.

table 1.9

RUN	A	B	AB	C	AC	BC	D	y_1	y_2	y_3	$\overline{Y}$	S
1	−	−	+	−	+	+	−	22.27	21.12	21.37	21.59	.60
2	−	−	+	+	−	−	+	14.22	15.40	10.46	13.36	2.58
3	−	+	−	−	+	−	+	22.49	23.15	22.08	22.57	.54
4	−	+	−	+	−	+	−	9.96	13.80	11.92	11.89	1.92
5	+	−	−	−	−	+	+	17.35	18.60	17.97	17.98	.62
6	+	−	−	+	+	−	−	27.08	24.54	24.57	25.40	1.46
7	+	+	+	−	−	−	−	18.36	17.63	17.04	17.68	.66
8	+	+	+	+	+	+	+	22.78	26.97	27.14	25.63	2.47

A graphical analysis for each effect in the design matrix is obtained by plotting the average response at the (−) and (+) settings and then connecting these two dots (see figure 1.13).

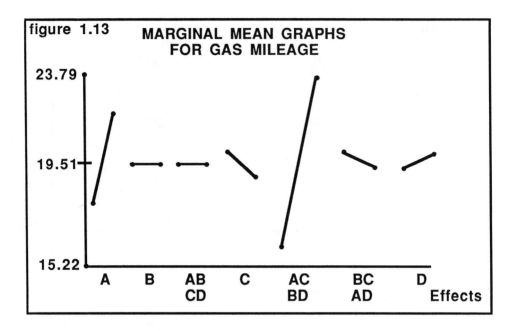

figure 1.13   MARGINAL MEAN GRAPHS
                FOR GAS MILEAGE

The steepness of the slope reflects the importance of each effect. From the graphical results in figure 1.13, it is apparent that the most important effects are factor A and the AC or BD interaction. Since AC is aliased with BD, we must go back to the factors to try and determine which interaction is most likely responsible for the steep slope in figure 1.13. Since it is highly unlikely to have a "tire pressure x type of oil" interaction (AC), it is decided that the steep slope in figure 1.13 is due to the BD, or "timing x type of gas" interaction. The BD interaction graph is shown in figure 1.14.

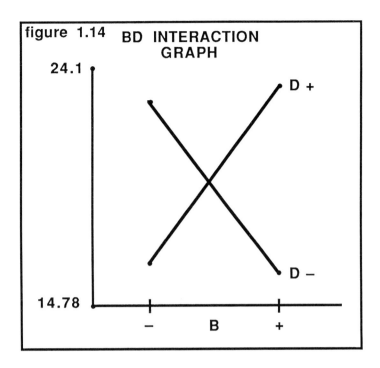

figure 1.14 BD INTERACTION GRAPH

A similar graphical analysis is used to determine if any of the factors shift the variability in the gas mileage. Figure 1.15 indicates that factor C has a big effect on variability.

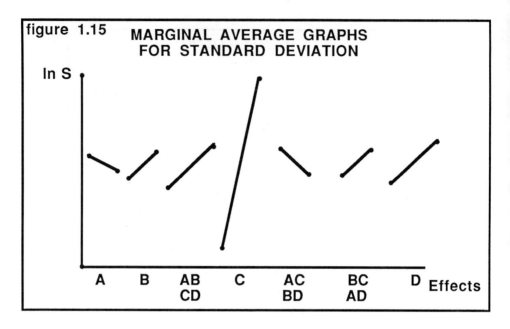

figure 1.15 **MARGINAL AVERAGE GRAPHS FOR STANDARD DEVIATION**

In S

A B AB C AC BC D Effects
        CD   BD AD

To determine the best factor settings for maximizing gas mileage while minimizing variability, we must return to figures 1.13, 1.14 and 1.15. Based on figure 1.13, set factor A at (+). Figure 1.14 indicates that factors B and D should be set at (+). Factor C is set at (–) to minimize variability. Thus, the final factor setting decision is as follows:

factor	–, + setting	true setting
A	+	35 psi
B	+	high timing
C	–	type 1 oil
D	+	type 2 gas

At this point, one should feel pretty good about solving a quality problem and thus satisfying the customer. Be careful, though, because we made the assumption that the customer wanted high gas mileage. When a survey was conducted in Madison, Wisconsin [8] as to what the customer (the driver) thought was important, their response was not mileage, but safety. The point is that designing experiments to solve problems which the customer may not think important will not always be a productive strategy. One should first survey the customer through the use of a procedure such as Quality Function Deployment (QFD). After assessing the customer needs and how to translate these needs into product/process design, then the designed experiments approach provides the type of profound knowledge we should really be striving for.

1.11 Steps for Quality Improvement: Getting Started

Across America, purchasers of products and services are demanding that their suppliers provide (1) **better** quality, (2) **faster** response times and (3) **cheaper** prices. To meet these demands, suppliers must be able to use a host of statistical and non-statistical tools. Other texts that address tools not included in this book are *Quality Through Leadership: TQM in Action* (1990) by Watson, F. and Schmidt, S.R. and *Basic Statistics: Tools for Continuous Improvement* (1990) by Kiemele, M. and Schmidt, S.R. (For more information on these texts, contact Air Academy Press at (719) 531-0777.) To improve quality, it is important to be familiar with all available tools; however, equally important is knowing when to use them. The following steps are suggested as an example of how to get started in the quality improvement cycle. We highly recommend that you use them as an outline for your company and that you make modifications as necessary.

QUALITY IMPROVEMENT STEPS

WHAT	**WHO**	**HOW**
1. Commitment	Everyone, from the top to the on-line worker	Cultural Change Strategic Planning Awareness Training
2. Define products/ processes	Crossfunctional Teams	Process Flow Diagrams Brainstorming
3. Mission Statement	Crossfunctional Teams	Brainstorming Nominal Group Technique (NGT)

4. Identify customers (Internal/External)	Crossfunctional Teams	Brainstorming NGT
5. Identify customer expectations	Crossfunctional Teams and the Customer	Surveys Brainstorming
6. Translate customer expectations into key quality characteristics	Crossfunctional Teams and the Customer	QFD Brainstorming NGT
7. Determine current state of quality for all products/processes	Process Action Teams (PATs) and the Customer	Benchmarking Loss Function Process Capability (C_{pk}) Defective Parts per Million (dpm) Surveys
8. Identify the best opportunity for improvement	PATs and the Customer	Loss Functions Cost Analysis Pareto Charts
9. For the best opportunity, identify causes of degraded quality	PATs	Brainstorming Cause and Effect Diagrams Historical Data
10. Characterize product/ process, i.e., find relationship of outputs (effects) and inputs (causes)	PATs	Designed Experiments
11. Find optimal settings for input factors	PATs	Designed Experiment Analysis Empirical Models
12. Determine input tolerances	PATs	Use empirical models in (11) as simulators for Designed Experiments using available tolerances

13. Set up control charts on critical inputs and outputs	PATs	Statistical Process Control (SPC)

14. Recycle the steps, implementing the philosophy of continuous process improvement (CPI).

EXPERIMENTAL DESIGN
INFORMATION SHEET

I. STATEMENT OF THE PROBLEM: _____
_____
_____

II. OBJECTIVE OF THE EXPERIMENT _____
_____

III. START DATE: _____   END DATE: _____

IV. Quality characteristics (also known as response, dependent variables or output variables.)

RESPONSE	TYPE *	ANTICIPATED RANGE	How will you measure the response?
1.			
2.			
3.			
4.			
5.			

V. Factors (also known as independent or input variables) which are anticipated to have an effect on the response.

FACTOR	TYPE *	CONTROLLABLE OR NOISE	RANGE OF INTEREST	LEVELS	ANTICIPATED INTERACTIONS WITH
1.					
2.					
3.					
4.					
5.					
6.					
7.					
8.					
9.					
10.					

VI.   DESIRED NUMBER OF RUNS _____
       MAX ALLOWABLE RUNS _____
       COST PER RUN _____
       TIME PER RUN _____

VII. CAN ALL RUNS BE RANDOMIZED? _____
       WHICH FACTORS ARE HARDEST TO RANDOMIZE? _____

VIII. WHAT EXPERIMENTAL DESIGNS ARE APPROPRIATE?

IX.   CONDUCT THE EXPERIMENT AND COLLECT DATA (monitor the data
       gathering process).

X.   Indicate which analysis techniques will be used

      a. Graphs of Main and Interaction Effects

      b. Pareto Diagrams

      c. Normal Probability Plots

      d. ANOVA

      e. Regression

      f. Signal-to-Noise

      g. Response Surface Methodology

XI.   Draw conclusions, make predictions and recommend further experimentation (if
       necessary).

XII. Make confirmiration runs to verify predicted results. Re-evaluate as in XI.

XIII. Predict your performance capability, using C_{pk}, based on the confirmation runs.

XIV. If your performance capability in XIII is unsatisfactory, discuss plans for further
       brainstorming and experimentation.

*   Put a Q for quantitative variables, an A for attribute variables, and a C for
      categorical or qualitative variables.

R&M 2000 VARIABILITY REDUCTION PROCESS

BRUCE A. JOHNSON, CAPTAIN, USAF

Office of the Special Assistant for Reliability and Maintainability
Headquarters, United States Air Force

THE PROCESS

Improving combat capability is a major Air Force objective. This is becoming increasingly difficult in the face of constrained manpower and fiscal resources. However, there is a solution. Substantial increases in combat capability are achievable through more reliable and maintainable weapon systems. Such systems are able to complete more missions with less spares, support equipment, facilities and maintenance personnel.

Weapon systems fail for many reasons. Some components, like tires, wear out. But most systems fail because of poor design, the use of defective parts and materials, or poor workmanship. The cause of these failures is *variability* in the design and manufacturing processes. The problem variability presents is that it exists in nearly all processes and it results in marginal or non-conforming products. The variability comes from the fact that conditions under which these items are produced change. Variability reflects the differences in raw material, machines, their operators and the manufacturing conditions. When process variation increases, the product's physical properties or functional performance can degrade, and the number of product defects increases. The significance variation has on a product's reliability and quality depends on the criticality of the manufacturing process and part characteristics.

There are two ways to reduce variability. Traditionally, the approach has been to tighten design tolerances and increase inspections. Costs climb as scrap and rework increase, and productivity drops. Inspections and tighter tolerances only treat the symptoms and do not resolve the actual problem.

The preferred method is to reduce the variability by improving the process. This can be done by eliminating the causes of variation through statistical techniques, and by developing more robust products which are insensitive to the causes of variation. The methods of reducing variability is aptly named the Variability Reduction Process (VRP).

VRP is a proven set of practices and technologies which yield more reliable and nearly defect-free products at lower cost. It is a structured, disciplined design and manufacturing approach aimed at meeting customer expectations and improving the development and manufacturing process while minimizing acquisition time and cost (Figure 1).

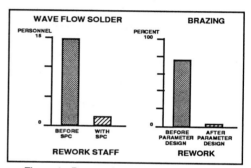

Figure 1. Process Improvements from VRP.

The objectives of VRP are to design robust products which are insensitive to the causes of failure; to achieve capable manufacturing processes that produce nearly defect-free products; and to adopt the managerial attitude of continuously improving all processes. The basic tools are teamwork, statistical process control (SPC), loss function, design of experiment (DOE), parameter design and quality function deployment (QFD). VRP must span all of engineering, manufacturing and management, and include the suppliers (Figure 2).

PURPOSE: MEET CUSTOMER EXPECTATIONS, IN MINIMUM TIME, AT LOWEST COST			
OBJECTIVES	ROBUST DESIGNS	CAPABLE PROCESSES	CONTINUOUS IMPROVEMENT
PRIMARY RESPONSIBILITY	ENGINEERS	MANUFACTURING	MANAGEMENT
TEAMS	INTERDISCIPLINARY	IMPROVEMENT	MULTI-FUNCTIONAL
TOOLS / TECHNOLOGIES	QFD / DOE / PARAMETER DESIGN	DOE / SPC	LOSS FUNCTION

Figure 2. The Elements of VRP.

CAPABLE MANUFACTURING PROCESSES

Capable manufacturing processes can only be achieved when the critical parameters are known, and the causes of variation are eliminated or minimized. For most processes, SPC is highly effective (Figure 3). It allows the operator to observe the process and distinguish between patterns of random and abnormal variation. It assists the operator in making timely decisions such as adjusting or shutting down the process before defects are produced. When combined with other statistical tools and problem solving techniques (Figure 4), the worker can isolate and remove the causes of abnormal variations.

When the abnormal variations are removed from the process, the process is said to be under statistical control. In many processes, this will not be sufficient. The random variations alone can result in

PART		PROCESS		TARGET: 95 FT·LBS					
OPERATOR		MACHINE		SPEC LIMITS: ±20 FT·LBS					
DATE									
TIME									
MEASUREMEN 1	102	106	104	89	95	100	96	107	92
2	94	97	91	94	105	95	85	103	100
3	97	102	105	92	103	92	104	100	93
4	106	107	98	96	92	93	90	105	90
AVERAGE, X	100	103	99	93	99	95	94	104	94

Figure 3. SPC Control Chart

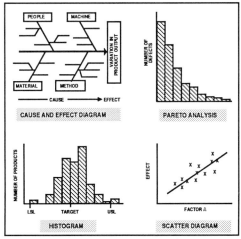

Figure 4. Tools Workers Can Use to Identify the Causes of Variation.

defective products, and their causes should be identified and removed until the process is capable of producing near defect-free products. However, causes of random variation are more difficult to identify, are usually systemic and normally require management action to remove.

The difference between VRP and traditional methods of quality control is that improvements in quality are achieved through improvements in the manufacturing processes. No longer is better quality to be achieved through tightening specifications and more inspections. In the case of SPC, the manufacturing processes are improved by eliminating the causes of variability. Usually, the process can be centered around the design target and variation reduced well within the specifications (Figure 5).

When implemented correctly, the results can be impressive. For Parlex Nevada Inc, a circuit card manufacturer, SPC was used to cut scrap cost by 90 percent in one year, and changed the company's losses into profits. Boeing used SPC to resolve a rivet flushness problem on the nose section of the 737 aircraft. The improvements saved a half-million dollars a year.

A more powerful method of resolving difficult or complex industrial problems is the statistical design of experiments (DOE). DOE methods have been around for 60 years and have been extensively used by the agricultural, pharmaceutical and chemical industries to advance their products. These techniques can greatly accelerate the rate of improving product designs and manufacturing processes. Such statistical experiments will aid the engineer in identifying the critical parameters for SPC, isolating the causes of variation, and improving the product's technical or operational characteristics.

DOE works by measuring the effects that different inputs have on a process. This is done by identifying a prospective set of input factors, varying

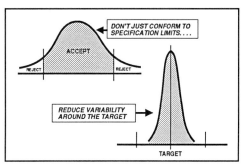

Figure 5. Design and Build to Target Values, Not Specification Limits.

the inputs over a series of experiments, collecting the data and analyzing the results. An input may be varied over a range of values such as can be done with an oven's curing temperature or conditionally, such as the decision to add or withhold a curing additive. The methods, whether they employ a full-factor, fractional-factor, orthogonal array or surface response technique, use a statistical approach that ensures accuracy and validity.

Well-planned experiments can have dramatic results. For example, a government-owned-company-operated (GOCO) munitions plant had a serious problem in producing the ADAM mine. Although SPC was in use and 12 of 13 processes were within their tolerances, 19 out of 25 lots were rejected. Aerojet Ordinance, the plant operator, decided to apply a Taguchi experiment to identify the critical parameters. They selected the 13 parameters used in the SPC program and tested parameters at three different levels. Only 27 experiments were conducted, firing 6 rounds each. The results were profound. Four parameters were found to be critical, and when set at their best levels, the process produced good lots without any rejects. The other nine parameters were less important and their tolerances can be relaxed. Results: production schedule met while achieving significant cost savings.

ROBUST DESIGNS

Having a capable manufacturing process is not enough. It may not be economical to remove or control some of the causes of variation. Therefore, it is necessary to develop robust manufacturing processes which are insensitive to the manufacturing conditions, materials, machines and operators. In most cases, *the greatest improvements come from robust designs*. These improvements are achieved through parameter design, a technique of selecting the optimum conditions (i.e. determining the ideal parameter settings) that minimize the variability without removing the causes of variation.

During parameter design, a set of parameters is identified to enhance the product, and a series of experiments is conducted to observe the effects of the parameters on the desired part characteristics. The results

could identify new parameter settings that improve the product and increase yield. For example, the problem an engineer may want to solve is the variability of ceramic parts. The source of variation is the uneven temperatures in the kiln. Because modifying the kiln is too expensive, the engineer conducts several experiments to identify a way to minimize the effects of the uneven temperature. For the experiment, he selects as parameters the amounts of the ingredients, their textures, blending procedures and firing temperatures. For validity, the engineer should use DOE procedures for conducting the experiment. An orthogonal array may be used to minimize experimental time and cost. The results will enable the engineer to fine new parameter settings that minimize the effects of uneven firing temperatures.

The problem with most design approaches is that parameter design is rarely done. Most engineers focus on the system design to develop the product, and immediately transition to tolerance design to establish the specification limits. Often, the results is a inferior product which is sensitive to variations in the manufacturing process. Parameter design should be done before tolerance design.

Parameter design can also be used to design and produce a more robust product that will perform better over a wider range of operating conditions and environments. It can be used to enhance a desired customer's need such as a smooth automobile ride, or to enhance an engineering requirement such as to lower the susceptibility of corrosion.

The success of parameter design and SPC hinge on the engineers' understanding of the customers' needs. Quality Function Deployment (QFD)

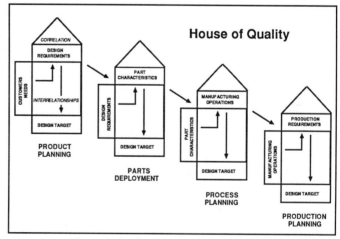

Figure 6. Quality Function Deployment.

is a systematic approach for developing and translating the customers' needs into the critical part characteristics and production requirements. The QFD requirements matrices are designed to minimize the chance of starting the design process with incomplete or erroneous requirements. They provide a methodology which assures an orderly translation of the customers' requirements throughout the product development process (Figure 6).

The basic approach used in QFD is conceptually similar to the practice employed by most companies. The difference lies in its structure. It compels the different disciplines and departments to communicate. QFD starts by defining the customers' requirements in the customers' terminology and translates these requirements into engineering requirements. These engineering requirements become the product characteristics which should be measurable and given target values. If properly executed, the product should fulfill the expectations of the customer.

The other matrices translate the engineering requirements into part characteristics, required manufacturing operations, and production requirements. Each matrix identifies the design targets, interrelationships and priorities. The end result should be a set of operating procedures which the factory can follow to consistently produce the critical part characteristics.

The design environment best suited to produce robust products is concurrent engineering (also referred to as simultaneous engineering). Concurrent engineering addresses all the customer, design and manufacturing issues up-front starting with concept exploration. The process employs good design practices, interdisciplinary teams and a structured requirements process, such as QFD, to concurrently develop the product and manufacturing processes. Its practice encourages communication between the design, product and production engineers. Concurrent engineering replaces the typical "sequential" approach to product design, which is more costly and time consuming. The effects concurrent engineering can have may be summed up by the following example. Using sequential engineering practices, the Allison Transmission Division estimated in 1982 it would cost $100 million in capital investment and $75,000 per unit to replace the transmission in the M-113 Armored Personnel Carrier. In 1987, using concurrent engineering, Allison's estimate dropped to $20 million for capital investment and $50,000 per unit (Figure 7).

CONTINUOUS IMPROVEMENT

For VRP to succeed, management from the top down must adopt new attitudes about reliability and quality, and must become *directly involved* in continuously improving the design and manufacturing processes. They must implement programs to foster improvement, teardown the barriers that inhibit change, instill teamwork, establish goals for improvement, and provide education and training for successful implementation.

Management's primary objective should be to satisfy the customer and serve the customer's needs if a company is to stay in business and make a profit. Reliability and quality must come first — not profit. If done smartly, reliability and quality will reduce cost and increase profit. For example, Hewlett-Packard's Yokogawa plant implemented many of the VRP techniques and, after eight years, they achieved 240 percent increase in profit, 120 percent increase in productivity, 19 percent increase in market share, 79 percent decrease in failure rate, and 42 percent decrease in manufacturing costs.

Management must become *process-oriented* and stimulate efforts to improve the way employees do the job. Teamwork is the foundation for continuous improvement. An important part of team building is the assignment of people to multifunctional management teams, interdisciplinary design teams and process improvement teams. Everyone should be involved in process improvement.

Management should implement programs to foster continuous improvement, use education to change attitudes and provide training. Change will be gradual and will require a long-term outlook. In Japan, most of the small improvements come from the workers' suggestion system called *Kaizen Teian*. Kawasaki Heavy Industries Aircraft Works has one of the more impressive programs. In 1987, each employee submitted an average of 229 suggestions

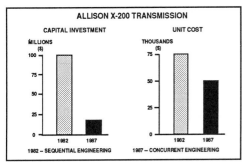

Figure 7. Concurrent Engineering.

and 92 percent were adopted. Savings were estimated at $35 million. At a Texas Instruments plant, they introduced an enhancement program, and over the past five years, the program has reduced defects by 2,300 percent (Figure 8).

Management must take responsibility for process improvement while giving the workers the responsibility of maintaining the process. Without the ability to maintain the existing process, there can be no improvement. This means management must give the worker ownership of his processes and allow the worker to improve or stop the process as necessary. In many progressive companies, workers are involved in the development of their own operating procedures, and in some cases, they write their procedures.

Management must change the accounting procedures. The notion that loss only occurs when the product is outside the specification limits is obsolete. Loss includes not only the cost of scrap and rework, but also the cost of warranties, excess inventory and capital investment, customer dissatisfaction, and eventual loss of market share. The traditional go/no-go approach to quality should be replaced with a powerful monetary loss function to better account for loss (Figure 9). A quadratic loss function allows management to better assess the true cost of production processes and the benefits derived from process improvements (Figure 10). Most important, the loss function supports continu-

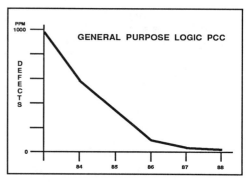

Figure 8. Continuous Improvement.

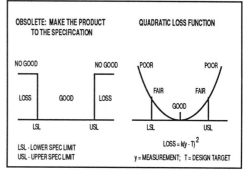

Figure 9. Quality Loss Functions.

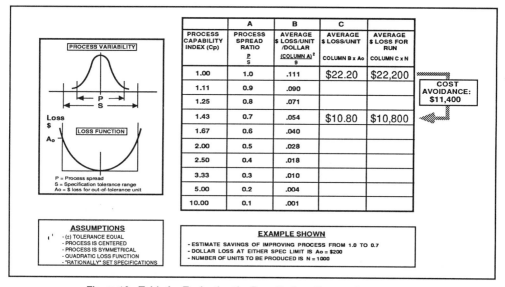

Figure 10. Table for Evaluating the Benefits from Process Improvement.

ous process improvement because minimizing the loss equates to reducing the variability around the target.

IMPLEMENTATION

The adoption of VRP begins with the conviction that change is necessary and beneficial. Implementation is an evolutionary process. VRP may start out in a single product line at one factory with a single group of suppliers. But it must be grown with the long-range goal that it will encompass the entire enterprise. The Air Force strategy for VRP is to encourage defense contractors and suppliers to (1) foster top-level commitment to VRP, (2) involve all levels, departments and vendors, (3) apply the VRP in a systematic approach, and (4) create a culture of continuous process improvement. Within the Air Force, the Vice Chief of Staff has directed all commands involved in weapon system acquisition and support to implement VRP by 1993. The Air Force acquisition regulations have been rewritten to incorporate VRP in the acquisition process. VRP will be an essential part of Air Force's Total Quality Management program (Figure 11).

	TQM TOOLS	R&M 2000 VRP	
		CAPABLE PROCESSES	ROBUST DESIGN
TEAM WORK		X	X
SPC		X	
LOSS FUNCTION		X	X
DESIGN OF EXPERIMENTS		X	X
PARAMETER DESIGN			X
HOUSE OF QUALITY (QFD)			X

(left side vertical labels: INCREASING — COMPLEXITY)

Figure 11. TQM / VRP Relationhhip

SUMMARY

The Variability Reduction Process makes two seemingly contradictory goals compatible: to produce highly reliable and maintainable weapon systems while reducing development time and costs. The method is to design robust systems, produce them with capable manufacturing processes, and achieve continuous improvement.

VRP provides a win-win situation. The Air Force obtains more combat capability with the available dollars. Industry is able to satisfy their customers, improve productivity and lower costs.

REFERENCES

Statistical Process Control:

E. L. Grant and R. S. Leavenworth, Statistical Process Control (5th ed), McGraw Hill, New York, 1979.
J. S. Oakland, Statistical Process Control, Wiley, New York, 1986.
D. J. Wheeler and D. S. Chambers, Understanding Statistical Process Control, Statistical Process Controls, Inc., Knoxville, TN, 1986.

Design of Experiment/Parameter Design:

G. C. P. Box, W. G. Hunter and J. S. Hunter, Statistics for Experimenters, Wiley, New York, 1978.
G. C. P. Box, S. Bisgaard and C. Fung, "An Explanation and Critique of Taguchi's Contribution to Quality Engineering," Quality and Reliability Engineering International, Vol. 4, 123-131, (1988).
C. Daniel, Applications of Statistics to Industrial Experimentation, Wiley, New York, 1976.
R. V. Hogg and J. Ledolter, Engineering Statistics, MacMillan Publishing, New York, 1987.
S. R. Schmidt and R. G. Launsby, Understanding Industrial Design of Experiments, Department of Mathematical Sciences, USAF Academy, CO, (pending publication, Summer 1989).
G. Taguchi and Y. Wu, Introduction to Off-Line Quality Control, Central Japan Quality Control Association, Nagoya, Japan. 1980.
G. Taguchi, System of Experimental Design — Engineering Methods to Optimize Quality and Minimize Costs, UNIPUB/Kraus International Publications, White Plains, NY, 1987.

Quality Function Deployment:

J. R. Hauser and D. Clausing, "The House of Quality," Harvard Business Review, (May-June 1988)
B. King, Better Designs in Half the Time: Implementing QFD Quality Function Deployment in America, GOAL/QPC, Methuen, MA, 1987.
L. P. Sullivan, "Quality Function Deployment," Quality Progress, (June 1986), pp 39-50.

Concurrent (Simultaneous) Engineering:

R. I Winner, J. P. Pennell, H. E. Bertrand and M. Slusarczuk, The Role of Concurrent Engineering in Weapon System Acquisition (IDA Report R-338), Institute for Defense Analyses, Alexandria VA, 1988.

Continuous Improvement:

M. Imai, Kaizen, Random House Business Division, New York, 1986.
W. W. Scherkenbach, The Deming Route to Quality and Productivity — Road Maps and Roadblocks, Mercury Press/Fairchild Publications, Rockville, MD, 1986.

Chapter 1 Problem Set

1. What is experimental design ?

2. Discuss how the use of experimental design can result in higher quality products at lower costs.

3. List the disadvantages of one factor at a time experiments.

4. List the disadvantages of full factorial (all possible combinations) experiments.

5. What is variability in the response and how can it be reduced ?

6. Describe what is meant by a robust design.

7. Assuming your process is in control, use the following information to calculate Cp and Cpk. Interpret these values and discuss the proper use of each.

$$
\begin{aligned}
\text{Upper Spec} &= 2100 \\
\text{Target} &= 2000 \\
\text{Lower Spec} &= 1900
\end{aligned}
$$

$$
\begin{aligned}
\bar{y} &= 2040 \\
\hat{\sigma} &= 20
\end{aligned}
$$

8. What is meant by randomization and why is it important ?

9. How important is the brainstorming session ? How much time should be devoted to brainstorming ?

10. Discuss the difference in the following two philosophies:
    1) Manufacture the product to meet specifications.

    2) Manufacture the product to the designed target value.

    What are the advantages/disadvantages of each philosophy ?

11. Your travel time to work tends to vary by several minutes. As a result of this variation problem you have been late on several occasions. Your boss is concerned and has asked that you brainstorm the factors that contribute to travel time vaiation. Use the fishbone diagram below to organize the brainstorming process.

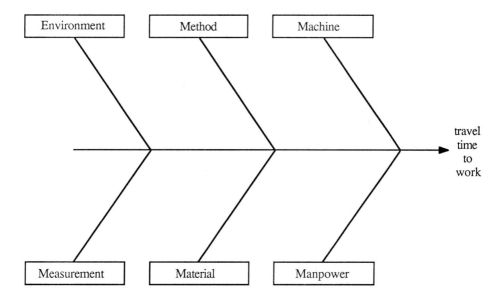

Chapter 1 Bibliography

1. Barker, Thomas B. (1985) *Quality by Experimental Design*, Marcel Dekker, Inc.

2. Deming, W. E. (1982) *Quality, Productivity and Competitive Position*, Cambridge, Mass: MIT Center for Advanced Engineering Study, 1982.

3. Hogg, Robert V. and Ledolter, Johannes (1987), *Engineering Statistics*, MacMillan Publishing Company, New York.

4. Prince, George M. (1972) *The Practice of Creativity*, Collier Books.

5. Taguchi, G. (1986) *Introduction to Quality Engineering*, Asian Productivity Organization.

6. Ott, Ellis R., (1975) *Process Quality Control* , McGraw-Hill, Inc., New York

7. Phadke, M.S. (1989) *Quality Engineering using Robust Design,* Prentice-Hall, New Jersey.

8. Hunter, W., O'Neill, J., and Wallen, C. (1988) *Doing more with Less in the Public Sector: A progress report from Madison, Wisconsin,* Center for Quality and Productivity Improvement, University of Wisconsin.

CHAPTER 2: STATISTICAL TECHNIQUES

2.1 Introduction

In order to fully comprehend subsequent discussions on experimental design, the reader should have working knowledge of inferential statistical techniques. This chapter covers the broad topics of Analysis of Variance (ANOVA), Simple Linear Regression (SLR) and Multiple Linear Regression (MLR) in enough detail to provide the required level of expertise for subsequent use. In no way does this chapter exhaust these three topics.

2.2 Analysis of Variance

In the way of an example, a company wants to investigate ways to improve a process measured by the variable y. It is decided to test the process at two different temperatures (70° and 90°) to determine if there is a temperature effect. In order to infer that temperature did or did not cause a difference in response, random assignments of experimental units to each temperature are conducted. The randomization procedure results in spreading all possible nuisance variable effects evenly over the two treatments. Thus, if there does exist a significant difference in group means, it can be concluded that temperature did in fact cause that difference, and not some uncontrolled factor. As stated in Chapter 1, randomization will not reduce the experimental error. This is only accomplished by including all measurable factors, whether controllable or noise, in the design. For the sake of simplicity, only temperature is initially included in this example.

To conduct the randomization procedure you could simply flip a coin where heads implies 70° and tails is 90°. Say the first coin flip resulted in a head; then run #1 would be made with the temperature at 70°. The process continues until a predetermined number of runs are made for each temperature. (Random numbers are typically used when the number of experimental conditions are greater than 2). When there are factors with levels that are difficult or costly to change, pseudo randomization techniques can be used. This procedure is accomplished by placing difficult to change factors in design columns which have the least number of level changes. The sample data for demonstrating 1-way ANOVA are shown in table 2.1.

---

table 2.1 Sample Data on the response y for 2 temperature settings

Treatment (Temperature)

$I_{(70°)}$	$II_{(90°)}$
2.2	4.6
2.8	5.0
3.2	5.4
3.6	5.8
$\bar{y}_I = 2.950$	$\bar{y}_{II} = 5.20$
$S_I^2 = .357$	$S_{II}^2 = .267$

The mean for the response at each temperature level, j, indicated by $\bar{y}_j = \sum_1^{n_j} \dfrac{y_{ij}}{n_j}$ is a measure of the center of the data values. The sample variance

$$S_j^2 = \sum_1^{n_j} \frac{\left(y_{ij} - \bar{y}_j\right)^2}{n_j - 1}$$

is a measure of the spread in the data. Assumptions made about the response values are:

(1) their distribution is mound shaped (Normal; also referred to as Gaussian)
(2) independence
(3) variability of the response in both treatment groups is approximately the same.

It is obvious from the sample data that $\bar{y}_{II} > \bar{y}_I$, but is this difference large enough, based on sample variation, to conclude that the population means μ_I and μ_{II} are different? To conduct the test, we define the following hypotheses:

H_0: $\mu_I = \mu_{II}$ (i.e., no difference in treatment means)

H_1: $\mu_I \neq \mu_{II}$ (i.e., there is a difference in treatment means)

Since variation exists within each sample and between each sample's mean response, we can obtain two estimates of the overall population variability, σ^2. Assumption (3) implies equal variability within samples, therefore both sample variances can be used to estimate σ^2. However, a better estimate than the separate S_j^2 values is a combination or pooling of the two estimates of within variability using the mean square error (MSE) such that

$$MSE = \left[\frac{(n_I - 1)S_I^2 + (n_{II} - 1)S_{II}^2}{n_I + n_{II} - 2} \right].$$

The combination of assumption (3) and the null hypothesis (H_0) true, implies both samples are from the same population [i.e., Normal (μ, σ^2)]. Based on these assumptions, $\bar{y}_I$ and $\bar{y}_{II}$ are both estimates of μ and vary about μ with variance $\sigma_{\bar{y}}^2 = \sigma^2/n$. In terms of the population variance, $\sigma^2 = \sigma_{\bar{y}}^2 \cdot n$. This indicates that if the sample size is the same for both groups ($n_I = n_{II} = n$) then another estimate of σ^2 would be called mean square between. See the formula below where k is the number of factor levels.

$$MSB = n S_{\bar{y}}^2 = n \sum_{j=1}^{k} \frac{\left(\bar{y}_j - \bar{\bar{y}}\right)^2}{k-1} .$$

It can be shown that these two estimates of σ^2 are independent because of assumptions (1) and (2) and thus their ratio has an F distribution. Basically, the ratio of $\dfrac{MSB}{MSE}$ would be expected to be approximately 1.0 if H_0 is true. If H_0 is not true, there should not be any discernable change in MSE; however, MSB should increase due to an expected increased difference in $\bar{y}_I$ and $\bar{y}_{II}$, causing $S_{\bar{y}}^2$ to increase. Therefore, values of $\dfrac{MSB}{MSE}$ larger than 1.0 are evaluated using the F distribution to assess their probability (p) of being that large due to chance alone or because H_1 is true. Low probabilities (i.e., less than .05) typically indicate that the ratio is large not due to chance when H_0 is true but rather because H_0 is not true. This leads to the conclusion that there is a treatment effect. The test is formalized by the following steps:

(1) define hypotheses H_0: $\mu_I = \mu_{II}$

H_1: $\mu_I \neq \mu_{II}$

(2) determine at which specific level p (called α) you would reject H_0 as true. This value of α is usually referred to as the probability of concluding H_1 when it is false, i.e., the risk you are willing to take in concluding H_1 when H_0 is true.

(3) Compute

$$\text{i) } MSB = \sum_{j=1}^{k} n_j \frac{\left(\bar{y}_j - \bar{\bar{y}}\right)^2}{k-1} \quad \text{where} \quad \bar{\bar{y}} = \frac{\sum_1^k n_j \bar{y}_j}{\sum_1^k n_j}$$

$$\text{ii) } MSE = \frac{\sum_1^k (n_j - 1) S_j^2}{\sum_1^k (n_j - 1)}$$

The denominators of MSB and MSE contain values referred to as degrees of freedom (df). They are simply explained as the number of independent observations minus the number of population parameters estimated. In MSB, there are k independent observations (one for each treatment group) and $\bar{\bar{y}}$ is an estimate of μ; therefore $df_B = k-1$. For MSE, remember $S_j^2 = \sum_1^{n_j} \dfrac{(y_{ij} - \bar{y}_j)^2}{n_j - 1}$ for which there are n_j independent observations for each j. Since each $\bar{y}_j$ estimates μ_j, each S_j^2 has $(n_j - 1)$ degrees of freedom and therefore MSE has $df_E = \sum_1^k (n_j - 1)$.

(4) In an F table, look up $F(\alpha, df_B, df_E) = F_c$.

(5) Compare the ratio $F_0 = \dfrac{MSB}{MSE}$ to F_c from (4).

   If $F_0 \le F_c$ fail to reject H_0

   If $F_0 > F_c$ reject H_0 with $(1-\alpha)$ 100% confidence of being correct.

For our example, the five steps appear as follows:

(1) $H_0: \mu_I = \mu_{II}$

   $H_1: \mu_I \ne \mu_{II}$

(2) select $\alpha = .05$

(3) $\bar{\bar{y}} = \dfrac{4(2.95) + 4(5.20)}{8} = 4.075$

   $MSB = \dfrac{4(2.95 - 4.075)^2 + 4(5.20 - 4.075)^2}{(2-1)} = 10.13$

   $MSE = \dfrac{(4-1)(.357) + (4-1)(.267)}{(4-1) + (4-1)} = .312$

   $F_0 = \dfrac{10.13}{.312} = 32.468$

(4) $F(.05, 1, 6) = 5.99$

(5) Since $F_o > F_c$ reject H_o and conclude H_1 with 95% confidence.

Notice that $F(.01, 1, 6) = 13.75$ implies we could be as high as 99% confident in concluding H_1. It is also important to note that in the case of $K = 2$ treatments a t test could have been used to test the same hypotheses.

Using the t test changes step 3 to $t_0 = \dfrac{\bar{y}_1 - \bar{y}_2}{\sqrt{MSE}\sqrt{\dfrac{1}{n_1} + \dfrac{1}{n_2}}}$ and step 4

becomes $t_c = t\left(\frac{\alpha}{2}, n_1 + n_2 - 2\right)$. Step 5 would be worded as follows:

If $|t_o| > t_c$, reject H_o with $(1 - \alpha)100\%$ confidence, otherwise fail to reject H_o. The conclusion, however, will be the same regardless of the choice of a t test or F test. In addition, it can be easily shown that $t_o^2 = F_o$ and $t_c^2 = F_c$.

Now consider the following one way ANOVA example where $K > 2$

Factor A

	A_1	A_2	A_3	A_4	A_5
$\bar{y}_j$	8	6	7	12	10
s_j^2	4.888	7.333	12.222	9.333	10.667
n_j	10	10	10	10	10

1) $H_0: \mu_1 = \mu_2 = \mu_3 = \mu_4 = \mu_5$

H_1: not all μ_i are equal

2) $\alpha = .01$

3) $\overline{\overline{Y}} = \dfrac{\Sigma n_j \overline{y}_j}{\Sigma n_j} = \dfrac{10(8+6+7+12+10)}{50} = 8.6$

$$\text{MSB} = \dfrac{\displaystyle\sum_{1}^{K} n_j \left(\overline{y}_j - \overline{\overline{y}} \right)^2}{K-1} = \dfrac{10(-.6)^2 + 10(-2.6)^2 + 10(-1.6)^2 + 10(3.4)^2 + 10(1.4)^2}{5-1} = 58$$

$$\text{MSE} = \dfrac{\displaystyle\sum_{1}^{K} \left(n_j - 1 \right) s_j^2}{\displaystyle\sum_{1}^{K} \left(n_j - 1 \right)} = \dfrac{9(4.88) + 9(7.333) + 9(12.222) + 9(9.333) + 9(10.667)}{9+9+9+9+9}$$

$$= \dfrac{400}{45} = 8.887$$

4) $F(.01;4,45) = 3.83$

5) Since $F_0 > 3.83$, reject H_0 with 99% confidence.

In this case the conclusion is H_1: μ_i are not all equal. The next question is "which levels are different and which are not?" To address this question, additional tests must be conducted. Using a regular t test or F test on each possible pair of treatments, i.e. H_0: $\mu_i = \mu_j$ for all $i \leq j$, would result in an increase in the overall α known as the experiment-wise α, α_{EW}. The α_{EW} is also referred to as the probability of committing at least one type one error after conducting all the individual pairwise tests. A procedure which allows for testing all pairwise levels at a specified

α_{EW} is the Tukey Post Hoc comparison test. Use of the test is only valid when the factor overall F_0 is significant. The test is fairly simple and requires the use of the studentized ranges provided in Chapter 8. The Tukey test will be illustrated using the previous example.

Since factor A is significant, it is desired to determine which levels differ and which do not. The procedure is as follows:

a) Use the Tukey comparison test and compute the critical Tukey distance $\bar{d}_T = q_T \sqrt{\frac{MSE}{n_i}}$ where q_T is found in chapter 8 and all n_i are assumed equal.

b) In chapter 8, p is the number of factor levels. Thus, for our example p = 5. Since df_E = 45 and given α = .05 then q_T = 3.171. Therefore,

$$\bar{d}_T = 3.171 \sqrt{\frac{8.887}{10}} = 2.989.$$

c) Compute all factor level mean pairwise differences and conclude those differences greater than $\bar{d}_T$ as significant at the chosen α level (See the chart shown below).

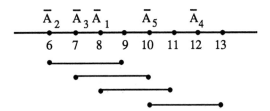

Conclude μ_4 is significantly different than μ_2, μ_3, and μ_1. Also conclude μ_5 is significantly different than μ_2 and μ_3; fail to reject H_0 for all other pairwise comparisons. What does this mean to the experiment? If your objective is to maximize the response then you are 95% confident that material 4 produces a larger response than materials 2, 3 and 1, and material 5 produces a larger response than 2 or 3. Therefore, the best material type is either 4 or 5. Unless 5 is substantially cheaper than 4, the most conservative decision is to select type 4 because you failed to reject μ_5 different from μ_1. However, there exists probabilities of error, α (rejecting H_0 incorrectly) and β (failing to reject H_0 when you should). Small sample sizes contribute to large β values. Our sample size of ten for each factor level may not provide enough overwhelming information to reject H_0: $\mu_5 = \mu_1$ when we should.

Often the question is asked, "How large should my sample be?" The answer to this question depends on the following:

1) The desired level for α

2) The desired level for $1-\beta$ (β is the probability of concluding H_0 given that H_1 is true, therefore $1-\beta$ is the probability of concluding H_1 correctly. The term $1-\beta$ is also referred to as the power of the test)

3) Estimated amount of experimental error, $\hat{\sigma}^2$

4) A specific alternative hypothesis deemed critical to detect with $(1-\beta)100\%$ confidence

The appropriate equation containing these terms is:

$$\Phi^2 = \frac{n' \left[\sum_{1}^{k} (\mu_i - \mu)^2 \div k \right]}{\hat{\sigma}^2}$$

where $\hat{\sigma}^2$, α, n' and degrees of freedom are used to enter the power charts in chapter 8. A step-by-step procedure for determining the approximate n' is as follows:

2-9

1) Determine k, the number of factor levels you intend to investigate

2) Determine the alternative hypothesis of interest either with specific values of μ_i or with the anticipated range, R, of the μ_i values and their suspected variability (low, medium, or high) over that range.

   Low variability implies that one μ_i is at the maximum, one μ_i is at the minimum and the other (K-2)μ_i values occur at a point in between.

   Medium variability implies that the μ_i are approximately normally distributed,

   i.e. $\sum_{1}^{k} \dfrac{(\mu_i - \mu)^2}{k} \doteq \dfrac{R^2}{16}$.

   High variability implies that half of the μ_i are at the maximum and the other half are at the minimum.

3) Estimate $\hat{\sigma}^2$ via

   a) historical data

   b) pilot study

   c) $\dfrac{R_i^2}{16}$ where R_i is the anticipated range of the raw data for some factor level i. This method assumes $Y_{ij} \sim N(\mu_i, \sigma_i^2)$ for each level i.

4) Select an estimated value of a factor level sample size, n'.

5) Compute Φ

6) Enter the appropriate chart from chapter 8 where υ_1 is the degrees of freedom for the factor effect and υ_2 is the degrees of freedom for error.

7) If $(1-\beta)$ is larger then desired, lower n' otherwise raise n'.

8) Continue until step 7 converges on the desired $(1-\beta)$.

An example follows:

Assume $\hat{\sigma}^2 = 20$, $k = 4$, $\alpha = .05$ and the desired power is 0.8 for detecting $\mu_1 = 16$, $\mu_2 = 21$, $\mu_3 = 23$, $\mu_4 = 20$.

$$\Phi^2 = \frac{\dfrac{n'\left[(16-20)^2 + (21-20)^2 + (23-20)^2 + (20-20)^2\right]}{4}}{20} = .325\,n'$$

Thus $\Phi = .57\sqrt{n'}$.

Using the table with $\upsilon_1 = 3$, you could iterate n' as follows:

n'	V_2	Φ	$1 - \beta$	Comments
5	16	1.28	.45	Too small, increase n'
7	24	1.51	.64	Too small, increase n'
9	32	1.71	.78	Slightly less than .8
10	36	1.80	.83 *	Use n = 10 for each level

2.3   2-Way ANOVA

Assume that in the example problem of section 2.2, there exists another variable of interest, pressure. The one factor at a time approach implies that we collect data on two temperatures as shown in section 2.2 and then collect more data on a desired number of pressure levels. In each case, we would hold the other variable at some constant level. Using this approach, our experimentation is grossly inefficient and it is possible to determine that neither temperature nor pressure has an effect when in fact the contrary may be true. The missing effect when conducting one at a time designs is the interaction of the two effects when analyzed together. To estimate whether or not there is a **main effect** and/or an **interaction effect**, a **factorial design** including all combinations of levels of temperature and pressure must be used. Assume that we are interested in only two levels of pressure, 100 and 200 psi. Then tables 2.2 (a) and (b) display the different combinations possible for temperature and pressure.

table 2.2   2 factor combinations of temperature and pressure

(a) Temperature (°)	70	90
Pressure (PSI) 100	1	2
Pressure (PSI) 200	3	4

(b) Combination Number	Temp	Pressure
1	70	100
2	90	100
3	70	200
4	90	200

If the average observed values in cells 1, 2, 3, and 4 are 1, 3, 3, and 5, they would appear graphically as shown in figure 2.1.

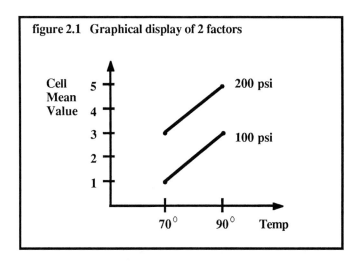

figure 2.1 Graphical display of 2 factors

The graph in figure 2.1 indicates a potential temperature effect (average response over pressure levels increases for temperature changes from 70° to 90°); a potential pressure effect (average response for 200 psi is consistently above that for 100 psi) and there appears to be no 2-way interaction because the change in response from 70° to 90° for 100 psi is equal to the change at 200 psi, (i.e., the lines are approximately parallel). An example of cell mean values with a potential 2–way interaction would be 2, 4, 5, 3 for cells 1, 2, 3, and 4, respectively. Graphically, these values appear as shown in figure 2.2. The non-parallel nature of these lines (whether they cross or not) is an indicator of a potential interaction. The non-horizontal nature of each pressure line over the two temperatures indicates a potential positive temperature effect when pressure is 100 psi and a potential negative temperature effect when pressure is 200 psi. The term potential is used because you must consider the error variance prior to establishing significance of any effect. Notice that if you average the response over levels of pressure there will appear to be no temperature effect; thus, the importance of the interaction term. The impact of not properly modelling interactions is illustrated by way of an example in Appendix 4C.

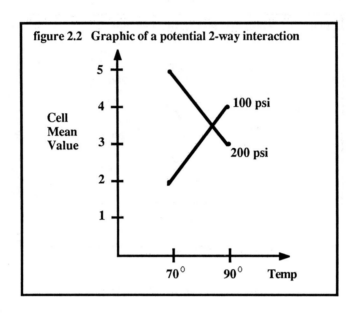

figure 2.2 Graphic of a potential 2-way interaction

To analyze the 2-way ANOVA, consider the following table with formulas, where factor A is temperature, B is pressure, and

a = # of levels of factor A

b = # of levels of factor B

n_j = # of observations in column j

n_i = # of observations in row i

n_{ij} = # of observations per ij experimental condition

SS = sum of squares; the numerator of a variance estimate

S_{ij}^2 = sample variance of observations in cell ij.

table 2.3	Formulas and table structure for 2-way ANOVA			
Source	**SS**	**df**	**MS**	**F_0**
Temp (A)	$SS_A = \sum_1^a n_i (\bar{A_i} - \bar{\bar{y}})^2$	$a - 1$	$\dfrac{SS_A}{df_A}$	$\dfrac{MS_A}{MS_E}$
Press (B)	$SS_B = \sum_1^b n_j (\bar{B_i} - \bar{\bar{y}})^2$	$b - 1$	$\dfrac{SS_B}{df_B}$	$\dfrac{MS_B}{MS_E}$
Inter-action (AxB)	$SS_{AxB} = $ $SS_T - SS_A - SS_B - SS_E$	$df_{AxB} = $ $df_T - df_A - df_B - df_E$	$\dfrac{SS_{AxB}}{df_{AxB}}$	$\dfrac{MS_{AxB}}{MS_E}$
Within Error (E)	$SS_E = \sum_1^a \sum_1^b (n_{ij} - 1)S_{ij}^2$	$\sum_1^a \sum_1^b (n_{ij} - 1)$	$\dfrac{SS_E}{df_E}$	
TOTAL	$SS_T = \sum_1^a \sum_1^b \left[\sum_1^{n_{ij}} (y_{ijk} - \bar{\bar{y}})^2 \right]$			

In this case, the data would appear similar to that of table 2.4.

table 2.4   Data structure for a 2-way ANOVA

FACTOR B

		1	2	- - - - - - - - -	b
FACTOR A	**1**	y_{111} y_{112} $\vdots$ $y_{11n_{11}}$	y_{121} y_{122} $\vdots$ $y_{12n_{12}}$		
	2	y_{211} y_{212} $\vdots$ $y_{21n_{21}}$	y_{221} y_{222} $\vdots$ $y_{22n_{22}}$		
	$\vdots$				
	a				y_{ab1} y_{ab2} $\vdots$ $y_{abn_{ab}}$

For an example, using the temperature and pressure factors, see the data in table 2.5 and the calculated values for the 2-way ANOVA table in table 2.6.

table 2.5   (a)   **Example data for 2-way ANOVA**

Press (B)

		100	200	$\bar{A}_i$
Temp (A)	70°	2.2 2.8	3.2 3.6	2.95
	90°	5.0 4.6	5.8 5.4	5.20
	$\bar{B}_j$	3.65	4.5	$\bar{\bar{y}}=4.075$

(b)   **Sample cell means and variance; mean (variance)**

Press (B)

		100	200
Temp (A)	70°	2.5 (0.18)	3.4 (0.08)
	90°	4.8 (0.08)	5.6 (0.08)

table 2.6   **Results of 2-way ANOVA example**

Source	SS	df	MS	F_0
Temp (A)	10.125	1	10.125	96.428 *
Press (B)	1.445	1	1.445	13.762 *
A x B	.005	1	.005	.0476
error	.420	4	.105	
TOTAL	11.995	7		

F (.05,1,4) = 7.7086

These results indicate a significant temperature and pressure effect with no significant interaction. This finding is verified by the graphic in figure 2.3.

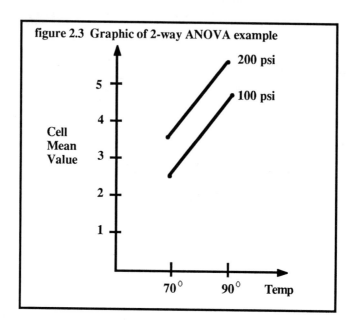

figure 2.3 Graphic of 2-way ANOVA example

If the second independent variable was something outside of the experimenter's control, it would have been called a **blocking variable**. As an example, consider materials of various hardness to be treated with one of three treatments. The population of materials can be divided (stratified) into 3 levels of hardness: high, medium and low. Since the experimenter cannot assign materials to these inherent characteristics, randomization occurs within each stratum to the various treatments instead of randomizing assignment to the 9 different combinations of treatment and hardness. The calculations would not differ from that of the previous 2-way ANOVA example provided levels of the strata are fixed.

2.4   Simple Linear Regression (SLR)

Consider the same example from section 2.3; however, assume that there are 4 levels of temperature: $70°$, $80°$, $90°$ and $100°$. Your objective is to develop a model which will allow you to estimate the response at levels other than those mentioned above and be able to place prediction intervals about these estimates. You also need a measure of model effectiveness. The technique to be discussed next, **Simple Linear Regression (SLR)**, will assist you in accomplishing these objectives. Assume the data you collected from the 4 temperature settings is that of table 2.7.

table 2.7	Data for SLR example		
	Temperature		
$70°$	$80°$	$90°$	$100°$
2.3	2.5	3.0	3.3
2.6	2.9	3.1	3.5
2.1	2.4	2.8	3.0

A graph of the data is given in figure 2.4. The line represents an "eyeball fit" or free-hand regression line. The closeness of all the observations to the line indicates the accuracy of the predicted values of y for any given temperature.

The objective in placing the line is to attempt to minimize the distance that observations are from the line. Using the slope-intercept formula $[E(y) = B_0 + B_1x]$, you can estimate "B_0" graphically as .1 using figure 2.4. The slope, B_1, is found by measuring the change in y (Δy) for some specific change in x (Δx), i.e.,

$$\frac{\Delta y}{\Delta x} = \frac{2.7 - 2.3}{80 - 70} = 0.04.$$

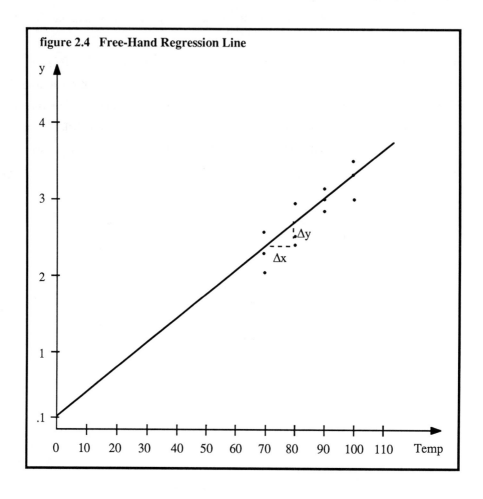

figure 2.4 Free-Hand Regression Line

Since all observations do not lie exactly on the line, there is obviously some error in our straight line estimate. To incorporate this _error_ in the formula for predicting y given any x value, we use $y = B_0 + B_1 x + \varepsilon$, where ε represents the error term which is assumed to be distributed normally about 0; have equal variability for all levels of x; and the error values are independent.

The relationship $y = B_0 + B_1 x + \varepsilon$ is applicable for the _population data_ (set of all possible x and y values). The associated regression line is the true regression line represented by $E(y) = B_0 + B_1 x$.

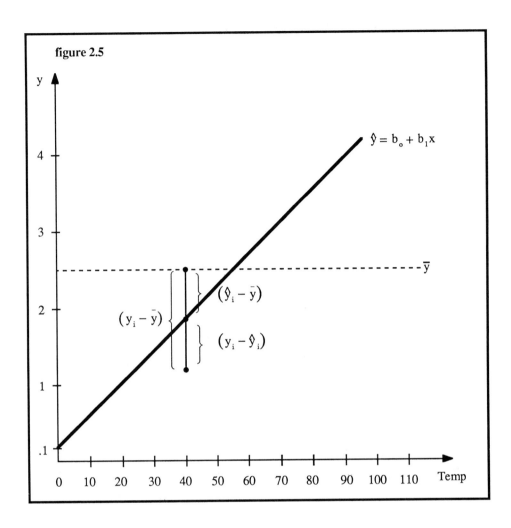

figure 2.5

Data used in experimentation and process control are almost always sample data (a subset of population data); therefore, use $\hat{y} = b_0 + b_1 x$ to approximate the regression line; i.e., $\hat{y}$, b_0 and b_1 estimate $E(y)$, B_0 and B_1 respectively. In addition, $e_i = y_i - \hat{y}_i$ is called the ith residual which estimates ε_i. These terms are displayed graphically in figure 2.5.

From figure 2.5, it can also be seen that each observations's deviation from $\bar{y}$ can be partitioned such that $(y_i - \bar{y}) = (y_i - \hat{y}_i) + (\hat{y}_i - \bar{y})$. Summing and squaring both sides can be simplified as

$$\sum_1^n (y_i - \bar{y})^2 = \sum_1^n (y_i - \hat{y}_i)^2 + \sum_1^n (\hat{y}_i - \bar{y})^2$$

where SST $= \sum_1^n (y_i - \bar{y})^2$ is the numerator of the variance of y and is called the sum of squares total (SST). The two terms partitioning SST are called the sum of squares regression, SSR $= \sum_1^n (\hat{y}_i - \bar{y})^2$ and the sum of squares error, SSE $= \sum_1^n (y_i - \hat{y}_i)^2$

If we don't use x to predict y, then the best prediction for some future y is $\bar{y}$. The variance of observations about $\bar{y}$ is

$$s_y^2 = \sum_1^n \frac{(y_i - \bar{y})^2}{n - 1}.$$

When information on x is used to predict y, then $\hat{y} = b_0 + b_1 x$ is the prediction and the variance of observations about $\hat{y}$ is

$$\sum_1^n \frac{(y_i - \hat{y}_i)^2}{n - 2} = \frac{SSE}{n - 2} = MSE$$

referred to as the mean square error or error variance. The objective of SLR is to develop a regression line such that MSE is much smaller than s_y^2. The size of the error variance determines how good a fit the regression line will be and it plays a key role in the prediction intervals to be developed later in this chapter. One mathematical method for finding b_0 and b_1 is the **method of least squares**. The name is derived from the property of minimizing the sum of squared deviations from the line, i.e., minimize $\sum_1^n e_i^2$. The appropriate equations are

$$b_1 = \cfrac{\displaystyle\sum_1^n xy - \cfrac{\displaystyle\sum_1^n x \displaystyle\sum_1^n y}{n}}{\displaystyle\sum_1^n x_i^2 - n\overline{x}^2} \qquad\qquad b_0 = \overline{y} - b_1\overline{x}$$

For the example data, the calculations are shown in table 2.8.

table 2.8   Simple Linear Regression Calculation

n	x	y	x^2	xy	y^2
1	70	2.3	4900	161	5.29
2	70	2.6	4900	182	6.76
3	70	2.1	4900	147	4.41
4	80	2.5	6400	200	6.25
5	80	2.9	6400	232	8.41
6	80	2.4	6400	192	5.76
7	90	3.0	8100	270	9.00
8	90	3.1	8100	279	9.61
9	90	2.8	8100	252	7.84
10	100	3.3	10000	330	10.89
11	100	3.5	10000	350	12.25
12	100	3.0	10000	300	9.00
	1020	33.5	88200	2895	95.47

$$\overline{x} = 85 \qquad\qquad \overline{y} = 2.79$$

$$b_1 = \cfrac{\displaystyle\sum xy - \cfrac{\displaystyle\sum x \displaystyle\sum y}{n}}{\displaystyle\sum x^2 - n\overline{x}^2} = \cfrac{2895 - \cfrac{1020(33.5)}{12}}{88200 - 12(85)^2} = \cfrac{47.5}{1500} = 0.032$$

$$b_0 = \overline{y} - b_1\overline{x} = 2.79 - 0.032(85) = 0.07$$

$\hat{y} = .07 + .032x$ The $\hat{y}$ indicates that an average value of y is a function of x derived from sample data .

As you can see, the estimates from the free-hand regression are close to those of least squares for this example. When the data are more spread out and more variables are added, the "eyeball fit" becomes an impossible task.

A measure of the strength of the linear relationship of y with temperature is the correlation coefficient,

$$R = \frac{\sum xy - \frac{\sum x \sum y}{n}}{\sqrt{(\sum x^2 - n\bar{x}^2)(\sum y^2 - n\bar{y}^2)}}$$

For our example,

$$R = \frac{2895 - \frac{1020\,(33.5)}{12}}{\sqrt{[88200 - 12(85)^2][95.47 - 12(2.792)^2]}} = \frac{47.5}{53.763} = .883$$

The value of R is limited to the interval [-1,1] where -1 indicates perfect negative correlation, 1 indicates a perfect positive correlation and 0 implies no linear relationship at all. The example value of .883 represents a moderate positive linear relationship between y and x. The proportion of variability in y which is explained by y's relationship with x is measured by R^2. The example data $R^2= .780$ indicates that about 78.0% of the variability in y is explained through y's linear relationship with x. The variance of y without regard to x is

$$S_y^2 = \frac{\sum y^2 - n\bar{y}^2}{n - 1} = \frac{95.47 - 12(2.792)^2}{11} = .1752$$

and since R^2=SSR/SST we can complete an ANOVA table for our regression analysis. The SST = $S_y^2 \cdot$ (n-1) = .1752(11) = 1.927 and $R^2 = \frac{SSR}{SST}$ implies SSR = 1.9271(.780) = 1.503. Thus, SSR is 78.0% of the SST (total variability) as discussed previously.

The degrees of freedom for the error term in a regression model are equal to n minus the number of parameters estimated. In SLR, b_0 and b_1 are both estimates obtained from sample data of the two population parameters B_0 and B_1, therefore, $df_E=n-2$. The degrees of freedom for regression in SLR are the number of parameters estimated minus 1, which will be 1. The Regression-ANOVA table appears in table 2.9.

table 2.9 SLR ANOVA Table				
Source	SS	df	MS	F_o
Regression	$SSR = R^2(SST)$	1	SSR/1	MSR/MSE
ERROR	$SSE = (1 - R^2)(SST)$	n-2	SSE/(n-2)	
TOTAL	$S_y^2 \cdot (n-1)$	n-1		

The F test in this table represents a test of the hypothesis H_0: $R^2 = 0$; H_1: $R^2 \neq 0$ which is the same in SLR as testing H_0: $B_1 = 0$; H_1: $B_1 \neq 0$. The MSE is an estimate of the variance of y at any level of x. Table 2.10 displays the example data in table 2.9 format.

table 2.10 SLR ANOVA Table for Example Data				
Source	SS	df	MS	F_o
Regression	1.503	1	1.503	35.448 *
ERROR	.424	10	.0424	
TOTAL	1.927	11		
F(.01, 1, 10) = 10.04				
*Since $F_o > 10.04$ we can conclude H_1: $R^2 \neq 0$ or H_1: $B_1 \neq 0$ with 99% confidence				

To obtain a point estimate for y given any x value (x_h) you simply insert the value for x_h in $\hat{y} = b_0 + b_1 x_h$. Point estimates, however, do not include any reference to precision. In order to indicate how precise our estimate is , a **prediction**

interval is formed in which the researcher is $(1-\alpha)\cdot 100\%$ confident where the true value of y will occur. Since our data is just one sample, it is reasonable to take another sample and compute different values of b_0 and b_1. This implies that there exists sample variability in the slope and intercept in the sample regression lines in addition to observation variability about the line. The slope and intercept variability will cause large deviations in predicted response the further you estimate from the center of the domain of x. For these reasons, a prediction interval for the response at any given value of x (labeled x_h) will appear as shown below.

$$\hat{y}|_{x_h} \pm t(\tfrac{\alpha}{2}, n-2) \sqrt{MSE\left(1 + \frac{1}{n} + \frac{(x_h - \bar{x})^2}{\sum x^2 - n\bar{x}^2}\right)}^{\,*}$$

MSE represents variability of observations about the regression line, MSE/n represents variability of the average response which is similar to variability of b_0, and

$$MSE\left[\frac{x_h - \bar{x}^2}{\sum x^2 - n\bar{x}^2}\right]$$

represents variability due to the variability in slope, b_1.

For our example, the prediction intervals based on 95% confidence are displayed in table 2.11 for all x values previously used to develop the model. Predicting values of y for $x_{min} \leq x_h \leq x_{max}$ is called **interpolation**, which is typically done. Predicting values of y for $x_h < x_{min}$ or $x_h > x_{max}$ is called **extrapolation**, which assumes the same relationship of y with x outside the region of the sampled x. Notice how wide the 95% prediction interval would be outside [80,100]. This lack of accurate predictability and uncertainty of the relationship of y with x outside the region of sampled x is why extrapolation is not recommended. If you need to predict outside the sampled region for x you should expand the sampled region.

---

    * When actually performing the calculations, $n\bar{x}^2$ is more prone to numerical error than

       its equivalent, $\dfrac{(\sum x_i)^2}{n}$.

table 2.11   Prediction Intervals for y based on 95% confidence

x	$\hat{y}$	y_L	y_U	Δ
70	2.31	1.80	2.82	1.02
* 80	2.63	2.15	3.11	.96
90	2.95	2.47	3.43	.96
100	3.27	2.76	3.78	1.02

$t(.025, 10) = 2.228$

$MSE = .0424$

$\sum x^2 - n\bar{x}^2 = 1500$

$\bar{x} = 85$

$$\hat{y}|x_h = .07 + .032(80) = 2.63$$

* for $x_1 = 80$

$$y_L = 2.63 - 2.228\sqrt{.0424\left(1 + \tfrac{1}{12} + \frac{(80-85)^2}{1500}\right)} = 2.15$$

$$y_U = 2.63 + 2.228\sqrt{.0424\left(1 + \tfrac{1}{12} + \frac{(80-85)^2}{1500}\right)} = 3.11$$

When prediction intervals are too wide for your application, there are ways to reduce these intervals. The researcher can

(1) decrease the confidence level $(1-\alpha)$; then $t(\alpha/2, n-2)$ will get smaller

(2) increase the sample size; then $1/n$ becomes smaller

(3) increase $\left[\sum x^2 - n\bar{x}^2\right]$.

It is not recommended that one alter the confidence level to reduce interval width. In fact, the subject of experimental design discussed in Chapter 3 addresses only items (2) and (3) as the major areas of interest. It is also apparent that in conducting industrial experiments the sample size is desired to be a minimum which puts even more emphasis on maximizing $\left[\sum x^2 - n\bar{x}^2\right]$. This is equivalent to maximizing the variance of x, which is accomplished by placing n/2 observations at both x_{min} and x_{max}. The problem with only sampling at two levels of x is that we are restricted to linear estimates. For this reason, you'll see in Chapter 3 that center points can be added to 2 level designs to allow for curvature examination. The idea of maximizing $\left[\sum x^2 - n\bar{x}^2\right]$ has another effect in that

$$\sigma^2(\hat{b}_1) = \frac{MSE}{\sum x^2 - n\bar{x}^2}.$$

Therefore, the variance of the slope is minimized if $\left[\sum x^2 - n\bar{x}^2\right]$ is maximized, creating more power in the F test for H_0: $R^2 = 0$ or H_0: $B_1 = 0$. This implies that if we minimize $\left[\hat{\sigma}^2(b_1)\right]$ we are more likely to find a real linear relationship if it in fact exists. Once multiple regression is covered, the reader will also see that maximizing $\left[\sum x^2 - n\bar{x}^2\right]$ is related to D-optimality to be discussed in Chapter 3.

2.5 Polynomial Regression and Optimization in SLR.

Suppose that the example data in section 2.4 included 3 more observations taken at $x=50°$, which when added to table 2.7 appear as shown in table 2.12.

table 2.12 Example data for Polynomial Regression				
Temperature				
50°	70°	80°	90°	100°
3.3	2.3	2.5	3.0	3.3
2.8	2.6	2.9	3.1	3.5
2.9	2.1	2.4	2.8	3.0

Graphically, the data are presented in figure 2.6. This graphic indicates the danger in extrapolation because the function takes on a different shape outside the original sampled space of x.

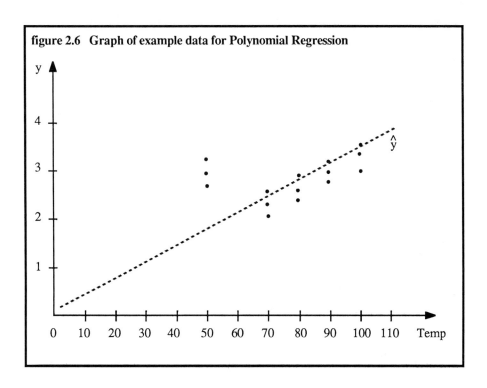

figure 2.6 Graph of example data for Polynomial Regression

As you can see, the linear dashed line is no longer a good fit. In fact, if we try SLR, the result is $b_1 = 0$ which implies $R^2 = 0$. However, the more appropriate thing to do is to try a quadratic term. Staying with SLR, the new model becomes $\hat{y} = b_0 + b_{11}x_1^2$ where b_{11} is not b eleven but b one-one to indicate that x_1 is squared. Table 2.13 has the computations for SLR.

table 2.13	Computations for $\hat{y} = b_0 + b_{11}x_1^2$			
(x_1^2)	y	$(x_1^2)^2$	y^2	$(x_1^2)y$
2500	3.3			
2500	2.8			
2500	2.9			
4900	2.3			
4900	2.6			
4900	2.1			
6400	2.5			
6400	2.9			
6400	2.4			
8100	3.0			
8100	3.1			
8100	2.8			
10000	3.3			
10000	3.5			
10000	3.0			
95700	42.5	710490000	122.61	276810

$$\hat{y} = 2.472 + 0.000056\,(x_1^2) \qquad R^2 = .146$$

The low R^2 indicates that something is wrong because we'd expect a better fit. What happened? Creating (x^2) as a new variable using the model above resulted in fitting a quadratic with the vertex on the y axis. Graphically, you can see the vertex is far from $x = 0$.

To properly conduct polynomial regression in the way we've stated the problem, we must have x_1 and x_1^2 both in the model, i.e., $\hat{y} = b_0 + b_1 x_1 + b_{11} x_1^2$. This second order polynomial regression model is similar to a multiple regression problem. The computer output shown in table 2.14 indicates that this model is a fairly good fit, i.e., the probability of an incorrect model is very small ($p = .001$).

table 2.14 Computer Output for Polynomial Regression Example		

Model $\hat{y} = 7.960 - .15374x_1 + .00108x_1^2$ $\quad R^2 = .673;$ $\quad P = .001$

Parameter	estimate	p-value
b_0	7.9600	.0001
b_1	-0.1537	.001
b_{11}	.00108	.001

In order to find the x value resulting in the minimum for y, differential calculus is used to find $\frac{dy}{dx} = -.15374 + .00216x_1 = 0$. Solving for $x_1 = \frac{.15374}{.00216} = 71.176$ is the point where y is a minimum. This seems reasonable from the graph of $\hat{y} = 7.96 - .15374x_1 + .00108x_1^2$ in figure 2.7.

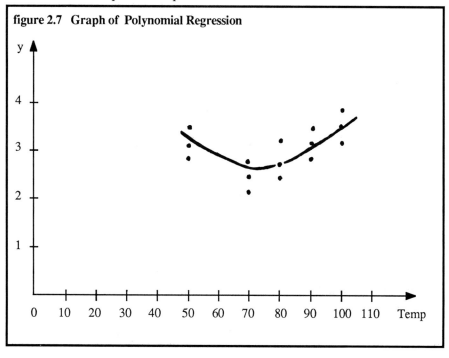

figure 2.7 Graph of Polynomial Regression

2.6 Multiple Regression

If a variable for pressure was also included in our model, the multivariable model would be referred to as a **Multiple Regression Model**. A new set of example data appears in table 2.15.

table 2.15 Example Data for MLR		
y	x_1 = Temp	x_2 = Press
3.1	50	100
3.3	50	100
2.6	50	200
2.4	50	200
2.5	70	100
2.6	70	100
3.0	70	200
3.3	70	200
2.4	80	100
2.3	80	100
3.2	80	200
3.5	80	200
2.8	90	100
2.6	90	100
3.1	90	200
3.0	90	200
3.2	100	100
3.4	100	100
2.5	100	200
2.4	100	200

The initial model to be tested is $y = B_0 + B_1 x_1 + B_2 x_2 + \varepsilon$ which is represented using a matrix notation as shown below.

$$\underline{Y} = X \underline{B} + \underline{\varepsilon} \quad \text{where}$$

$$
\underline{Y} = \begin{bmatrix} y_1 \\ y_2 \\ y_3 \\ \vdots \\ y_n \end{bmatrix} \qquad X = \begin{bmatrix} 1 & x_{11} & x_{12} \\ 1 & x_{21} & x_{22} \\ 1 & x_{31} & x_{32} \\ \vdots & \vdots & \vdots \\ 1 & x_{n1} & x_{n2} \end{bmatrix} \qquad \underline{B} = \begin{bmatrix} B_0 \\ B_1 \\ B_2 \end{bmatrix} \qquad \text{and} \qquad \underline{\varepsilon} = \begin{bmatrix} \varepsilon_1 \\ \varepsilon_2 \\ \varepsilon_3 \\ \vdots \\ \varepsilon_n \end{bmatrix}
$$

As in SLR, $\underline{b}$ is used to estimate $\underline{B}$ and $\underline{e}$ is an estimate of $\underline{\varepsilon}$. In general, $\underline{Y}$ is an (n x 1) vector of the response values, X is an (n x p) matrix where the first column is all 1's to estimate b_0 and the rest are the levels of the k independent variables; $\underline{b}$ is a (p x 1) vector of (k + 1) regression coefficients; and $\underline{e}$ is an (n x 1) vector of random error values (one for each of the n observations). For our example data, $\underline{Y}$, X, $\underline{b}$ and $\underline{e}$ appear as

$$
\underline{Y} = \begin{bmatrix} 3.1 \\ 3.3 \\ 2.6 \\ 2.4 \\ 2.5 \\ \vdots \\ 2.4 \end{bmatrix} \qquad X = \begin{bmatrix} 1 & 50 & 100 \\ 1 & 50 & 100 \\ 1 & 50 & 200 \\ 1 & 50 & 200 \\ 1 & 70 & 100 \\ \vdots & \vdots & \vdots \\ 1 & 200 & 200 \end{bmatrix}
$$

$$
\underline{b} = \begin{bmatrix} b_0 \\ b_1 \\ b_2 \end{bmatrix} \qquad \text{and} \qquad \underline{e} = \begin{bmatrix} e_1 \\ e_2 \\ e_3 \\ \vdots \\ e_n \end{bmatrix}
$$

Values of $\underline{b}$ and $\underline{e}$ are determined from the least squares multiple regression which has the objective of minimizing $[\underline{e}'\underline{e}]$. To find $\underline{b}$ such that $\hat{Y} = X\underline{b}$ we use $\underline{b} = (X'X)^{-1}X'\underline{Y}$ and then after computing $\hat{Y}$ we can compute $\underline{e} = \underline{Y} - \hat{Y}$. The variance-covariance matrix of the regression coefficients is

$$\text{cov}(\underline{b}) = \sigma^2 (X'X)^{-1} = \begin{bmatrix} \sigma^2_{b_0} & \sigma_{b_0 b_1} & \sigma_{b_0 b_2} \\ & \sigma^2_{b_1} & \sigma_{b_1 b_2} \\ & & \sigma^2_{b_2} \end{bmatrix}$$

$\sigma^2_{b_0}$ is the variance of b_0 and $\sigma_{b_1 b_2}$ is the covariance of b_1 and b_2. The estimate for σ^2 is based on the variance of the residuals which have n-p degrees of freedom, i.e.,

$$\hat{\sigma}^2 = \frac{\sum_{1}^{n}(y_i - \hat{y}_i)^2}{n - p}$$

which is referred to as the regression error variance, or MSE = SSE/(n-p). For our example data, the computer output is displayed in table 2.16.

---

table 2.16   Output for $\hat{y} = b_0 + b_1 x_1 + b_2 x_2$

Dep Var: Y   N: 20   Multiple R: .108   Squared Multiple R: .012
Adjusted Squared Multiple R: .000   Standard Error of Estimate: 0.410

Variable	Coefficient	STD Error	STD Coef	Tolerance	T	P(2 Tail)
Constant	2.695	0.506	0.000	1.0000000	5.322	0.000
x1	0.001	0.005	0.026	1.0000000	0.108	0.915
x2	0.001	0.002	0.105	1.0000000	0.437	0.668

Analysis of Variance

Source	Sum-of-Squares	DF	Mean-Square	F-Ratio	P
Regression	0.034	2	0.017	0.101	0.904
Residual	2.854	17	0.168		

Since $R^2 = .012$ with $p = .904$, the model is insufficient in predicting the response. In order to calculate lack of fit for the model, the SSE is partitioned into pure error and lack of fit error. The pure error SSE_p is acquired from repeated observations of y at the same levels of x. Since the example data has 2 observations of y for the 10 different combinatons of x_1 and x_2, the pure error is estimated with

$$SSE_p = \sum_{1}^{10} (n_{ij} - 1) S_{ij}^2 = 0.19.$$

The $MSE_p = .19/10 = .019.$ [*] The sum of squares for lack of fit, SSE_{LF}, is calculated using $SSE_{LF} = SSE - SSE_p$. For our example, $SSE_{LF} = 2.854 - .19 = 2.664$ and the degrees of freedom for testing lack of fit are $df_{LF} = df_E - df_P$ where $df_P = \sum_{1}^{10} (n_{ij} - 1)$. Therefore, the example data has a mean square for lack of fit of

$$MS_{LF} = \frac{SSE_{LF}}{df_{LF}} = \frac{2.664}{17 - 10} = 0.3806 .$$

The F test for lack of fit is $F_0 = MS_{LF}/MSE_p = .3806/.019 = 20.032$ compared to $F_C = F(\alpha, df_{LF}, df_P) = F(.05, 7, 10) = 3.14$. Since the lack of fit is significant, it is necessary to investigate a better model using interactions and quadratic effects. If we test a 2nd order model, it will appear as

$$y = b_0 + b_1 x_1 + b_2 x_2 + b_{12} x_1 x_2 + b_{11} x_1^2 + e$$

Notice that we do not estimate a quadratic effect for x_2 because it only has 2 levels. The computer output is shown in table 2.17.

If the design did not contain any replicated points at designated levels of x, replications can be made at the center point (for our example, this would be 75° and 150 psi) and the SSE_p would then be calculated for this one combination of independent variables. If using replicated center points, then $df_P = n_C - 1$; where n_C is the number of center points.

table 2.17 **Computer Output for** $\hat{y} = b_0 + b_1 x_1 + b_2 x_2 + b_{12} x_1 x_2 + b_{11} x_1^2$

Dep Var: Y   N: 20   Multiple R: .115   Squared Multiple R: .013

Adjusted Squared Multiple R: .000   Standard Error of Estimate: 0.436

Variable	Coefficient	STD Error	STD Coef	Tolerance	T	P(2 Tail)
Constant	2.973	2.371	0.000	1.0000000	1.254	0.229
x1	-0.004	0.057	-0.203	.0100250	-0.079	0.938
x2	-0.000	0.009	-0.054	.0463950	-0.046	0.964
x1 x2	0.000	0.000	0.194	.0327289	0.137	0.893
x1 x1	0.000	0.000	0.124	.0110192	0.051	0.960

Analysis of Variance

Source	Sum-of-Squares	DF	Mean-Square	F-Ratio	P
Regression	0.038	4	0.010	0.50	0.995
Residual	2.850	15	0.190		

To test for lack of fit once more, $SSE_{LF} = SSE - SSE_p = 2.850 - .19 = 2.66$ (about the same as before). The F test for lack of fit is $F_0 = \frac{2.66/(15-10)}{.019} = 28.0$ which is still significant at the .05 level. What's wrong? The problem is that 2nd order models can also have quadratic interaction terms. Therefore, the new model to be tested will include a quadratic interaction of x_1^2 with variable x_2:

$$y = b_0 + b_1 x_1 + b_2 x_2 + b_{12} x_1 x_2 + b_{(11)2} x_1^2 x_2 + b_{11} x_1^2 + e$$

The results are shown in table 2.18. It should be of concern that adding $x_1^2 \cdot x_2$ increased R^2 to .901 which is significant, but the coefficient for $x_1^2 \cdot x_2$ is 0 (The problem here is due to the use of only 3 decimal place accuracy. Formating the output to 4 or more decimal places will produce non-zero coefficients.). There is another more insidious problem occuring in this analysis. The tolerance column measures $1-R_j^2$, where R_j^2 is the squared correlation coefficient obtained by regressing predictor variable j on all other predictor variables. Tolerance values other than 1.0 indicate that a phenomena referred to as multicollinearity exists. Tolerance values close to zero imply strong multicollinearity conditions which can grossly affect the accuracy of the estimated coefficients. Since the vectors for x_1, x_1^2 and $x_1^2 \cdot x_2$ are intercorrelated, the predictor variable matrix is not orthogonal and therefore the

---

table 2.18 Computer Output for $\hat{y} = b_0 + b_1 x_1 + b_2 x_2 + b_{12} x_1 x_2 + b_{(11)2} x_1^2 x_2 + b_{11} x_1^2$

Dep Var: Y   N: 20   Multiple R: .949   Squared Multiple R: .901

Adjusted Squared Multiple R: .866   Standard Error of Estimate: 0.143

Variable	Coefficient	STD Error	STD Coef	Tolerance	T	P(2 Tail)
Constant	23.815	2.015	0.000	1.0000000	11.818	0.000
x1	-0.596	0.056	-26.984	.0011019	-10.656	0.000
x2	-0.139	0.013	-18.337	.0025125	-10.934	0.000
x1, x2	0.004	0.000	49.539	.0003606	11.192	0.000
x1 x1	0.004	0.000	27.053	.0006840	10.683	0.000
x1 x1 x2	-0.000	0.000	-36.030	.0006840	-11.210	0.000

Analysis of Variance

Source	Sum-of-Squares	DF	Mean-Square	F-Ratio	P
Regression	2.602	5	0.520	25.504	0.000
Residual	0.286	14	0.020		

estimates of the b_is can be very ambiguous. In fact, sometimes a variable with a known positive effect on the response will have a negative coefficient due to high multicollinearity. In addition, multicollinearity increases the off-diagonal values of $(X'X)^{-1}$ causing inflated variance of the coefficients which is estimated by $\hat{\sigma}^2(X'X)^{-1}$. This can result in significant F values without significant regression weights (See Appendix 5.A for an example). When including terms for a second order model, multicollinearity is difficult to avoid, unless a transformation of the predictor variables is used. Chapter 3 contains information on how to orthogonally design and code the data. A final note is that a MLR prediction interval for a future observaton at some independent level $\underline{x}'_h = (x_{1h}, x_{2h}, ... x_{kh})$ is written as

$$\hat{y}_{\underline{x}_h} \pm t(1 - \alpha/2, n - p)\sqrt{\hat{\sigma}^2(1 + \underline{x}'_h(X'X)^{-1}\underline{x}_h)}$$

This interval is greatly increased as values in $(X'X)^{-1}$ increase. To measure the effect of $(X'X)^{-1}$ using a single value, most experimental designers consider the determinant of $(X'X)$ denoted by $|X'X|$. The larger $|X'X|$ the smaller the prediction interval. It can be shown that if vectors of X are orthogonal or close to orthogonal (i.e., low or zero multicollinearity) then $|X'X|$ is large. The D value of a design (to be discussed in detail in Chapter 3) is found using $D = |X'X|$. If a design has the maximum value of $D = |X'X|$ for a specific n and number of levels, that design is referred to as D-optimal.

A mathematical explanation behind maximizing $|X'X|$ is based on the following procedure. Let the element of the ith row and j^{th} column of the matrix $A^{-1} = (X'X)^{-1}$ be $\tilde{a}_{ij}$. Then,

$$\tilde{a}_{ij} = \frac{A_{ji}}{\det A}$$

2-38

where A_{ji} is the cofactor of the element a_{ji} of the matrix $A=X'X$. Thus, to find A^{-1} we replace each element a_{ij} of A by its cofactor A_{ij}, form the transpose of the resulting matrix and divide each element by $|A|$. Therefore, maximizing $|A|$ will minimize the elements of $(X'X)^{-1}$ and thus minimize $\sigma^2(b) = \sigma^2(X'X)^{-1}$.

2.7   Crossvalidation

When conducting SLR or MLR, researchers typically use R^2 as a measure of the strength of the model. It should be noted that the measured R^2 is model and sample specific. It is possible to continue to add variables up to $k = n-1$ where R^2 is monotonically increasing up to 1.0. This results in an overfit condition which inaccurately predicts the strength of the model. A more accurate measure of model strength is to use half of the sample data to build the model and the other half for cross validation. Another strategy is to build the model on sample data and obtain a crossvalidated R^2 from confirmation experiments. To get the crossvalidated estimate of R^2, calculate $e_i = y_i - \hat{y}_i$ for all observations in the crossvalidated sample data, then

$$R^2 = 1 - \frac{\sum e_i^2}{SST_2}$$

where SST_2 is the total sum of squares for the data used for crossvalidation. It is also possible to correlate the y_i with $\hat{y}_i$ from the second data set to obtain $R_{y\hat{y}}$. Squaring $R_{y\hat{y}}$ also results in the crossvalidation R^2.

One way to reduce overfit is to ensure that for any regression model, n is sufficiently large compared to the number of parameters estimated. A good rule of thumb is for the final prediction model to have n be greater than or equal to 10 times the number of parameter estimates.

2.8 Notes of Caution

When using SLR or MLR on "happenstance data", you should be cautious of high correlations because they may be significant due only to a spurious condition (see case 1 in Chapter 1). This problem can be averted by designing proper experiments instead of analyzing non-experimental data. Furthermore, there are times when a researcher has designed an experiment properly but obtains unanticipated results due to improper levels for independent variables. For example, consider the relationship between y and x in figure 2.8. If the levels of x are a and d, you will find a significant linear relationship between y and x, as you should. However, if the range on x is b to c, this relationship will not be discovered.

figure 2.8

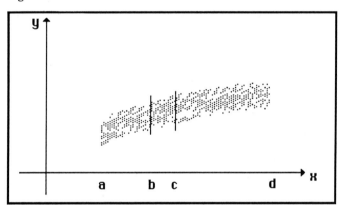

A less likely occurrence, but certainly a potentiality, appears in figure 2.9. Due to the quadratic relationship, if the sampled levels of x are a and b for a SLR, a non-significant linear relationship will be found even though a significant quadratic exists.

figure 2.9

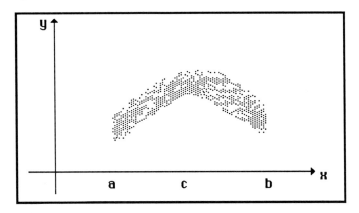

To find a significant linear relationship, the sampled region of x must be [a,c] or [c,d]. Obviously, using a quadratic (or 2nd order model) over [a,b] you could detect the significant quadratic relationship. Thus, experimenters should not only carefully plan which variables to be in a model, but also which levels are necessary to detect anticipated relationships.

Finally, it will be briefly stated that outliers in the data can significantly alter regression results. Before any regression is conducted or any other data analysis takes place, you should explore the data distribution for each variable, identify outliers, ascertain their accuracy and take the appropriate action. You can also use sophisticated techniques to identify outliers in a multivariable setting. [1]

APPENDIX 2A SIMPLE APPROACH TO ANOVA WHEN ALL
FACTORS ARE SET AT 2 LEVELS

Using the data given on page 2-17, you can form a design matrix where each row represents an experimental condition (i.e. a specific combination of input factors). The resulting design matrix with response values is shown below.

RUN	TEMP A	PRESS B	RESPONSE y_1	y_2	$\bar{y}$	s^2
1	70	100	2.2	2.8	2.5	.18
2	70	200	3.2	3.6	3.4	.08
3	90	100	5.0	4.6	4.8	.08
4	90	200	5.8	5.4	5.6	.08

Replacing the low factor settings with a (-1) and the high factor settings with a (+1), the above design matrix with the response values will be:

RUN	A	B	A · B	RESPONSE y_1	y_2	$\bar{y}$	s^2
1	-1	-1	+1	2.2	2.8	2.5	.18
2	-1	+1	-1	3.2	3.6	3.4	.08
3	+1	-1	-1	5.0	4.6	4.8	.08
4	+1	+1	+1	5.8	5.4	5.6	.08

Notice that the A • B column is generated by multiplying the (+1) and (-1) values in columns A and B for each row. This new column represents the 2 way linear interaction between A and B.

To analyze the significance of the A, B, and A • B effects, you need only complete the following table based on the formulas indicated below.

	A	B	A • B
(-1) Level response Avg	2.95	3.65	4.10
(+1) Level response Avg	5.20	4.50	4.05
(+1) Avg - (-1) Avg	2.25	.85	-.05
SS_{effect}	10.125	1.445	.005
F_{effect}	96.428	13.762	.0476

FORMULAS

a) $SS_{effect} = \frac{N}{4}[AVG(+1) - AVG(-1)]^2$ where N is the total number of observations (i.e., the number of runs multiplied by the number of replicates).

EXAMPLE: $SS_A = \frac{8}{4}[5.20 - 2.95]^2 = 10.125$

b) df_{effect} is the number of factor levels minus 1 (which is 1 for all effects in a 2- level factor design).

c) $MS_{effects} = SS_{effects} / df_{effects}$

d) $SS_{error} = \sum_{1}^{\#of\ rows}(n_r - 1)S_r^2$ where n_r is the number of replicated response values in row r and S_r^2 is the variance of those values.

EXAMPLE: $SS_{error} = (2-1)(.18) + (2-1)(.08) + (2-1)(.08) + (2-1)(.08)$
$$= .42$$

e) $df_{error} = \sum(n_r - 1)$

EXAMPLE: $df_{error} = (2-1) + (2-1) + (2-1) + (2-1) = 4$

f) $MS_{error} = SS_{error} / df_{error}$

EXAMPLE: $MS_{error} = \dfrac{.42}{4} = .105$

g) $F_{effect} = MS_{effect} / MS_{error}$

h) $F_c = F(\alpha, df_{effect}, df_{error})$

EXAMPLE: $F(.05, 1, 4) = 7.71$

Using the method just demonstrated, it now becomes a simple task to perform ANOVA for any number of factors provided they all have 2 levels. For example, consider the following 3 way ANOVA.

RUN	A	B	-A•B	C	-A•C	-B•C	A•B•C	y_1 y_2 y_3	$\bar{y}$	s^2
1.	1	1	1	1	1	1	1	5 7 6	6.0	1.00
2.	1	1	1	2	2	2	2	20 21 23	21.33	2.33
3.	1	2	2	1	1	2	2	6 4 5	5.0	1.00
4.	1	2	2	2	2	1	1	23 19 18	20	7.00
5.	2	1	2	1	2	1	2	21 21 20	20.67	.33
6.	2	1	2	2	1	2	1	4 5 6	5.0	1.00
7.	2	2	1	1	2	2	1	24 21 19	21.33	6.33
8.	2	2	1	2	1	1	2	5 8 7	6.67	2.33

	A	B	-A•B	C	-A•C	-B•C	A•B•C	
AVG (-)	13.08	13.25	13.83	13.25	5.66	13.33	13.08	$\bar{y} = 13.25$
AVG (+)	13.42	13.25	12.67	13.25	20.83	13.16	13.42	
Δ = AVG (+) − AVG (-)	.34	0.0	-1.16	0.0	15.17	-.17	.34	
SS effect	.6936	0.0	8.0736	0.0	1380.77	.1734	.6936	
F effect	.260	0.0	3.029	0.0	518.11	.065	.260	

The reason for using the negative of the two-way interaction columns is to comply with the Taguchi design matrix on page 3-25.

$$\text{MS}_{error} = \frac{2(1) + 2(2.33) + 2(1) + 2(7) + 2(.33) + 2(1) + 2(6.33) + 2(2.33)}{2+2+2+2+2+2+2+2} = \frac{42.64}{16} = 2.665$$

Because each factor had 2 levels which were coded (+1) or (-1), the multiple regression approach for building the prediction equation can be simplified as follows:

$$\hat{y} = \overline{Y} + \frac{(\overline{A}_+ - \overline{A}_-)}{2}A + \frac{(\overline{B}_+ - \overline{B}_-)}{2}B + \frac{(\overline{C}_+ - \overline{C}_-)}{2}C$$

$$+\frac{(\overline{AB}_+ - \overline{AB}_-)}{2}(-A \cdot B) + \frac{(\overline{AC}_+ - \overline{AC}_-)}{2}(-A \cdot C) + \frac{(\overline{BC}_+ - \overline{BC}_-)}{2}(-B \cdot C)$$

$$+\frac{(\overline{ABC}_+ - \overline{ABC}_-)}{2}A \cdot B \cdot C$$

However, since most of the effects are not significant, the only terms which should be used in the prediction equation are the grand mean and the A • C interaction. The exception would be if you follow the rule of hierarchy, i.e., if interactions or quadratic terms are significant you must also include the linear terms in the prediction equation. Using the rule of hierarchy, our prediction equation becomes:

$$\hat{y} = 13.25 + .17A + 0 \cdot C + 7.585(-A \cdot C) = 13.25 + .17A - 7.585(A \cdot C)$$

In order to use this equation to predict the response for various factor settings you must ensure the levels for A and C are in the coded form, i.e., somewhere between -1 and +1.

If your objective is to minimize or maximize the response, you can determine the optimum settings by using the prediction equation. For example, consider minimizing the average predicted response, $\hat{y}$. Since the coefficient for A is positive, setting A at (-1) will move $\hat{y}$ in the minimum direction. The negative coefficient of A•C and the previously determined negative setting for A will dictate that factor C be set to (-1). The resulting predicted average response will be:

$$\hat{y} = 13.25 + .17(-1) - 7.585(-1)(-1) = 5.495$$

If your objective is to hit a target value for $\hat{y}$, say 16, then the factor settings are determined as follows.

1) Set $\hat{y}$ = 16 as shown below.

$$16 = 13.25 + .17A - 7.585\ A \cdot C$$

2) Select a coded setting for either A or C based on economics.

e.g., Assume A is twice as expensive to set at (+1) than it is at (-1). Therefore, you would want to set A at (-1) to minimize cost.

3) Solve the equation for C

$$16 = 13.25 + .17(-1) - 7.585(-1)(C)$$

$$C = \frac{2.92}{7.585} = .385$$

4) The true values for A and C are easily decoded using the following equation from page 3-1.

$$f_i = \frac{x_i d_i}{2} + \bar{f}$$

where f_i is the true value for factor i

$\bar{f_i}$ is the true average for factor i

d_i is the range for factor i true values (i.e., $f_{max} - f_{min}$)

x_i is the coded value for factor i

As an example, assume the true low and high values for factor C are 100 and 400. Converting the coded C = .385 the true factor C value becomes:

$$f = \frac{385\,(300)}{2} + 250 = 307.75$$

Graphically, the linear interpolation appears as shown below.

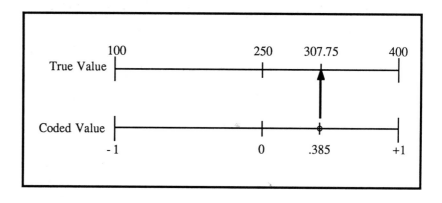

APPENDIX 2B COMPARISON OF CONFIDENCE INTERVAL, t–TEST, ANOVA, AND REGRESSION APPROACHES FOR ANALYZING A FACTOR WITH 2 LEVELS

Given some factor A measured at 2 levels, A_1 and A_2, this section will compare four approaches which could be used to test the following hypotheses: $H_0 : \mu_1 = \mu_2$ versus $H_1 : \mu_1 \neq \mu_2$

The following sample data will be used to make the comparison.

SAMPLE DATA			
	A_1	A_2	
	4	14	
	10	8	
	2	16	
	12	6	
Mean $\left(\bar{A}_j\right)$	7	11	Grand Mean, $\bar{Y} = 9$
Variance $\left(s_j^2\right)$	22.667	22.667	
Sample size $\left(n_j\right)$	4	4	

For the sake of this comparison α will be set to .05.

1. Confidence Interval Approach

(1-α) 100% confidence interval for $\mu_2 - \mu_1$ is:

$$\left(\bar{A}_2 - \bar{A}_1\right) \pm t\left(\tfrac{\alpha}{2}, n_1 + n_2 - 2\right) \sqrt{MSE\left(\tfrac{1}{n_1} + \tfrac{1}{n_2}\right)}$$

where $MSE = \dfrac{\sum(n_j - 1) s_j^2}{\sum(n_j - 1)}$. The sample data results are:

$$\left[(11 - 7) \pm t(.025, 6) \sqrt{22.667\left(\frac{1}{4} + \frac{1}{4}\right)} \right]$$

$$= [(4) \pm (2.447)(3.366)]$$

$$= [-4.236, 12.236]$$

Since the 95% confidence interval for $\mu_2 - \mu_1$ contains zero, your conclusion will be to fail to reject H_0.

2. t-Test Approach

The 5 step method for the t-test is:

(1). $H_0 : \mu_1 = \mu_2$

   $H_1 : \mu_1 \neq \mu_2$

(2). $\alpha = .05$

(3). $t_0 = \dfrac{\bar{A}_2 - \bar{A}_1}{\sqrt{MSE\left(\dfrac{1}{n_1} + \dfrac{1}{n_2}\right)}} = \dfrac{4}{\sqrt{22.667\left(\dfrac{1}{4} + \dfrac{1}{4}\right)}} = 1.188$

(4). $t_{CRITICAL} = t\left(\dfrac{\alpha}{2}, n_1 + n_2 - 2\right) = t(.025, 6) = 2.447$

(5). Since $[|t_0| = 1.188]$ is less than $[t_{CRITICAL} = 2.447]$ your conclusion will be to fail to reject H_0.

2-50

3. Analysis of Variance (ANOVA) Approach

The 5 step method for ANOVA is:

(1). $H_0 : \mu_1 - \mu_2 = 0$
$\quad\;\; H_1 : \mu_1 - \mu_2 \neq 0$

(2). $\alpha = .05$

(3). $F_0 = \dfrac{MSB}{MSE} = \dfrac{32}{22.667} = 1.412$

```
                    ANOVA TABLE

    SOURCE    SS     df      MS        F
    BETWEEN   32      1       32      1.412
    ERROR     136     6     22.667

    TOTAL     168     7
```

	ANOVA TABLE			
SOURCE	SS	df	MS	F
BETWEEN	32	1	32	1.412
ERROR	136	6	22.667	
TOTAL	168	7		

(4). $F_{CRITICAL} = F(.05, 1, 6) = 5.99$

(5). Since $[F_0 = 1.412]$ is less than $[F_{CRITICAL} = 5.99]$ your conclusion will be to fail to reject H_0.

Before completing a regression approach to this problem it is worth reviewing the following:

- All previous conclusions were the same
- $t_0^2 = F_0$
- $\left[t\left(\frac{\alpha}{2}, 6\right) \right]^2 = F(\alpha, 1, 6)$

4. Regression Approach

To solve a two group problem by way of regression you must form a group membership column. Although there are several ways to code this column, the most common would be using effect coding as shown below.

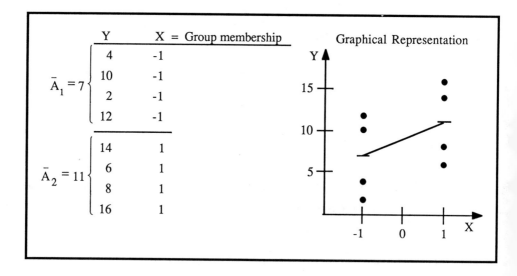

	Y	X = Group membership
	4	-1
$\bar{A}_1 = 7$	10	-1
	2	-1
	12	-1
	14	1
$\bar{A}_2 = 11$	6	1
	8	1
	16	1

The regression equation is built based on the slope intercept equation, i.e.,

$$\hat{y} = \text{intercept} + \text{slope } X = b_0 + b_1 X$$

$$= 9 + \frac{11 - 7}{2} X$$

$$= 9 + 2 X$$

Using the formula on page 2-24, R^2 is calculated to be .1904.

The 5 step hypothesis test would be:

1a) $H_0 : R^2 = 0$ or 1b) $H_0 : B_1 = 0$

 $H_1 : R^2 \neq 0$ $H_1 : B_1 \neq 0$

In either case, H_0 implies no group differences, i.e., $\mu_1 = \mu_2$.

2) $\alpha = .05$

3a) To test $H_0 : R^2 = 0$,

$$F_0 = \frac{\dfrac{R^2}{1}}{\dfrac{(1 - R^2)}{(N - 2)}} = \frac{.1904}{\dfrac{(1 - .1904)}{(8 - 2)}} = 1.412$$

4a) $F_{CRITICAL} = F(.05, 1, 6) = 5.99$

5a) Since $F_0 < F_{CRITICAL}$, you fail to reject H_0.

3b) To test $H_0 : B_1 = 0$

$$t_0 = \frac{b_1}{\sqrt{\dfrac{MSE}{\left(\sum x^2 - n\,\bar{x}^2\right)}}} = \frac{2}{\sqrt{\dfrac{22.667}{8}}} = 1.188$$

4b) $t_{CRITICAL} = t(.025, 6) = 2.447$

5b) Since $\left|t_0\right| < t_{CRITICAL}$, you fail to reject H_0.

The conclusion is that all 4 methods produced identical results. In addition, the t_0 values were all the same and equal to $\sqrt{F_0}$. Likewise, the $t_{CRITICAL}$ values were all the same and equal to $\sqrt{F_{CRITICAL}}$.

Chapter 2 Problem Set

1. Given 2 different soft bake methods for a photoresist process, determine if the mean critical dimension in microns for the feature of interest is the same for each process. (Conduct this test using Confidence Intervals, t-test and ANOVA)

Setup #	Convection Bake	Hot Plate Bake
1	4.1	3.8
2	3.9	3.5
3	4.3	4.1
4	4.1	4.1
5	4.1	3.8

All data is in microns, and the exposure was constant in all cases.

2. A precision bearing manufacturer who wants to determine a minimum sample size to check the accuracy of two groups of bearings manufactured 2 months apart. The manufacturer wants to insure that he can detect a specific $H_1: \mu_1 - \mu_2 = 10$ microns with 90% confidence given: $\sigma^2 = 81$ and $\alpha = 0.05$.

What is the minimum sample size ?

3a) Given the following etch rate data as a function of DC bias voltage, determine which levels of DC bias voltage are significantly different.

Level
A1 = –220 Volts
A2 = –240 Volts
A3 = –260 Volts
A4 = –280 Volts
A5 = –300 Volts

Etch Rate Data: Angstroms/Min.

A1	A2	A3	A4	A5
====	====	====	====	====
1280	1285	1301	1345	1401
1277	1296	1310	1350	1390
1275	1291	1298	1340	1395
1282	1287	1316	1356	1402
1272	1294	1300	1351	1391
1289	1281	1285	1360	1405
1271	1285	1305	1347	1406

alpha = 0.01

b). If the objective is to maximize the etch rate, which level(s) should be selected ?

4. A manufacturer of a proprietary plating additive wants to understand the reaction kinetics of the additive. The manufacturer assumes that the following reaction represents his process:

Concentration of Additive C_A —> Rate of Reaction $-r_A$.

Given this Data:

n	1	2	3	4	5
C_A	5.6	9.2	15.6	22.5	26.2
$-r_A$	1.5	3.2	7.3	13.9	22

a). Calculate the slope and intercept of the linear model between C_A and $-r_A$ where $-r_A$ is the dependent variable.

b). Find the correlation coefficient, R and the proportion of variability explained by the regression.

c). Graph C_A vs. $-r_A$. Is a linear relationship a good assumption ?

d). The manufacturer now thinks the relationship is:

$$-rA = k\, C_A^N, \quad k = \text{rate constant}$$
$$N = \text{order of reaction}$$

Taking the natural log of this equation:

$$\ln(-r_A) = \ln(k) + N * \ln(C_A)$$

Which is a linear model. Take the natural log of the raw data and repeat a) and b) to find the order of the reaction, N.

5. A graduate student was asked to determine if a relationship existed between sulfuric acid concentration (wt. %) and temperature (deg. C), and their effect on a 1st order rate constant (k) with respect to the hydrolysis of cellulose.

   a). Using the data he derived, generate a ANOVA table.
   b). Is there an interaction between Wt.% H_2SO_4 and temperature ?
   c). Is there a significant effect by temperature or Wt.% H_2SO_4 ?

Data:

T (deg C)	Wt.% H_2SO_4	k1	k2
170	0.5	0.0196	0.0201
190	0.5	0.0376	0.0355
210	0.5	0.871	0.900
170	1.0	0.0808	0.0799
190	1.0	0.166	0.150
210	1.0	0.742	0.721
170	2.0	0.0736	0.0691
190	2.0	0.163	0.170
210	2.0	0.448	0.435

6. Given the following data:

Run	A	B	–AB	C	–AC	D	E	Y_1	Y_2	$\bar{Y}$	S
1	–1	–1	–1	–1	–1	–1	–1	66	62	64	2.83
2	–1	–1	–1	+1	+1	+1	+1	68	63	65.5	3.54
3	–1	+1	+1	–1	–1	+1	+1	88	80	84	5.66
4	–1	+1	+1	+1	+1	–1	–1	63	65	64	1.41
5	+1	–1	+1	–1	+1	–1	+1	73	71	72	1.41
6	+1	–1	+1	+1	–1	+1	–1	37	42	39.5	3.54
7	+1	+1	–1	–1	+1	+1	–1	38	39	38.5	0.71
8	+1	+1	–1	+1	–1	–1	+1	57	48	52.5	6.36

2-58

a). Complete the following table:

	A	B	–AB	C	–AC	D	E
Avg –							
Avg +							
delta							
delta/2							
SS							
F							

b) Complete a standard ANOVA table.

c) Determine the prediction equation: $\hat{Y}$

d) What are the optimal settings to minimize $\hat{Y}$?

e) What is the prediction response for the settings in c) above for $\hat{Y}$?

7. If only one variable is being tested at 2 levels, is there any difference between using ANOVA or a t-test ?

8. What is effect of using pairwise t-test if multiple μ_i are found not to be equal ? Why is it bad to increase α_{EW} ?

9. What does R^2 from a regression model measure ?

10. $\left[t\left(\dfrac{\alpha}{2}, n \right) \right]^2 = F\,(\underline{\quad}\,,\,\underline{\quad}\,,\,\underline{\quad})$

11. What is an example of a blocking variable ? How can an experimenter deal with a blocking variable ?

12. What is a prediction interval and how would you interpret it ?

Chapter 2 Bibliography

1. Myers, Raymond H. (1986) *Classical and Modern Regression with Applications.* Duxbury Press, Boston, Mass.

2. Walpole, R.E. and Myers, R.H. (1985) *Probability and Statistics for Engineers and Scientists,* MacMillian, Inc. New York.

CHAPTER 3: DESIGN TYPES

3.1 Introduction

A design matrix displays the level of each factor (independent variable) at which each experimental run is made. For example, table 3.1 contains 3 factors and 4 runs where the first run is made with factors A, B and C set at values of 20, 800 and 50, respectively.

table 3.1		factor	
Run	A	B	C
1	20	800	50
2	20	600	30
3	10	800	30
4	10	600	50

For convenience, researchers will typically recode the factor values using the transformation

$$x_i = 2\frac{f_i - \overline{f}_i}{d_i}$$

where f_i is the actual reading for factor i

$\overline{f}_i$ is the mean value for factor i

d_i is the distance between the largest and smallest values of factor i.

Since $\overline{f}_i$ and d_i are known, it is easy to untransform x_i back to f_i whenever necessary. Transforming f_i values in table 3.1 to x_i yields the design matrix shown in table 3.2a. Researchers will sometimes abbreviate "1" and "-1" with "+" and "-" as shown in table 3.2b.

table 3.2								
(a)		factors			(b)		factors	
Run	A	B	C		Run	A	B	C
1	1	1	1		1	+	+	+
2	1	-1	-1		2	+	-	-
3	-1	1	-1		3	-	+	-
4	-1	-1	1		4	-	-	+

Each column of factor values is referred to as a vector in the design matrix.

A design is considered to be **balanced** when the transformed values, x_i, for each factor sum to zero, i.e. $\sum_{u=1}^{n} x_{iu} = 0$ for all factors i. Balanced designs are desirable because they simplify the calculations during analysis and under certain conditions they lend themselves to orthogonal designs. A design is said to be **orthogonal** if the sum of every possible dot product for all possible variable pairs is zero, i.e. $\sum_{u=1}^{n} x_{iu} x_{ju} = 0$ for all combinations of factors i and j, i≠j. (Notice that the design in table 3.2 is balanced and orthogonal.) The reason for orthogonality is that it allows the desired effects to be measured independently of each other. Non-orthogonal designs result in a dependency of effects which is particularly undesirable. For example, if effects A and B are dependent and only A is important, B may appear important due to the dependency with A. Even worse, if A and B are both important but one in a positive direction and the other in a negative direction, their dependency may cause both to appear unimportant.

Designs can also be balanced and non-orthogonal as shown in table 3.3. In this case, factors B and C have the same set of values and are referred to as **aliased**. Since not all factor dot products in table 3.3 sum to zero, the design is non-orthogonal. However, it is not necessary to have identical vectors in order for a design to be non-orthogonal. For example, the design in table 3.4 is a balanced, non-orthogonal design without any two vectors alike.

table 3.3	factors		
Run	A	B	C
1	1	1	1
2	1	-1	-1
3	-1	1	1
4	-1	-1	-1

table 3.4	factors			
Run	A	B	C	D
1	1	-1	-1	-1
2	-1	1	-1	-1
3	-1	-1	1	-1
4	-1	-1	-1	1
5	1	1	1	1
6	-1	1	1	1
7	1	-1	1	1
8	1	1	-1	1
9	1	1	1	-1
10	-1	-1	-1	-1

In this design, the sum of each possible dot product is not equal to zero, and factors are not aliased; instead, all factors are intercorrelated with each other (i.e., aliased factors have an intercorrelation of 1 or -1 which can be referred to as perfect positive or negative confounding; intercorrelations $\neq 0$ and $\neq \pm 1$ represent partial confounding).

The designs in tables 3.1 through 3.4 all represent factors measured at only two levels. For that reason, they are typically referred to as **2 level designs** which are particularly useful for factor screening and in estimating first order models with or without interactions. The inclusion of a few center points in a 2-level design will create a third level to allow for curvature analysis. However, there exist several different types of 3 level designs, some of which are discussed in detail later in this chapter. For now, an example of a balanced orthogonal 3 level design is shown in table 3.5.

table 3.5		factors			
Run	A	B	C	D
1	1	1	1	1
2	1	0	0	0
3	1	-1	-1	-1
4	0	1	0	-1
5	0	0	-1	1
6	0	-1	1	0
7	-1	1	-1	0
8	-1	0	1	-1
9	-1	-1	0	1

3.2 Full Factorial Designs

Given that the experimenter desires to examine 3 factors (A, B and C) each at two levels, the design used to estimate all possible effects would be a full factorial design. The set of possible effects are described in table 3.6

table 3.6		
main effects	2 way interactions	3 way interactions
A	A x B	A x B x C
B	A x C	
C	B x C	

In this case, a total of 7 effects can be analyzed. The minimum number of runs in any orthogonal design is equal to the number of effects to be analyzed plus 1 (the extra run allows for estimation of the population mean response). Therefore, when using 2 level designs and estimating all possible interactions, 2^k runs are required (k = number of factors). The design matrix for a 2^3 full factorial is shown in table 3.7. Notice that all effects are balanced and mutually orthogonal. The experimenter need only set factors A, B and C at the various levels shown in columns A, B and C; the remaining columns are crossproducts generated during the analysis phase.

table 3.7							
Run	A	B	C	AB	AC	BC	ABC
1	1	1	1	1	1	1	1
2	1	1	-1	1	-1	-1	-1
3	1	-1	1	-1	1	-1	-1
4	1	-1	-1	-1	-1	1	1
5	-1	1	1	-1	-1	1	-1
6	-1	1	-1	-1	1	-1	1
7	-1	-1	1	1	-1	-1	1
8	-1	-1	-1	1	1	1	-1

Graphically, these runs appear as shown in figure 3.1.

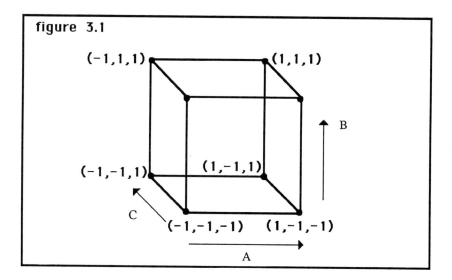

figure 3.1

For 3 factors, each with 3 levels, a full factorial of 27 runs is displayed in table 3.8. It can easily be verified that the design contains effects that are all balanced and

3-5

mutually orthogonal. A visual check for orthogonality of the design in table 3.8 would be to verify that every level of any one factor contains an equal number of runs for all 3 levels of any other factor.

table 3.8

	factors				factors				factors		
Run	A	B	C	Run	A	B	C	Run	A	B	C
1	1	1	1	10	0	1	1	19	-1	1	1
2	1	1	0	11	0	1	0	20	-1	1	0
3	1	1	-1	12	0	1	-1	21	-1	1	-1
4	1	0	1	13	0	0	1	22	-1	0	1
5	1	0	0	14	0	0	0	23	-1	0	0
6	1	0	-1	15	0	0	-1	24	-1	0	-1
7	1	-1	1	16	0	-1	1	25	-1	-1	1
8	1	-1	0	17	0	-1	0	26	-1	-1	0
9	1	-1	-1	18	0	-1	-1	27	-1	-1	1

This 3^3 full factorial is graphically displayed in figure 3.2.

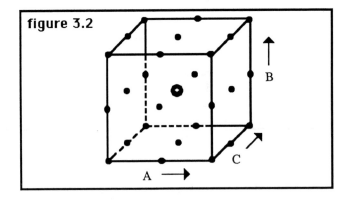

figure 3.2

3.3 Fractional Factorials for 2-Level Designs

Often a researcher does not have the time, funds or supplies to complete a full factorial design. Obviously it doesn't take too many factors, k, before 2^k or 3^k runs are infeasible. A type of design requiring fewer runs and which is balanced and orthogonal is a **fractional factorial**. Given any number of factors, you can build a type of 2 level fractional factorial design with 2^{k-q} runs by completing the following:

(1) Write the full factorial design for k-q variables. (q is some positive integer < k)

(2) Associate the extra factors with the higher order interactions which must be assumed to be nonsignificant.

Consider the 2^{4-1} design in table 3.9 as a simple example. The full factorial for k = 4

table 3.9		factors		
Run	A	B	C	D = ABC
1	1	1	1	1
2	1	1	-1	-1
3	1	-1	1	-1
4	1	-1	-1	1
5	-1	1	1	-1
6	-1	1	-1	1
7	-1	-1	1	1
8	-1	-1	-1	-1

implies $2^4 = 16$ full factorial runs. If we can assume an insignificant ABC interaction (the column of signs for ABC is found by multiplying together the respective row coefficients for A, B and C), then a fractional factorial can be accomplished in $2^{4-1} = 8$ runs, referred to as a half fraction. Since ABC is used to generate the D main effect, the defining relation for this design is found by multiplying both sides of D = ABC by D such that $D^2 = ABCD$. It is obvious that in 2-level designs if any factor is multiplied by itself the result is an identity vector, I, which is a column of 1's. (Verify this using table 3.9). Therefore, I=ABCD which is called the **defining**

relation. Since there is only one defining relation for this design, the alias pattern is easily constructed by multiplying both sides of the defining relation by the desired effect. The right side then becomes the alias of the effect on the left. See table 3.10.

table 3.10	
effect	alias
A	BCD
B	ACD
C	ABD
D	ABC
AB	CD
AC	BD
AD	BC
BC	AD
BD	AC
CD	AB
ABC	D
ABD	C
ACD	B
BCD	A
ABCD	I

In the event you want to de-alias the above design with respect to the ABC interaction, you need only run the other half fraction using the negative of the generator ABC = D, i.e., build another 8 runs with factors A, B and C and let $D = {}^-ABC$. Then augment these 8 runs with the previous 8 runs to make a 2^4 full factorial. The D factor will be de-aliased from the ABC interaction; however, the ABC interaction will be aliased with the potential block effect.

A more complex example would be to have 7 factors each at 2 levels. Not wanting $2^7 = 128$ runs, you choose to fractionate the design so that n = 16. The first step is to build a $2^{7-3} = 2^4$ full factorial as shown in table 3.11 where higher order interactions are aliased with E, F and G main effects.

table 3.11				factors			
Run	A	B	C	D	E=ABC	F=BCD	G=ABD
1	+	+	+	+	+	+	+
2	+	+	+	-	+	-	-
3	+	+	-	+	-	-	+
4	+	+	-	-	-	+	-
5	+	-	+	+	-	-	-
6	+	-	+	-	-	+	+
7	+	-	-	+	+	+	-
8	+	-	-	-	+	-	+
9	-	+	+	+	-	+	-
10	-	+	+	-	-	-	+
11	-	+	-	+	+	-	-
12	-	+	-	-	+	+	+
13	-	-	+	+	+	-	+
14	-	-	+	-	+	+	-
15	-	-	-	+	-	+	+
16	-	-	-	-	-	-	-

The 1/8 fraction for k = 7 contains the generators E = ABC, F = BCD and G = ABD which result in the following defining words: I = ABCE = BCDF = ABDG. Multiplying any 2 defining words at a time results in I^2 = I = ADEF = CDEG = ACFG and multiplying all 3 at one time yeilds I^3 = I = BEFG. Therefore, the complete defining relation consists of

I = ABCE = BCDF = ABDG = ADEF = CDEG = ACFG = BEFG.

This defining relation will allow the easy computation of the aliasing pattern for the entire design. For example, if you want to determine all effects aliased with factor A, just multiply the defining relation by A which then appears as

A = BCE = ABCDF = BDG = DEF = ACDEG = CFG = ABEFG.

Similarly, the aliasing pattern for EF is

EF = ABCF = BCDE = ABDEFG = AD = CDFG = ACEG = BG.

In all, there exist 8 different 1/8 fractional designs, represented by "+" or "-" the previous set of generators, i.e.,

E = ± ABC, F = ± BCD and G = ± ABD.

3-9

To de-alias any particular factor from its 3-way interaction generator, simply add a different 1/8 fraction to the first one. For example, to de-alias factor E from ABC, take two 1/8 fractions, one with E = ABC and the other E = –ABC.

At this time the reader is ready to understand the term **resolution** of a design. In general, the resolution of a 2-level design is the length of the shortest word in the defining relation. The previous 2^{7-4} fractional factorial has a resolution of IV. The meaning of different resolution levels is as follows:

R_V: a design that does not alias the main effects with each other or with 2-way interactions. In addition, 2-way interactions are not aliased with one another. Thus, this design provides for unconfounded main and 2-way interaction effects.

R_{IV}: a design which does not alias mains with other 2-way interactions but does alias 2-way interactions with other 2-way interactions.

R_{III}: a design which does not alias main effects with one another but does alias mains with 2-way interactions.

R_{II}: a design which contains main effects aliased with other main effects.

A R_{II} design is referred to as **supersaturated** and is not recommended unless n has to be less than k. When n is approximately equal to k and 2-way interactions are assumed insignificant, a R_{III} design which is **saturated** is used. If 2-way interactions are of interest, **unsaturated** designs of R_{IV} or R_V are suggested. In the event you need a R_{IV} design with minimum n you can fold over any R_{III} design. A **foldover** design of resolution IV consists of two R_{III} designs augmented where the second R_{III} design has all the signs "+" or "-" reversed. The new design will have twice as many runs as the original R_{III}. An additional factor can be added by placing a column of "+" values for the first R_{III} runs and a column of "-" values for the second R_{III} runs. As an example, consider the 2^{3-1} design of resolution III in table 3.12.

table 3.12

factors

Runs	A	B	C = A · B
1	+	+	+
2	+	-	-
3	-	+	-
4	-	-	+

The defining relation is I = ABC. By folding over this R_{III} design, you obtain the design in table 3.13.

table 3.13

factors

Run	A	B	C	D = new factor
1	+	+	+	+
2	+	-	-	+
3	-	+	-	+
4	-	-	+	+
5	-	-	-	-
6	-	+	+	-
7	+	-	+	-
8	+	+	-	-

For the first set of 4 runs D = ABC and for the second set of 4 runs –D = –ABC. Therefore, both sets produce I = ABCD which is the only defining relation and the resulting design resolution is R_{IV}. Notice that factor D could also be used as a blocking variable to account for different days, weeks, operators, shifts, etc. for the different sets of 4 runs.

In the very recent past, classical statistics and short courses typically spent chapters on the concept of alias patterns, fractionalization, and resolution. With the advent of powerful software, computers can quickly determine what the student was once asked to spend many hours on. Since this is the case, we shall forge on into additional useful design matrices.

3-11

3.4 Plackett-Burman Designs

The disadvantage of a fractional factorial design is that the number of runs is a function of powers of 2, i.e., n = 4, 8, 16, 32, 64, 128 .... A balanced, 2-level, orthogonal design which does not have this problem is the Plackett-Burman (P-B). The Plackett-Burman design matrix originates from what is known as a Hadamard matrix. Each element is 1 or -1 and all $n-1$ columns are balanced and pairwise orthogonal (see John [5]). The number of runs for Plackett-Burman designs is a multiple of 4, i.e., n = 4, 8, 12, 16, 20 ... . These designs were developed for evaluating up to $n-1$ main effects with few or no interactions of interest. When $n = 2^{k-q}$, the design is referred to as a geometric P-B design which is the same as the 2 level fractional factorial discussed previously. All other 2 level P-B designs, such as n = 12, 20, 24, 28, 36, 40 etc, are non-geometric and have a different confounding structure of the mains and 2-way interactions. For geometric P-B designs of R_{III}, it can be shown that each 2-way interaction will be positively or negatively aliased with a main effect (see table 3.14). However, the non-geometric P-B designs contain main effects which are partially confounded (intercorrelated) with 2-way interactions (see table 3.15). In fact, each 2-way interaction is confounded with each of the other main effects (i.e., the BC interaction of table 3.15 is intercorrelated with the A, D, E, F and G main effects). The biggest disadvantage of these non-geometric designs is that the amount of intercorrelation is not easily determined. The simplest way to obtain the intercorrelations of effects is to use the correlation coefficient discussed in Chapter 2 for each pair of design vectors.

To build a Plackett-Burman design, take a single vector for a particular size n design and generate the remaining vectors. For example, if n = 8, the generating vector consists of the following $n-1$ values (+ + + − + − −). The design matrix is completed as follows: (1) use the generating vector as column A, (2) build column B by making the last value in A the first value in B, then add the rest of A below that value, (3) vector C is made by using the last value in B as the first value in C with the rest of B below it, (4) continue until all k columns are complete, and (5) add the nth row as all "−" values as shown in table 3.14. For more information on P-B designs, see reference article by Plackett and Burman [8].

table 3.14			factors				
Run	A	B	C	D	E	F	G
1	+	-	-	+	-	+	+
2	+	+	-	-	+	-	+
3	+	+	+	-	-	+	-
4	-	+	+	+	-	-	+
5	+	-	+	+	+	-	-
6	-	+	-	+	+	+	-
7	-	-	+	-	+	+	+
8	-	-	-	-	-	-	-

For n = 12, the design matrix appears as shown in table 3.15.

table 3.15				factors							
Run	A	B	C	D	E	F	G	H	I	J	K
1	+	-	+	-	-	-	+	+	+	-	+
2	+	+	-	+	-	-	-	+	+	+	-
3	-	+	+	-	+	-	-	-	+	+	+
4	+	-	+	+	-	+	-	-	-	+	+
5	+	+	-	+	+	-	+	-	-	-	+
6	+	+	+	-	+	+	-	+	-	-	-
7	-	+	+	+	-	+	+	-	+	-	-
8	-	-	+	+	+	-	+	+	-	+	-
9	-	-	-	+	+	+	-	+	+	-	+
10	+	-	-	-	+	+	+	-	+	+	-
11	-	+	-	-	-	+	+	+	-	+	+
12	-	-	-	-	-	-	-	-	-	-	-

The entire effect confounding pattern can be found by calculating all pairwise intercorrelations.

3.5 Non-Orthogonal 2-Level Designs

There are also three balanced non-orthogonal 2-level designs worthy of discussion; the random balanced, foldover Koshal, and D-optimal designs. The **random balanced** design is typically used as a supersaturated or saturated design and is formed by randomly generating the first $n/2$ runs for k factors. In order to balance the design, you fold it over, generating an equal number of "+" and "-" values in each column. In this case, n can be determined independently of k, which can be advantageous. The primary disadvantage is that the confounding structure for main effects and interactions is totally random and outside of the experimenter's control. However, a design can be generated and intercorrelations calculated for all possible pairs of effects. If any pairs are too highly intercorrelated, they can be avoided or the design can be regenerated randomly until satisfactory levels of intercorrelation are achieved.

A balanced non-orthogonal design which is closely related to the one–at–a–time design is a **Foldover Koshal**. Basically, the Foldover Koshal is a one–at–a–time design for 2 levels which is folded over to achieve balance. Except for $n = 8$, the design is far from orthogonal, which is why simulation studies indicate it is only about 60% effective in discriminating significant main and 2-way interaction effects when the number of factors $k < 20$. Intercorrelations will remain below the .6 level provided $k \leq 20$. Due to increased main effect intercorrelations as k increases, the effectiveness deteriorates for larger k. There are two reasons for including this design in this book: 1) some consultants are using it unwisely even though better alternatives exist and 2) if a one-at-a-time or Koshal design has already been run as shown in table 3.16, then folding over the design results in the Foldover Koshal of table 3.17, which is an attempt to salvage a bad design. One-at-a-time designs and Foldover Koshals are strongly discouraged due to their inefficiency, lack of orthogonality and lack of robustness to interactions.

table 3.16				
	factors			
Run	A	B	C	D
1	+	-	-	-
2	-	+	-	-
3	-	-	+	-
4	-	-	-	+
5	-	-	-	-

table 3.17				
	factors			
Run	A	B	C	D
1	+	-	-	-
2	-	+	-	-
3	-	-	+	-
4	-	-	-	+
5	-	-	-	-
6	-	+	+	+
7	+	-	+	+
8	+	+	-	+
9	+	+	+	-
10	+	+	+	+

The last and most important non-orthogonal design to be discussed is the **D–optimal** design. If we were to call our design matrix X, then premultiplying X with its transpose is X'X. D-optimality refers to a design that maximizes the determinant of X'X which is called the D value, i.e., $D = |X'X|$. Optimal values of D vary with n (the number of runs) and for some n values, D-optimal designs will be orthogonal. When n is other than a value associated with orthogonal designs, the columns of "+" and "–" values must be generated so as to maximize $|X'X|$. Good software packages such as RS/Discover, sold by BBN Software Products

Corporation, will do this for you, i.e., generate a D–optimal or close to D–optimal design for any n. It should be noted that less than orthogonal designs ought to be avoided when possible. The lack of orthogonality will confound effects and complicate the analysis. However, in situations where it is too costly or impossible to generate a fractional design which is orthogonal, the D–optimal alternative is very helpful. Consider the following three examples where the D–optimal design is useful. In example one, assume we are interested in 4 factors at 2 levels with all 2-way interactions and we want the number of runs to be 11. First of all, the odd number of runs will prevent the design from being balanced or orthogonal. The D–optimal design will generate 11 runs where the corresponding |X'X| is maximized.

As a second example, consider a design with 5 factors A, B, C, D and E. Factor A is qualitative with 3 levels, B is qualitative with 4 levels, C is qualitative with 2 levels, D and E are quantitative with 3 levels. Assume the C x D and C x E interactions are also of interest and the number of runs must be less than 25. In this case, standard designs will not be of much assistance. However, the D–optimal alternative will provide a design with |X'X| close, if not equal, to its maximum for some specific n < 25.

The last example consists of 4 quantitative factors each at 3 levels where the (+,+,+,+) combination is infeasible. Using the D–optimal option of the appropriate software package, you can generate a near orthogonal design excluding the infeasible point.

3.6 Fractional Factorials and Latin Squares for 3-Level Designs

Fractional factorials in designs of 3 levels, 3^{k-q} designs, are a bit more complicated. They are designs which include a mixture of corner points and mid-level points. Because of this combination of experimental points, these designs can be used to test all linear effects, all 2nd order (quadratic) effects, and some or all linear and quadratic interactions. This is the only type of 3 level design discussed in this book which can handle quadratic interactions. An example of a quadratic interaction term is $x_1 x_2^2$, which was discussed in Chapter 2. The only drawback of the 3^{k-q} design is the number of runs required to estimate all possible effects. Even if quadratic interactions and simple two-way interactions are not estimated, n must still be large to estimate 2nd order terms for all factors. The aliasing pattern for a 3^{k-q} design is difficult to obtain and will not be discussed in this book. Several software packages will do this for you, else you can use reference [7] or [1].

One method of formulating a 3^{k-1} fractional design or 1/3 fraction is through the use of Latin Squares. For example, given the Latin Square design in table 3.18, a list of all combinations of the 3 factors results in a balanced orthogonal 3^{k-1} design shown in table 3.19.

table 3.18		factor B			
		-1	0	1	
	-1	-1	0	1	
factor C	0	0	1	-1	factor A
	1	1	-1	0	

table 3.19		factors	
Run	A	B	C
1	-	-	-
2	0	0	-
3	+	+	-
4	0	-	0
5	+	0	0
6	-	+	0
7	+	-	+
8	-	0	+
9	0	+	+

When k = 4, a Latin cube is constructed to build a 3^{4-1} design and Latin hypercubes are the basis for designs with k > 4. As with the use of regular Latin Square designs, interactions of the treatment and blocking factors are not assumed to be significant and cannot be estimated.

3.7 Box-Behnken Designs

An efficient and frequently used 3 level design is the **Box–Behnken** (B–B). Through the use of embedded 2^k designs while holding certain factors at their centerpoint, the B–B designs are much more efficient than 3^k full factorials and are easily blocked in order to compensate for some uncontrolled factor. In their 1960 article, Box and Behnken [2] provide tabled B–B designs for k up to 16. When k = 4, the design appears as shown in table 3.20.

table 3.20

FACTORS

A	B	C	D
±1	±1	0	0
0	0	±1	±1
0	0	0	0
±1	0	0	±1
0	±1	±1	0
0	0	0	0
±1	0	±1	0
0	±1	0	±1
0	0	0	0

This design is divided into 3 orthogonal blocks of 9 runs where each ±1 refers to alternating columns of (+ + – –) and (+ – + –). The B–B design does not contain any corner points in the design space, which may or may not be of concern. In some experiments where corner points are infeasible, the B–B may be the only alternative. If only 3 factors were used, the design would appear graphically as shown in figure 3.3. The B–B design does allow for estimating main effects, quadratic effects and simple interactions, but cannot estimate quadratic interactions. The primary disadvantage of B–B designs is that the number of runs is large enough to estimate all factor 2nd order effects and all 2-way interactions, whether you want to or not. Typically, the B–B design will be less efficient with regard to the number of runs than the 3^{k-q} designs unless several 2-way interactions are of interest.

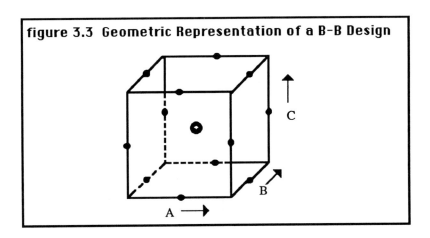

figure 3.3 Geometric Representation of a B-B Design

3.8   Box-Wilson (Central Composite) Designs

A 5-level design that focuses mostly on corner points is the Box–Wilson or Central Composite Design (CCD). For 3 factors, the CCD appears as shown in tables 3.21 a and b.

table 3.21

(a) CCD k=3, using 2^3

Run	A	B	C	
1	+	+	+	
2	+	+	-	
3	+	-	+	
4	+	-	-	F
5	-	+	+	
6	-	+	-	
7	-	-	+	
8	-	-	-	
9	α	0	0	
10	$-\alpha$	0	0	
11	0	α	0	
12	0	$-\alpha$	0	A
13	0	0	α	
14	0	0	$-\alpha$	
15	0	0	0	C

(b) CCD k=3, using 2^{3-1}

Run	A	B	C	
1	+	+	+	
2	+	-	-	
3	-	+	-	F
4	-	-	+	
5	α	0	0	
6	$-\alpha$	0	0	
7	0	α	0	
8	0	$-\alpha$	0	A
9	0	0	α	
10	0	0	$-\alpha$	
11	0	0	0	C

Graphically, the design from table 3.22a will appear as indicated in figure 3.4.

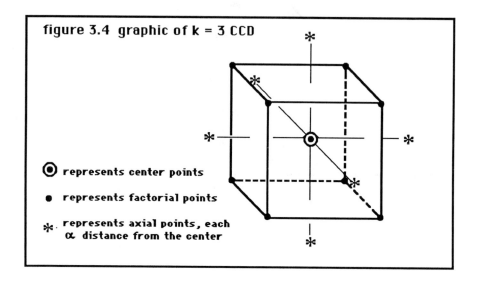

figure 3.4 graphic of k = 3 CCD

⊙ represents center points

● represents factorial points

⁎ represents axial points, each α distance from the center

If 2-way interactions are of interest, you can use a R_{IV} design in the factorial or F portion of the design; whereas if they are not you can save runs by using a R_{III}. The number of axial points will typically be 2k unless some factors don't lend themselves to such levels. The value for α is usually set at $(n_F)^{\frac{1}{4}}$ where n_F is the number of runs in the factorial portion of the design. These values are chosen to produce a design characteristic of **rotatability** which simply implies that the response is capable of being estimated with equal variance regardless of the direction from the center of the design space. You should notice that, in order to satisfy rotatability, α will take on values greater than 1 which means that ±1 no longer represents the factor min and max. An easy way to resolve the transformation to or from the design notation of 0, ±1, ±α is as follows:

(1) set up a number line as shown below

design values: -α   -1   0   1   α

3-20

(2) place the real factor value min and max (use 100 and 300 as an example) under $-\alpha$ and α, respectively. Place $\frac{max + min}{2}$ under 0. The values under -1 and 1 will be computed as the value under 200 $\pm$ some Δ.

design values:	$-\alpha$	-1	0	1	α
real values:	100	$(200 - \Delta)$	200	$(200 + \Delta)$	300

(3) to find the values for Δ you need to know your α value, (assume it is 1.5) then form the following ratio:

$$\frac{300 - 200}{1.5} = \frac{\Delta}{1}, \text{ therefore, } \Delta = 66.\overline{66}$$

implying $200 + \Delta = 266.\overline{66}$ and $200 - \Delta = 133.\overline{33}$.

The final result appears as follows:

design values:	$-\alpha$	-1	0	1	α
real values:	100	133.33	200	266.67	300

which indicates the real factor values for all 5 possible design levels. The number of center points, n_c, will vary but basically, enough are needed to get a good estimate of pure experimental error. According to John [5], $n_c = 4(\sqrt{n_F + 1}) - 2k$ will produce orthogonality and rotatability. For more information, see Box and Draper [3] or Myers [7]. Depending on which resolution you use in the 2^{k-q} portion of a CCD, it can be up to 47% more efficient in experimental runs than the B–B design [1]. In addition, the CCD is rotatable for $\alpha = (n_F)^{\frac{1}{4}}$ whereas the B–B is only rotatable for specific values of k. The biggest advantage of the CCD will be further discussed in Chapters 4 and 5 on factor screening and optimization.

Basically, the CCD allows you to screen with a 2 level design of R_{III} or R_{IV} and then add center points to check for curvature effects. If there are potential quadratic effects, you can sequentially add axial points for building a 2nd–order model until there no longer exists a quadratic lack of fit.

It is not necessary to always add the axial points. Typically, a fractional factorial of 2 levels is used to generate the vectors representing each effect. Replicated center points are then added to allow for testing curvature. The design may then be analyzed without the axial points. These points are only included when the optimal lies within the sampled region and tests for lack of fit indicate that second order terms are appropriate for the model. Through inclusion of the axial points, a 2nd order model is obtained which is used to locate the optimum.

3.9   Comparison of 3 Level Designs

The information in table 3.22 is provided as a comparison of the number of runs required for the 3^{k-q}, B-B, CCD and the Full Factorial given 3 level designs with various numbers of factors and 2–way interactions of interest. Numbers in parentheses indicate the number of replicated center points included in the design. The 3^{k-q} designs are similar to the Taguchi 3 level designs discussed in section 3.10.

table 3.22

	3^{k-q}	B-B	CCD**	D-Opt	Full Factorial
k = 3 with no interactions	9 (0)	15(3)	20(6) *	7(0)	27(1)
k = 5 with 1 2-way linear interaction	27(0)	46(6)	20(2) *	12(0)	243(1)
k = 7 with 7 2-way linear interactions	81(0)	62(6)	33(3) *	22(0)	2187(1)

* $n_c = 4\sqrt{n_F + 1} - 2k$ to attain rotatability

** The total number of runs for the CCD include 2k axial points. In many cases all the factors will not necessarily have a significant 2nd order term. Sequentially adding axial points in the CCD until there is no more quadratic lack of fit will typically lead to less than 2k axial points.

As you can see, the CCD is far more efficient than the 3^{k-q}, B-B and Full Factorial in all but one of the situations presented. Keep in mind, though, the objective for each of the designs shown in table 3.23.

table 3.23	
Design	Objective
3^{k-q}	• Used for qualitative or quantitative factors • Estimate all linear and quadratic effects and, when higher order resolutions are designed, you can also estimate linear and quadratic 2-way interactions.
B-B	• Used only for quantitative factors • Estimate all linear, quadratic, and 2-way linear interactions plus experimental error.
CCD	• Used only for quantitative factors • Estimate all linear effects, selected quadratics and 2-way linear interactions plus experimental error.
Full Factorial	• Used for qualitative or quantitative factors • Estimate all linear and quadratic effects plus all possible simple and higher order interactions.
D-Opt	• Used for quantitative, qualitative or a mix of both • Can estimate any desired effect with reduced n • Minimizes the non-orthogonality of the design

3.10 Taguchi Designs

The last set of designs to be discussed are those of Taguchi [9]. These designs are typically orthogonal with respect to main effects and contain either aliased or confounded mains and 2-way interactions. They are essentially in an R_{III} category. The designs used by Taguchi are not new but simply a form of Plackett-Burman, Fractional Factorial or Latin Square design. The fact that they are tabled provides for easy use. The most frequently used Taguchi designs are shown in table 3.24. These designs all appear in Appendix 3A. Taguchi has also published tabled designs for mixed levels of factors.

3-23

The most important contribution of Taguchi is not the orthogonal arrays which have been around for some time. Instead, it is his development of a loss function and robust designs which have made important contributions. These robust designs allow the investigation of factors which affect the response value and the variability or dispersion of the response. The result is a design which determines factor settings which optimize the response and minimize response variability. A more detailed discussion on Taguchi methods appears in Chapter 6.

table 3.24

TAGUCHI DESIGN DESIGNATION	MAXIMUM NUMBER OF VARIABLES ASSUMING NO INTERACTIONS	
	2 LEVELS	3 LEVELS
L4	3	
L8	7	
L9		4
L12	11	
L16	15	
L18	1	7
L27		13

$$L_4(2^3)$$

No.	1	2	3
1	1	1	1
2	1	2	2
3	2	1	2
4	2	2	1
	a	b	-a•b

$$L_8(2^7)$$

No.	1	2	3	4	5	6	7
1	1	1	1	1	1	1	1
2	1	1	1	2	2	2	2
3	1	2	2	1	1	2	2
4	1	2	2	2	2	1	1
5	2	1	2	1	2	1	2
6	2	1	2	2	1	2	1
7	2	2	1	1	2	2	1
8	2	2	1	2	1	1	2
	a	b	-a•b	c	-a•c	-b•c	a•b•c

$$L_9(3^4)$$

No.	1	2	3	4
1	1	1	1	1
2	1	2	2	2
3	1	3	3	3
4	2	1	2	3
5	2	2	3	1
6	2	3	1	2
7	3	1	3	2
8	3	2	1	3
9	3	3	2	1
	a	b	-a•b	ab^2

$$L_{16}(2^{15})$$

Column No	1	2	3	4	5	6	7	8	9	10	11	12	13	14	15
1	1	1	1	1	1	1	1	1	1	1	1	1	1	1	1
2	1	1	1	1	1	1	1	2	2	2	2	2	2	2	2
3	1	1	1	2	2	2	2	1	1	1	1	2	2	2	2
4	1	1	1	2	2	2	2	2	2	2	2	1	1	1	1
5	1	2	2	1	1	2	2	1	1	2	2	1	1	2	2
6	1	2	2	1	1	2	2	2	2	1	1	2	2	1	1
7	1	2	2	2	2	1	1	1	1	2	2	2	2	1	1
8	1	2	2	2	2	1	1	2	2	1	1	1	1	2	2
9	2	1	2	1	2	1	2	1	2	1	2	1	2	1	2
10	2	1	2	1	2	1	2	2	1	2	1	2	1	2	1
11	2	1	2	2	1	2	1	1	2	1	2	2	1	2	1
12	2	1	2	2	1	2	1	2	1	2	1	1	2	1	2
13	2	2	1	1	2	2	1	1	2	2	1	1	2	2	1
14	2	2	1	1	2	2	1	2	1	1	2	2	1	1	2
15	2	2	1	2	1	1	2	1	2	2	1	2	1	1	2
16	2	2	1	2	1	1	2	2	1	1	2	1	2	2	1
	a	b	-ab	c	-ac	-bc	abc	d	-ad	-bd	abd	-cd	acd	bcd	-abcd

$$L_{18}(2^1 \times 3^7)$$

Column No	1	2	3	4	5	6	7	8
1	1	1	1	1	1	1	1	1
2	1	1	2	2	2	2	2	2
3	1	1	3	3	3	3	3	3
4	1	2	1	1	2	2	3	3
5	1	2	2	2	3	3	1	1
6	1	2	3	3	1	1	2	2
7	1	3	1	2	1	3	2	3
8	1	3	2	3	2	1	3	1
9	1	3	3	1	3	2	1	2
10	2	1	1	3	3	2	2	1
11	2	1	2	1	1	3	3	2
12	2	1	3	2	2	1	1	3
13	2	2	1	2	3	1	3	2
14	2	2	2	3	1	2	1	3
15	2	2	3	1	2	3	2	1
16	2	3	1	3	2	3	1	2
17	2	3	2	1	3	1	2	3
18	2	3	3	2	1	2	3	1

$$L_{27}(3^{13})$$

No \ Column	1	2	3	4	5	6	7	8	9	10	11	12	13
1	1	1	1	1	1	1	1	1	1	1	1	1	1
2	1	1	1	1	2	2	2	2	2	2	2	2	2
3	1	1	1	1	3	3	3	3	3	3	3	3	3
4	1	2	2	2	1	1	1	2	2	2	3	3	3
5	1	2	2	2	2	2	2	3	3	3	1	1	1
6	1	2	2	2	3	3	3	1	1	1	2	2	2
7	1	3	3	3	1	1	1	3	3	3	2	2	2
8	1	3	3	3	2	2	2	1	1	1	3	3	3
9	1	3	3	3	3	3	3	2	2	2	1	1	1
10	2	1	2	3	1	2	3	1	2	3	1	2	3
11	2	1	2	3	2	3	1	2	3	1	2	3	1
12	2	1	2	3	3	1	2	3	1	2	3	1	2
13	2	2	3	1	1	2	3	2	3	1	3	1	2
14	2	2	3	1	2	3	1	3	1	2	1	2	3
15	2	2	3	1	3	1	2	1	2	3	2	3	1
16	2	3	1	2	1	2	3	3	1	2	2	3	1
17	2	3	1	2	2	3	1	1	2	3	3	1	2
18	2	3	1	2	3	1	2	2	3	1	1	2	3
19	3	1	3	2	1	3	2	1	3	2	1	3	2
20	3	1	3	2	2	1	3	2	1	3	2	1	3
21	3	1	3	2	3	2	1	3	2	1	3	2	1
22	3	2	1	3	1	3	2	2	1	3	3	2	1
23	3	2	1	3	2	1	3	3	2	1	1	3	2
24	3	2	1	3	3	2	1	1	3	2	2	1	3
25	3	3	2	1	1	3	2	3	2	1	2	1	3
26	3	3	2	1	2	1	3	1	3	2	3	2	1
27	3	3	2	1	3	2	1	2	1	3	1	3	2

Appendix 3-B Plackett–Burman Designs Generating Vectors [8]

N=8   + + + − + − −

N=12 + + − + + + − − − + −

N=16 + + + + −   + − + + −   − + − − −

N=20 + + − − +   + + + − +   − + − − −   − + + −

N=24 + + + + +   − + − + +   − − + + −   − + − + −   − − −

N=32 − − − − +   − + − + +   + − + + −   − − + + +   + + − − +
      + − + − −   +

N=36 − + − + +   + − − − +   + + + + −   + + + − −   + − − − −
      + − + − +   + − − + −

BOX-BEHNKEN DESIGNS [2]

Number of factors	Design Matrix	Number of Points	Blocking and Association Schemes
3	$\begin{bmatrix} \pm1 & \pm1 & 0 \\ \pm1 & 0 & \pm1 \\ 0 & \pm1 & \pm1 \\ 0 & 0 & 0 \end{bmatrix}$	12 3 N= 15	No orthogonal blocking BIB (one associate class)
4	$\begin{bmatrix} \pm1 & \pm1 & 0 & 0 \\ 0 & 0 & \pm1 & \pm1 \\ 0 & 0 & 0 & 0 \\ \hline \pm1 & 0 & 0 & \pm1 \\ 0 & \pm1 & \pm1 & 0 \\ 0 & 0 & 0 & 0 \\ \hline \pm1 & 0 & \pm1 & 0 \\ 0 & \pm1 & 0 & \pm1 \\ 0 & 0 & 0 & 0 \end{bmatrix}$	8 1 ------ 8 1 ------ 8 1 N = 27	3 blocks of 9 BIB (one associate class)
5	$\begin{bmatrix} \pm1 & \pm1 & 0 & 0 & 0 \\ 0 & 0 & \pm1 & \pm1 & 0 \\ 0 & \pm1 & 0 & 0 & \pm1 \\ \pm1 & 0 & \pm1 & 0 & 0 \\ 0 & 0 & 0 & \pm1 & \pm1 \\ 0 & 0 & 0 & 0 & 0 \\ \hline 0 & \pm1 & \pm1 & 0 & 0 \\ \pm1 & 0 & 0 & \pm1 & 0 \\ 0 & 0 & \pm1 & 0 & \pm1 \\ \pm1 & 0 & 0 & 0 & \pm1 \\ 0 & \pm1 & 0 & \pm1 & 0 \\ 0 & 0 & 0 & 0 & 0 \end{bmatrix}$	20 3 ------ 20 3 N = 46	2 blocks of 23 BIB (one associate class)

Chapter 3 Problem Set

1. Describe the difference between a Taguchi L_8 design and a) a fractional factorial A 2^{7-4} b) a fractional factorial A 2^{4-1}.

2. What are the similarities/differences between fractional factorial, geometric Plackett-Burman, non-geometric Plackett-Burman designs, and Taguchi designs ?

3. Define the following terms:
   - Orthogonal designs

   - Resolution of a design

   - Aliased

   - Confounded

   - Generator

   - Defining word

   - Defining relationship

4. Discuss when you would use the following designs:

   - 3 level fractional factorial (Taguchi)

   - Box-Behnken

   - Box-Wilson (CCD)

   - D-Optimal

5. How can you use sequential experimentation to increase resolution from III to IV ?

6. Given a 2^{7-2} design where F=ABCD and G=BCDE, determine the defining relationship.

7. For the example on page 3-9, let E=ABCD. How does this affect the defining relationship ?

8. Why should you use balanced and orthogonal designs ?

9. What is a saturated design ?

10. Given the generating line for a n=16 Plackett-Burman design (see page 3-30), write the entire design matrix. Compare this design with a 16 run fractional factorial and a L_{16}.

11. What 3 level designs can you use for qualitative factors ? Quantitative factors ? Mixtures of both ?

12. If Taguchi designs are not new, what has been his major contribution to experimental design ?

13. A chemical engineer with a major oil refinery wanted to try a number of different additives to boost the octane level of the unleaded fuel produced at the refinery. For the sake of secrecy his additives were uniquely coded: A, B, C, D, E. All additives were varied from their lowest to their highest. The engineer knew from prior experience that the following additives would interact with each other: AxB, AxE, BxD, DxE, CxE, and AxD. To make matters worse, his boss told him that if his

experiment did not produce positive results, he could forget about any futher raises.
The engineer ran the following screening design:

Note: D=AC, E=BC

Runs	A	B	C	AC	BC
1	+1	+1	+1	+1	+1
2	+1	+1	−1	+1	−1
3	+1	−1	+1	−1	−1
4	+1	−1	−1	−1	+1
5	−1	+1	+1	−1	+1
6	−1	+1	−1	−1	−1
7	−1	−1	+1	+1	−1
8	−1	−1	−1	+1	+1

a) What is the defining relation ?

b) Are any important interactions aliased with main effects ? If you answer "yes", list the alias problems you found.

c) Generate a foldover of the design generated for this problem.

d) What is the defining relation for the foldover design ?

e) Would a foldover design salvage this experiment ?

f) Should the engineer forget about his future raises ?

14. Given the following factor conditions:

a) 3 quantitative factors (3 levels for each factor).

b) 3 quantitative factors (3 levels for each factor).
   1 qualitative factor (3 levels for each factor).

c) 3 quantitative factors (3 levels for each factor).

2 qualitative factors (3 levels for each factor).

2 qualitative factors (4 levels for each factor).

For each scenerio, determine the minimum number of runs for a D-optimal Design. (Hint: you will have to count up the total degrees of freedom of the effects to be estimated)

15. A scrubber is used to remove particles from a wafer surface in semiconductor processing. An engineer would like to determine the most important variables that optimize the scrubber's ability to remove particles. There are 8 factors that can be manipulated. The engineer does not want to spend too much time doing this experiment. Given the following data:

				Factor
Pre-Rinse Speed (PRS)	500 – 1500	RPM	—	A
Pre-Rinse Speed (PRT)	10 – 25	Sec.	—	B
Scrub Speed (SRS)	500 –1500	RPM	—	C
Scrub Time (SRT)	10 – 50	Sec.	—	D
Post–Rinse Speed (PSRS)	500 –1500	RPM	—	E
Post–Rinse Time (PSRT)	10 – 25	RPM	—	F
Dry Speed (DRS)	4000 – 5000		—	G
Dry Time (DRT)	25 – 50	Sec.	—	H

Note: The interactions of interest   AxB, CxD, ExF, GxH.

a) Generate the most effective screening experiment.

b) What is the defining relationship ?

c) Are any of your important interactions aliased with any main effects ?

d) What is the resolution of your design ?

16. A researcher in immunology wanted to determine what effect various concentrations of nitrogen dioxide would have on the ability of the lungs to fight off airborne infections such as pneumonia. Her factors were as follows:

	Absolute Low and High	Factor
Nitrogen Dioxide Concentration:	.5% to 2%	A
Environmental Humidity:	10% to 80%	B
Environmental Temperature:	50 F to 90 F	C

She wants to create a CCD design.

a) What is the α distance for this experiment ?

b) What are the design and real values for each factor ?

c) How many center points should this design have in order to be rotatable with uniform precision ?

d) How many total runs are needed for this experiment ?

17. In the following design matrix, determine the degree of confounding between factors (i.e. factor intercorrelations).

Runs	A	B	C
1	+1	+1	+1
2	+1	+1	−1
3	−1	+1	+1
4	−1	−1	−1
5	+1	−1	+1
6	+1	−1	−1
7	−1	+1	+1
8	−1	−1	−1

Hint: To find the correlation between A and B, regress A on B and look at R^2.

18. If an engineer was running a 8 factor screening experiment, how many possible 2 factor interactions are there ?

Chapter 3 Bibliography

1. Barker, Thomas B. (1985), *Quality By Experimental Design*, Marcel Dekker Inc.

2. Box, George E. P. and Behnken, D.W. (1960), "Some New Three Level Designs for the Study of Quantitative Variables", *Technometrics, 2*, pp 455-475.

3. Box, George E. P. and Draper, Norman R. (1987), *Empirical Model-Building and Response Surfaces,* John Wiley and Sons, Inc.

4. Box, George E. P.; Hunter, William G.; and Hunter, J. Stuart (1978), *Statistics for Experimenters,* John Wiley and Sons, Inc.

5. John, Peter W. M. (1971), *Statistical Design and Analysis of Experiments,* MacMillan Company.

6. Montgomery, Douglas C. (1984), *Design and Analysis of Experiments,* John Wiley and Sons, 2nd Edition.

7. Myers, Raymond H. (1976), *Response Surface Methodology,* Virginia Polytechnic Institute.

8. Plackett, R. L. and Burman, J. P. (1946), "The Design of Optimum Multifactorial Experiments", *Biometricka, 33,* pp305 - 325.

9. Taguchi, G. and Konishi, S. (1987), *Taguchi Methods: Orthogonal Arrays and Linear Graphs,* American Supplier Institute, Inc.

CHAPTER 4: DESIGN EVALUATION

4.1 Overview and Objectives

Once you have selected which k independent variables need to be considered in the experiment, it will frequently be realized that designs for examining k main effects and all possible interactions will consume unreasonable amounts of resources. In general, not all k variables which initially go into an experiment will have a substantial effect on the response measure. For this reason, it is important to proceed in a sequential manner where you first screen out the important effects, thereby reducing the original k factors to a subset of $q < k$ factors. During this phase you want to expend only a small amount of the total available experimental resources for the project. Both Barker [1] and Box, Hunter and Hunter [2] recommend that only 25% of available resources be used to screen out important variables. The remaining 75% of experimental resources are used to optimize the process and make confirmation runs. Taguchi, on the other hand, does not advocate sequential experiments. He is of the opinion that you should adequately brainstorm all the factors and levels such that only one experiment is necessary. Since most analyses result in only a few significant effects, building one design to test all potential factors may not always be the most efficient approach. Furthermore, each experimental design provides new information which may require more testing to confirm or deny.

In designing a screening procedure, you should keep in mind that failure to find all important factors leads to model bias. Consequently, you will not be able to accurately proceed in optimizing the process. In addition, not eliminating unimportant factors will needlessly consume resources during future experimentation and increase the noise level which could lead to inappropriate findings. Thus, the screening process is a critical aspect of experimentation and should be conducted in the most efficient and accurate way possible.

4.2 Factor Screening Strategies

Since the previous section established the need for efficiency in factor screening, most researchers will agree in using 2-level unreplicated designs such as Plackett-Burman (P-B) or Fractional-Factorials (F-F). The non-geometric P–B designs tend to be more efficient than geometric P–B (also referred to as F–F). However, the aliasing structure, especially for 2-way interactions, is less complex in a 2^{k-q} F–F design. It is also possible to foldover a 2^{k-q} fractional factorial design or use sequential fractions with a blocking variable to allow the researcher to de-alias certain factors while making use of previous experimental runs. Therefore, if your objective is to go beyond factor screening and look at empirical model building and response optimization, it is recommended to start with a fractional factorial design, i.e. a geometric P–B instead of the non-geometric P–B [2].

Since you are only looking at two levels for each factor, it is important to choose these levels far enough apart to identify an important effect. If the difference in levels is such that you could expect a non-linear effect, you should add replicated center points to test for potential quadratic effects.

There are 3 strategies for designing factor screening experiments: (1) super-saturated designs where n must be less than k, (2) saturated designs where n is approximately equal to k, and (3) unsaturated designs where n can be much larger than k. The reasons for choosing either of these strategies and the limitations of each strategy are topics for the rest of this section.

Supersaturated designs are desired when you are considering large numbers of input factors and you have a severe limitation on the number of screening runs available. In this situation, you could generate a random balanced (RB) design (discussed in Chapter 3) or utilize a group screening (GS) procedure. Due to random confounding in RB designs, the discussion that follows will be restricted to group screening. To construct a GS design, a P-B should be used with the desired n. In supersaturated designs, interactions are typically not able to be estimated and thus either geometric or non-geometric P–B designs can be used.

You also need to partition all potential input factors into 3 groups: (1) factors which are anticipated to have a positive relationship with the response, (2) factors which are anticipated to have a negative relationship with the response, and (3) factors whose relationship with the response is currently unknown. In groups (1) and (2), these factors should be further partitioned into a group of factors with potentially strong effects and another group with potentially weak or no effects. Ideally, the strong effect factors should remain as individual factors (i.e., each having its own vector in the design matrix). The factors anticipated to have a weak or no effect are then paired as necessary. Pairing factors is equivalent to aliasing in that settings for both factors are based on the same design vector. All paired and individual factors are assigned to a vector in the P-B design matrix. Factors with unknown effects can either be paired or left individually depending on the number of vectors available in the design matrix. Due to factor aliasing, significance of a paired vector will require further investigation to determine which one (or if all) factors are significant. If the pairing process does not provide one vector per paired or individual factor, 3 or more factors in the same group can be aliased with each other. The reason for the complicated pairing procedure is to ensure that a factor with a negative effect is not paired with a positive factor resulting in two significant effects cancelling each other out. The obvious advantage of GS designs is the limited number of runs required; however, a major limitation is the inability to estimate factor 2-way interactions. Some researchers will argue that rarely will a factor interact without also having a main effect. You simply have to weigh this limitation with resources available when you consider supersaturated designs.

When n can be about as large as k, then **saturated designs** are available for factor screening. In this case, you would choose a P–B or F–F design such that n is approximately equal to k. The F–F design is desired because (1) the confounding or aliasing patterns are easier to obtain, (2) after the screening process you can remove unimportant factors and be left with a higher resolution F–F and (3) if empirical model building or response optimization is desired, the screening runs can be easily incorporated into a CCD design which is discussed further in Chapter 5. To construct the saturated design, select the appropriate F-F design matrix such that $n \geq k + 1$ and assign a factor to each vector. The first limitation of saturated designs is the aliasing of main effects with 2-way interactions. However, if your initial assumption about no 2–way interactions becomes untenable, you can fold over the R_{III} F–F design to obtain a R_{IV} design. The second limitation is the apparent lack of non-linear effect testing. This limitation can be overcome by adding a few replicated center points to test for curvature and at the same time estimate pure error. The pure error estimate will allow you to test for lack of fit as discussed initially in Chapter 2.

Unsaturated designs will de-alias main effects from 2–way interactions. These designs can be constructed by folding over the R_{III} saturated designs or simply using a F–F design where n » k. The foldover R_{III} design will be of resolution IV which de-aliases all main effects and 2–way interactions. Designs between R_{III} and R_{IV} will only have partial de-aliasing of mains and 2–way interactions. The unsaturated design will provide more information than the other two alternatives but the cost in resources is higher. To decide on which strategy to implement, you must consider if factor screening is the final phase of the experiment or if response optimization is of interest. When response optimization is to be pursued, factor screening should only consume 25% of experimental resources.

4.3   Analysis

The analysis of experimental data has 4 objectives: (1) identify the important factors which contribute to the response, (2) to determine which factor settings will produce the best response, (3) to build an empirical mathematical model , and (4) conduct response surface analysis and optimization. Only objectives (1), (2) and (3) are discussed in this chapter. Objective (4) is covered in Chapter 5.

To determine factor settings, the simplest analysis technique is to look at the response values for the experimental runs and select the response that best satisfies the objective (i.e. minimum, maximum, or closest to a target). This method, referred to as "pick the winner" has a severe limitation in that many other combinations of these factors are not given any consideration. As an example, consider the data in table 4.1. In this table, we are using a 1 to represent a low factor setting and a 2 to represent a high setting.

table 4.1

	Factors							Response
Run	A	B	C	D	E	F	G	Y
1	1	1	1	1	1	1	1	18
2	1	1	1	2	2	2	2	20
3	1	2	2	1	1	2	2	12
4	1	2	2	2	2	1	1	10
5	2	1	2	1	2	1	2	12
6	2	1	2	2	1	2	1	14
7	2	2	1	1	2	2	1	19
8	2	2	1	2	1	1	2	21

If a small response is desired then the optimal factor settings correspond to the highs (2) and the lows (1) of run #4. However, the 7 factors at 2 levels produce a total of 128 different combinations and we've only considered 8. To look at the remaining 120 possibilities we can analyze the marginal effects through marginal mean plots, Pareto diagrams of effects, normal probability plots, ANOVA and/or regression. Some of these procedures will not provide tests of significance of the data in table 4.1 unless there is some replication. This problem will be discussed as we examine each of these methods.

To analyze the data by way of marginal mean plots, consider figure 4.1. The appropriate factor settings which minimize the response are: $A_1B_2C_2D_1E_2F_1G_1$. The predicted average response based on these factor settings is

$$\hat{y} = \overline{Y} + (\overline{A}_1 - \overline{Y}) + (\overline{B}_2 - \overline{Y}) + (\overline{C}_2 - \overline{Y}) + (\overline{D}_1 - \overline{Y}) + (\overline{E}_2 - \overline{Y}) + (\overline{F}_1 - \overline{Y}) + (\overline{G}_1 - \overline{Y})$$
$$= 15.75 - .75 - .25 - 3.75 - .5 - .5 - .5 - .5$$
$$= 9$$

This is a tremendous improvement over run #4; however, this estimate could be overly optimistic if some of the included effects occurred only due to chance and in reality are not significantly different from zero.

One way to assess the significance of effects is to look at a Pareto diagram of the absolute value of all half effect values found in the equation for $\hat{y}$. For example, figure 4.2 contains a Pareto diagram of the absolute value of half effects for data set 4.1. The half effect for 2 level designs is defined as the difference in the factor marginal means divided by 2.

It appears that the only factor which stands out beyond what might be considered the noise level is factor C. The remaining effects could be considered to be different from zero due only to experimental error. This being the case, the best factor settings would consist of C_2 and the remaining factors set at levels which are based on: (1) economics, (2) convenience, (3) status quo, and /or (4) providing less variability in the response. The new predicted average response would be:

$$\hat{y} = \overline{Y} + (\overline{C}_2 - \overline{Y}) = 12$$

which is only slightly different from run #4. Since only factor C appears to be significant, production costs may be reduced through more flexibility in selecting levels for the remaining factors.

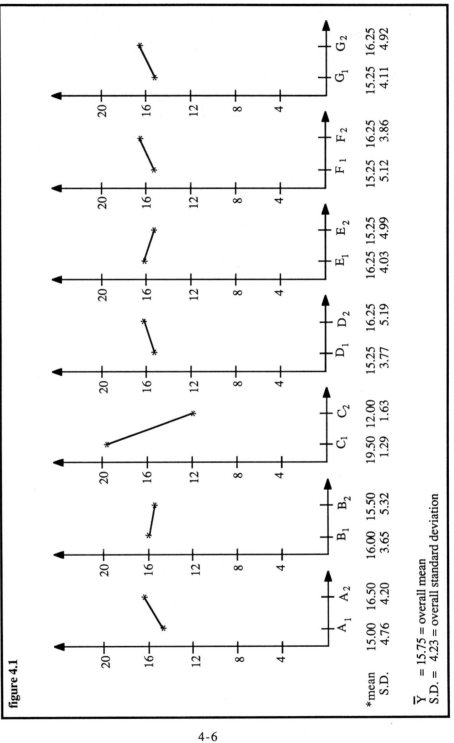

figure 4.1

4-6

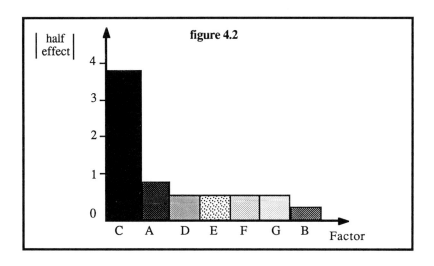

figure 4.2

Not all Pareto diagrams provide as clear a picture of effect significance as figure 4.2. Another approach would be the normal probability plot (NPP). This method is especially useful when designs are saturated and there are no degrees of freedom for error. Since the design in table 4.1 has n=8 runs and k=7 factors, there are no degrees of freedom for estimating error and therefore ANOVA and regression results with significance tests are not possible without some sort of modification. In these situations, the NPP is the recommended technique. To complete a NPP:

(1) for each factor and interaction, calculate the difference in mean response for high and low levels

e.g. $\overline{A}_2 - \overline{A}_1, \overline{B}_2 - \overline{B}_1, ----, \overline{G}_2 - \overline{G}_1, \overline{A \times B}_2 - \overline{A \times B}_1$, etc

(2) assign a rank, i, to the differences in (1) starting from lowest to highest (i=1,2,---M). M = # of effects; i.e., # of factor main effects and 2–way interactions.

(3) convert each factor rank to a percentile using $p = \dfrac{100(i - 0.5)}{M}$

(4) scale the horizontal axis of NPP paper to satisfy the values found in (1)

(5) on NPP paper, plot p versus values found in (1)

(6) draw a straight line through the majority of points which appear to line up (discounting any obvious "outliers")

(7) any point which deviates considerably from the straight line represents a factor which has a potential significant effect on the response.

Using the data set in table 4.1, steps (1) through (7) result in the following:

(1) Factor	A	B	C	D	E	F	G
Mean$_2$ - Mean$_1$	1.5	-.5	-7.5	1.0	-1.0	1.0	1..0
*(2) rank	7	3	1	5	2	5	5
(3) p	92.9	35.7	7.16	4.32	1.46	4.36	4

(4) (5) and (6) see figure 4.3

(7) the only factor which appears significant is C

To use ANOVA or regression, there must exist some degree of freedom for error. This can be accomplished by replicating the entire design or, more efficiently, replicating at one point. If runs are costly, replicating at one point is more economical but the question is, at which point? To answer this question, you must first determine if there is any need to address non-linear effects. If so, you should replicate at the center and thus be able to graphically test for curvature and estimate error. If non-linear effects are not reasonable to test, then replicate at the level chosen for the best response. This replication serves as a way to estimate error and also provides confirmation runs. The only problem with replicating at one point versus at all points is that standard ANOVA formulas are inappropriate if interaction terms are included. Because of the unbalanced nature of a design where replication occurs at only one point, regression is the preferred method for analysis. For balanced designs, the results from ANOVA and regression will always be the same.

We will use example data set 4.1, replicated at each of the 8 runs to demonstrate the use of ANOVA and regression. The new data set appears as shown in table 4.2 where level 1 refers to (-) and level 2 is (+).

---

* When two or more effects have equal values their ranks become the average of what would otherwise be assigned. For example, the ranks for D, F and G would be 4, 5 and 6, but since all effect values are the same, each receives an average rank of 5.0.

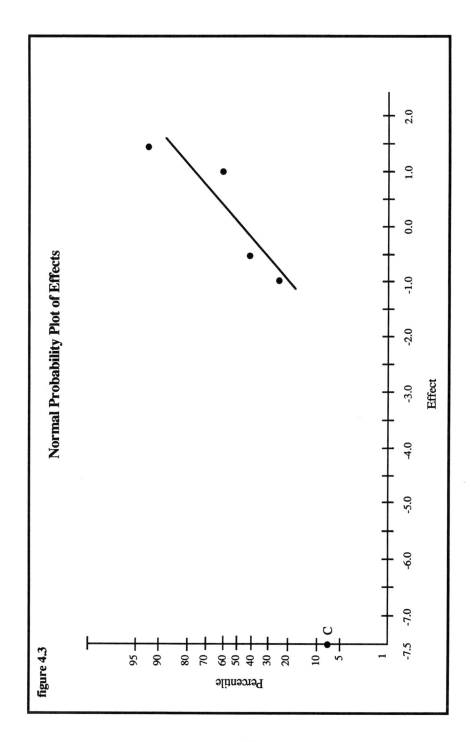

figure 4.3

Normal Probability Plot of Effects

table 4.2

Run	A	B	C	D	E	F	G	Response Y_1	Y_2	Mean $\overline{Y}$	** Variance S^2
	Factors										
1	1	1	1	1	1	1	1	18	23	20.5	12.5
2	1	1	1	2	2	2	2	20	18	19.0	2.0
3	1	2	2	1	1	2	2	12	13	12.5	0.5
4	1	2	2	2	2	1	1	10	14	12.0	8.0
5	2	1	2	1	2	1	2	12	11	11.5	0.5
6	2	1	2	2	1	2	1	14	9	11.5	12.5
7	2	2	1	1	2	2	1	19	25	22.0	18.0
8	2	2	1	2	1	1	2	21	20	20.5	0.5

The marginal means and variances are

Factor	A	B	C	D	E	F	G	
Mean 1	16.0	15.625	20.5	16.625	16.25	16.125	16.5	$\overline{\overline{Y}} = 16.1875$
Mean 2	16.375	16.75	11.875	15.75	16.125	16.25	15.875	
* STD Dev 1	4.444	4.868	2.449	5.423	4.950	4.998	5.782	$S_y^2 = 24.167$
STD Dev 2	5.655	5.230	1.808	4.683	5.222	5.175	4.257	

For 2 level orthogonal designs, the Chapter 2 formula for sum of squares between, is simplified as $SSB = \frac{N}{4}\left[(\overline{y}_2 - \overline{y}_1)^2\right]$. This formula is used to compute each factor's SS, where N = total # of observations, $\overline{y}_1$ is the mean for all observed values at level 1 and $\overline{y}_2$ is the mean at level 2. As an example, consider factor A. The sum of squares between for A is SS_A where

$$SS_A = \frac{16}{4}(16.375 - 16)^2 = .563$$

* STD Dev for A_1 is the STD Dev of (18,23,20,18,12,13,10,14)

** In practice, you should use more than 2 replications to estimate run variance or use the unreplicated approach described in Appendix 4.B

Computing all main effect SS results in the following:

Factor	A	B	C	D	E	F	G
SS	.563	5.063	297.563	3.063	.063	.063	1.563

The error sum of squares are found using the formula from Chapter 2,

$$SSE = \sum_{1}^{k}(n_r - 1)S_r^2 = 12.5 + 2.0 + .5 + 8.0 + .5 + 12.5 + 18 + .5$$
$$= 54.5$$

where n_r is the number of observations in row r

The completed ANOVA table appears as shown in table 4.3.

table 4.3

ANOVA TABLE

Source	SS *	df	MS	F	P
A	.563	1	.5625	<1	.781
B	5.063	1	5.0625	<1	.414
C	297.563	1	297.5625	43.68	.000
D	3.063	1	3.0625	<1	.521
E	.063	1	.0625	<1	.926
F	.063	1	.0625	<1	.926
G	1.563	1	1.5625	<1	.645
Error	54.5	8	6.8125	- -	
Total	362.438	15			

Using ANOVA, factor C is the only significant factor.

---

* For orthogonal 2 level designs, $SS_A = N \cdot b_A^2$ where N is the total number of runs (in this case N = 16) and b_A is the regression coefficient for factor A obtained when you regress the N response values on the design matrix of (+)s and (–)s.

After changing the 1 and 2 values in table 4.2 to -1 and +1, the regression results are shown in table 4.4.

```
┌─────────────────────────────────────────────────────────────────────────────┐
│  table 4.4                                                                    │
└─────────────────────────────────────────────────────────────────────────────┘
```

table 4.4

DEP VAR: Y   N: 16   MULTIPLE R: .922   SQUARED MULTIPLE R: .850

ADJUSTED SQUARED MULTIPLE R: .718   STANDARD ERROR OF ESTIMATE: 2.610

VARIABLE	COEFFICIENT	STD ERROR	STD COEF	TOLERANCE	T	P (2 TAIL)
CONSTANT	16.188	0.653	0.000	1.0000000	24.808	0.000
A	0.188	0.653	0.039	1.0000000	0.287	0.781
B	0.563	0.653	0.118	1.0000000	0.862	0.414
C	-4.313	0.653	-0.906	1.0000000	-6.609	0.000
D	-0.438	0.653	-0.092	1.0000000	-0.670	0.521
E	-0.063	0.653	-0.013	1.0000000	-0.096	0.926
F	0.063	0.653	0.013	1.0000000	0.096	0.926
G	-0.313	0.653	-0.066	1.0000000	-0.479	0.645

ANALYSIS OF VARIANCE

SOURCE	SUM-OF-SQUARES	DF	MEAN-SQUARE	F-RATIO	P
REGRESSION	307.938	7	43.991	6.457	0.009
RESIDUAL	54.500	8	6.813		

For an orthogonal 2 level design, the coefficients for each factor are equal to [response mean at +1 − response mean at −1] ÷ 2, thus regression analysis is very similar to marginal mean analysis. As in ANOVA and the previous test, only factor C is found to be significant. Since our objective is to minimize the response, the negative coefficient for factor C indicates that the orthogonal coded value for C must be +1. This implies factor C should be set at the high level (level 2) and the remaining factors can be set based on economics, convenience and/or status quo.

The prediction equation derived from table 4.4 is:

$$\hat{y}=16.188-4.313C.$$

Setting C at the high level (+1 orthogonal coded value) results in a predicted average response of 11.875. The next step would be to make as many confirmation runs as possible

possible (4 to 20) and determine the confirmation run average $\overline{CR}$. To test if the confirmation runs produced anticipated results you could conduct a t–test where H_0:

$\mu = 11.875$ and $H_1: \mu \neq 11.875$ using $t_0 = \dfrac{\overline{CR} - 11.875}{S_{CR} / \sqrt{n_{CR}}}$ and $t_C \left(\dfrac{\alpha}{2}, n_{CR} - 1 \right)$ (see

chapter 2 for steps required for a t–test). Assuming you fail to reject H_0, you have concluded the experiment and are ready to set the process. If you reject H_0, i.e. conclude $H_1: \mu \neq 11.875$, you must conduct an organized investigation of what may have caused the difference between the confirmation run average and the predicted average response. The following is a partial list of what could have caused the difference.

1). Factors not included in the experiment varying in an uncontrolled and non-random fashion.

2). Poor experiment discipline (i.e. settings were not made correctly, response values read incorrectly, etc).

3). Interactions aliased with main effects.

4). Model inadequacy, i.e. a second order model is appropriate, but only a first order model is estimated due to a two level type design.

Provided all factors are set at their highs or lows (i.e. no intermediate settings), the confirmation run problem is unlikely to be due to a missing 2nd order term. In our example, seven factors were placed in an eight run design which results in a resolution of III. Recall from chapter 3 that resolution III implies that main effects are aliased with two-way interactions. Therefore, there is a strong possibility that any problem with confirmation runs is associated with a strong two-way interaction. The next section will discuss the importance of including two-way interactions in the design matrix.

4.4 Importance of Modeling Interactions

How important is it to model interactions? This question can be addressed by way of a simple example. Consider the design below.

Run	A	B	A * B
1	+	+	+
2	+	-	-
3	-	+	-
4	-	-	+

Assume the researcher de-emphasizes interactions and he is also interested in evaluating factor C. The 4 run design above could be used to test the A, B and C main effects by aliasing A*B with factor C. Consider the following results:

A	B	C	Y
+	+	+	9
+	-	-	6
-	+	-	6
-	-	+	9

The marginal mean plots below indicate no A or B effect but a substantial C effect.

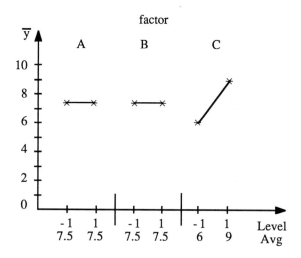

If "larger is better" then the researcher would suggest setting C at the high level and A and B settings to be based on economic considerations. For the sake of discussion, let's set A low and B high. Thus, the recommended settings are $A_{(-)}B_{(+)}C_{(+)}$ with a predicted average response of 9. We now make 5 confirmation runs with the following results.

Run	Y
1	6
2	4
3	5
4	5
5	7
Avg	5.4

What has gone wrong? The problem in this hypothetical example is that we assumed the A*B effect was not applicable and we therefore aliased factor C with A*B. In our example, the A and B effects appear unimportant; therefore, setting their levels independently and based on economic consideration should not be a problem. However, if A*B is important and C is not, then a true graphical representation will appear as shown below.

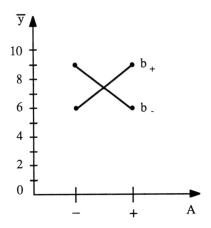

In this case, the settings for A and B cannot be made independently. For example, if A is set at + then y is maximized by setting B+. However, if A is set at − then y is maximized by setting B−.

The point to be made is that one should not ignore factor interactions just to simplify the design and to minimize the number of experimental runs. Even though it is true in most applications that interactions occur infrequently, the researcher should carefully consider whether or not to analyze interactions (see case study page 7.86 on the importance of identifying interactions). If you choose not to, at least be aware of ways to recover from problems similar to the one presented above; i.e., if you use a R_{III} design which provides confusing results, you can fold it over to get an R_{IV}, separating 2−way interactions from the main effects. See Chapter 3 for a detailed description of a foldover design.

4.5   Identifying Dispersion Factors (Factors which shift the Response Variability)*

Thus far, this analysis chapter has only discussed the identification of factors that shift the average response (i.e. location effects). Recent publicity of the Taguchi approach to parameter design for robust products and processes has emphasized the need to also identify which factors contribute to changes in variability of the response. The use of Taguchi's loss function described in chapters 1 and 6 provide the motivation for experiments to focus on target values instead of engineering tolerances or specifications. The average loss for any product is based upon the product variability and product average deviation from a designed target. This being the case, a set of objectives for a designed experiment should be as follows:

---

* Much of this section is taken from the SCHMIDT, S.R. and BOUDOT, J.R. paper presented at the 1989 Rocky Mountain Quality Conference and at the 1989 ORSA/TIMS Annual Conference [8].

(1) Identify factors which shift the mean (location effects) and then select those factor settings which minimize the difference between the average response and the desired target value. (see figure 4.4a)

(2) Identify factors which contribute to changes in the response variability (dispersion effects) and then select the factor settings which minimize response variability. (see figure 4.4b)

(3) Identify factors which shift the mean response and minimize response variability and then select the factor settings which best increase the quality of the product or process. (see figure 4.4c)

(4) Identify factors which have no effect on the response and then set these factors at levels that result in lower costs and/or faster through–put. (see figure 4.4d)

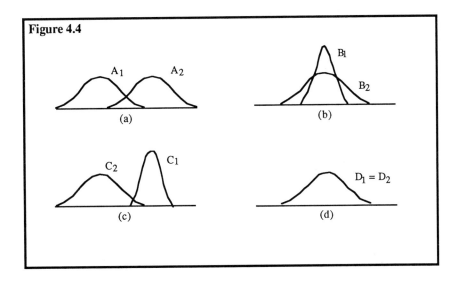

Figure 4.4

The bottom line is that the engineer needs to know which factors or knobs he/she can use to adjust the mean, which ones will reduce the variance and which ones do not make a difference.

To accomplish the four previously stated objectives, competing strategies have emerged. Since methods for finding location effects have already been discussed, the remainder of this section will concentrate on the following three commonly used methods for identifying despersion effects: 1) signal-to-noise ratios, 2) ln S and 3) residual analysis for replicated or unreplicated designs.

Taguchi's 3 signal-to-noise ratios are derived from loss functions associated with the following experimental goals: 1) maximize the response, 2) minimize the response and 3) adjust the response to a specified target or nominal value. The resulting signal-to-noise ratios are shown below.

(1) $S/N_L = -10 Log_{10} \frac{1}{n_r} \sum_1^{n_r} \left(\frac{1}{y_i^2} \right)$ for maximizing the response.

(2) $S/N_S = -10 Log_{10} \frac{1}{n_r} \sum_1^{n_r} \left(y_i^2 \right)$ for minimizing the response.

(3) $S/N_N = 10 Log_{10} \frac{1}{n_r} \left(\frac{S_m - V_e}{V_e} \right)$ for a target response.

where $S_m = n_r \bar{y}^2$

and $V_e = \dfrac{\sum y_i^2 - \frac{1}{n} \left(\sum y_i \right)^2}{n-1}$

(S/N_N is also proportional to $10 \, Log_{10} \left(\frac{\bar{y}^2}{s^2} \right)$)

The ln S method is based on the logarithm of the standard deviation associated with responses for each experimental condition or run. To demonstrate the use of signal-to-noise ratios and ln S consider the simple example presented in table 4.5.

table 4.5		FACTORS						
RUN	A	B	C	y_1	y_2	$\bar{Y}$	S	
1	1	1	2	20	30	25	7.071	
2	2	1	1	9	11	10	1.414	
3	1	2	1	26	24	25	1.414	
4	2	2	2	15	5	10	7.071	

If you look at the row changes in $\bar{Y}$ and S and how these changes are correlated with certain design columns, you should be able to see that the A knob adjusts the mean response, the C knob adjusts the variability of the response and the B knob has no effect at all. A graphic display of each factor's effect on y appears in figure 4.5 .

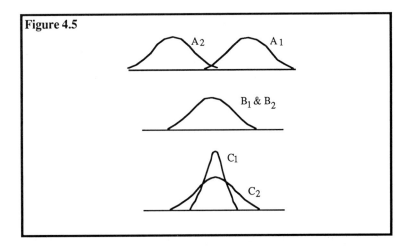

Figure 4.5

If you find the S/N_L values for each row in table 4.5 and then calculate marginal averages of S/N_L for each factor you should obtain results similar to those in table 4.6.

table 4.6	Analysis Using S/N_L		
	FACTOR		
	A	B	C
AVG 2	18.20	22.24	21.98
AVG 1	27.69	23.65	23.90
Δ	-9.49	-1.41	-1.92

Using the analysis results in table 4.6, the S/N_L strategy identified the correct location effect but not the dispersion effect.

Had we analyzed the data using S/N_S the marginal averages are displayed in table 4.7.

table 4.7	Analysis Using S/N_S		
	FACTOR		
	A	B	C
AVG 2	-20.51	-24.47	-24.55
AVG 1	-28.05	-24.09	-24.00
Δ	7.54	-.38	-.55

The marginal mean analysis for S/N_S also identified the location effect but not the dispersion effect.

Using S/N_N to analyze the data, the marginal averages are shown in table 4.8. In this case, the marginal mean analysis for S/N_N identified both A and C as important factors but not until the subsequent use of $10 \, Log_{10}(S_m)$ in another marginal table can you determine which factor is a location effect and which is a dispersion effect.

table 4.8	Analysis Using S/N_N		
	FACTOR		
	A	B	C
AVG 2	10.0	13.98	6.99
AVG 1	17.96	13.98	20.97
Δ	-7.96	0	-13.98

The marginal mean analysis using ln S is shown in table 4.9.

table 4.9	Analysis Using ln S		
	FACTOR		
	A	B	C
AVG 2	1.15	1.15	.35
AVG 1	1.15	1.15	1.95
Δ	0	0	-1.60

The analysis with ln S clearly indicates that factor C is a dispersion factor. Thus, given the type of data presented, the use of ln S appears to be an easier and more effective method for locating dispersion effects. Using ln S as the preferred approach for finding dispersion effects can also be verified by referring back to the example data in table 4.2. To analyze the variability effects of factors A through G, you can perform an analysis similar to the regression used for identifying location effects; however, the S^2 column of table 4.2 is used as the response. (As previously demonstrated you could also use S, ln S or ln S^2 as the response). Using an orthogonally coded design matrix (i.e. 1=-1 and 2=+1) the regression results appear in table 4.10.

table 4.10

DEP VAR:　　S2　　　N:　8　　MULTIPLE R:　1.000　　　SQUARED MULTIPLE R:　1.000

ADJUSTED SQUARED MULTIPLE R:　　1.000　　　STANDARD ERROR OF ESTIMATE: 0.000

VARIABLE	COEFFICIENT	STD ERROR	STD COEF	TOLERANCE	T	P(2 TAIL)
CONSTANT	6.813	0.000	.	1.0000000	.	.
A	1.063	0.000	.	1.0000000	.	.
B	-0.063	0.000	.	1.0000000	.	.
C	-1.438	0.000	.	1.0000000	.	.
D	-1.063	0.000	.	1.0000000	.	.
E	0.313	0.000	.	1.0000000	.	.
F	1.438	0.000	.	1.0000000	.	.
G	-5.938	0.000	.	1.0000000	.	.

The R^2 value of 1.0 and the 0 values for standard error are due to a perfect overfit (i.e., the number of effects are one less than the number of runs). This model is obviously inappropriate for significance tests but because of the orthogonal design, the coefficients can be used to estimate effects. The coefficients for each factor are computed from average variance at the factor high level minus the average variance at the low level. i.e. these coefficient values are equivalent to factor half effects using S_2 as the response $\left(\left(\overline{S}_2^2 - \overline{S}_1^2\right) \div 2\right)$. For factor A, the average run variance for levels 1 and 2 are 5.75 and 7.875 respectively. Thus, $\left(\overline{S}_{A2}^2 - \overline{S}_{A1}^2\right) \div 2 = 1.063$ which is the same as the regression coefficient. You can now use a Pareto diagram or NPP to determine which factors (if any) have substantial dispersion effects. Based on the coefficients in table 4.10, there appears to only be a G dispersion effect. The negative sign indicates that there is lower dispersion at the G_2 setting.

If you prefer to use ln S as the dependent variable, the results appear in table 4.11. The same conclusion is made either way; however, the distribution of ln S is less skewed than that of S^2. Since ln S analysis is less sensitive to occasional extreme values, it is the preferred response.

table 4.11

VARIABLE	COEFFICIENT	STD ERROR	STD COEF	TOLERANCE	T	P(2 TAIL)
CONSTANT	0.540	0.000	.	1.0000000	.	.
A	-0.036	0.000	.	1.0000000	.	.
B	-0.092	0.000	.	1.0000000	.	.
C	-0.137	0.000	.	1.0000000	.	.
D	-0.036	0.000	.	1.0000000	.	.
E	0.082	0.000	.	1.0000000	.	.
F	0.137	0.000	.	1.0000000	.	.
G	-0.713	0.000	.	1.0000000	.	.

DEP VAR: LNS   N: 8   MULTIPLE R: 1.000   SQUARED MULTIPLE R: 1.000

ADJUSTED SQUARED MULTIPLE R: 1.000   STANDARD ERROR OF ESTIMATE: 0.000

In conclusion, the analysis of example data set 4.2 indicates that you set factor C at level 2 for optimizing the location effect and factor G at level 2 for minimizing the dispersion effect. The remaining factors are set as dictated by economics, status quo or convenience. At this point you would want to make several (4 to 20) verification runs to determine a crude estimate of the capability index, C_{pk}, of the product based on the settings discussed above. If the verification runs did not produce desired results you would want to see section 5.1 to find ways to improve.

4.6   Using Taguchi S/N$_S$ on the Data in table 4.2.

Had we approached the sample data in table 4.2 using the "smaller is better" Taguchi signal to noise analysis the results would be as follows:

$$S/N_S = -10\log_{10}\left[\frac{1}{n_r}\sum y_i^2\right]$$

Run #1 from table 4.2 has a signal to noise of

$$S/N_S = -10\log_{10}\left[\frac{1}{2}(18^2 + 23^2)\right] = -26.30$$

The S/N$_S$ for all 8 runs are displayed below.

4-23

Run	y_1	y_2	S/N_S
1	18	23	-26.30
2	20	18	-25.59
3	12	13	-21.95
4	10	14	-21.70
5	12	11	-21.22
6	14	9	-21.41
7	19	25	-26.93
8	21	20	-26.24

The marginal S/N_S means for factors A through G are

Factor	A	B	C	D	E	F	G
Lo Mean	-23.884	-23.631	-26.263	-24.099	-23.974	-23.865	-24.086
High Mean	-23.951	-24.203	-21.571	-23.735	-23.860	-23.969	-23.748
Hi-Lo	-.067	-.572	4.692	.364	.114	-.104	.338

Using regression analysis with S/N_S as the dependent variable will provide similar information shown in table 4.12. Notice that the regression coefficients for each factor are $\left[\left(\overline{S/N}_{HI} - \overline{S/N}_{LO}\right) \div 2\right]$.

---

table 4.12

DEP VAR: S/N N: 8 MULTIPLE R: 1.000 SQUARED MULTIPLE R: 1.000

VARIABLE	COEFFICIENT	STD ERROR	STD COEF	TOLERANCE	T	P(2 TAIL)
CONSTANT	-23.917	0.000	.	1.0000000	.	.
A	-0.034	0.000	.	1.0000000	.	.
B	-0.286	0.000	.	1.0000000	.	.
C	2.346	0.000	.	1.0000000	.	.
D	0.182	0.000	.	1.0000000	.	.
E	0.057	0.000	.	1.0000000	.	.
F	-0.052	0.000	.	1.0000000	.	.
G	0.169	0.000	.	1.0000000	.	.

The S/N_S analysis identified factor C as an important factor but at this point it is still unclear if C adjusts the mean or the variance. As anticipated, the use of S/N_S, did not identify the large variability associated with factor G at its low level. Thus, it is again demonstrated that ln S is the more preferred method for identifying dispersion effects.

4.7   Residual Analysis for Replicated or Unreplicated Design

A third strategy for detecting dispersion effects allows for a reduction in experimental runs or increase in design resolution through utilization of an unreplicated design. It is presented as a modification of the Box-Meyer method [4]. A modified version of their procedure applied to a 2 level design is presented below. This procedure can be used for replicated or unreplicated designs.

(1) Use an R_{IV} or better design to avoid confounding interaction and dispersion effects.

(2) Fit the best model to the data.

(3) Compute the residuals.

(4) For each level of each factor, compute the standard deviation of the residuals.

(5) Compute the difference of the largest and smallest standard deviation for each factor.

(6) Rank order the differences found in (5).

(7) Assuming the Pareto Principle applies, use a Pareto chart or NPP to distinguish the important dispersion effects.

(8) For factors with important dispersion effects, determine the levels of least variance.

(9) If a factor setting for minimizing variance differs for the setting that optimizes the location effect, use subjective judgement as to which is most crucial to the product quality.

(10) Set all other factors based on economics, status quo or convenience.

The example used to illustrate this method contains data generated from the following simulation model

$$Y_i = 80 + 2A + 6B + 4A \cdot B + (z_i\sigma)$$

(where $\sigma = 3 + 2C$ and $z_i \sim N(0,1)$). Therefore, you anticipate location effects for A, B, and A * B as well as a dispersion effect for C. The generated data appear in table 4.13.

table 4.13					
RUN	A	B	C	D	Y
1	+	+	+	+	94.50
2	+	+	+	-	102.15
3	+	+	-	+	95.91
4	+	+	-	-	95.66
5	+	-	+	+	76.45
6	+	-	+	-	65.85
7	+	-	-	+	72.79
8	+	-	-	-	72.75
9	-	+	+	+	83.75
10	-	+	+	-	76.15
11	-	+	-	+	79.91
12	-	+	-	-	80.15
13	-	-	+	+	85.00
14	-	-	+	-	74.25
15	-	-	-	+	77.38
16	-	-	-	-	78.08

Using least squares regression, the predicted model is

$$\hat{Y} = 81.921 + 2.591A + 6.606B + 5.941A \cdot B$$

The computer output from the least squares regression procedure is shown in table 4.14.

table 4.14

DEP VAR: Y N: 16 MULTIPLE R: .939 SQUARED MULTIPLE R: .881

ADJUSTED SQUARED MULTIPLE R: .851 STANDARD ERROR OF ESTIMATE: 3.927

VARIABLE	COEFFICIENT	STD ERROR	STD COEF	TOLERANCE	T	P(2 TAIL)
CONSTANT	81.921	0.982	0.000	1.0000000	83.438	0.000
A	2.591	0.982	0.262	1.0000000	2.635	0.022
B	6.606	0.982	0.670	1.0000000	6.724	0.000
A*B	5.941	0.982	0.603	1.0000000	6.056	0.000

ANALYSIS OF VARIANCE

SOURCE	SUM-OF-SQUARES	DF	MEAN-SQUARE	F-RATIO	P
REGRESSION	1370.034	3	456.678	29.609	0.000
RESIDUAL	185.032	12	15.423		

The location effects of A, B and A $*$ B were correctly identified. Next, the predicted values $(\hat{Y})$ and the residuals $(Y - \hat{Y})$ were calculated and are displayed in table 4.15.

table 4.15

Run	Y	$\hat{Y}$	$Y - \hat{Y}$
1	94.50	97.06	-2.56
2	102.15	97.06	5.09
3	95.91	97.06	-1.15
4	95.66	97.06	-1.4
5	76.45	71.96	4.49
6	65.85	71.96	-6.11
7	72.79	71.96	.83
8	72.75	71.96	.79
9	83.75	79.99	3.76
10	76.15	79.99	-3.84
11	79.91	79.99	-.08
12	80.15	79.99	.16
13	85.00	78.68	6.32
14	74.25	78.68	-4.43
15	77.38	78.68	-1.30
16	78.08	78.68	-.60

The standard deviation of the residuals for each factor level appears in table 4.16.

table 4.16				
			Factor	
	A	B	C	D
$\sigma(+)$	3.675	3.039	5.035	3.190
$\sigma(-)$	3.596	4.148	.905	3.530
$\sigma(+) - \sigma(-)$	.079	-1.109	4.130	-.340

Eyeballing $\sigma_{(+)} - \sigma_{(-)}$, it is evident that the only dispersion effect which stands out from the rest is factor C. This can be verified with a Pareto diagram or NPP. To complete the NPP, plot the values of $\sigma_{(+)} - \sigma_{(-)}$ and p shown in table 4.17.

table 4.17				
Factor	A	B	C	D
$\sigma(+) - \sigma(-)$	.079	-1.109	4.130	-.340
Rank (i)	3	1	4	2
p	62.5	12.5	87.5	37.5

The NPP is displayed in figure 4.6. Given that "larger is better", the optimal settings for this problem are $A_{(+)}B_{(+)}$ and $C_{(-)}$. Factor D can be set based on economic or other considerations.

figure 4.6

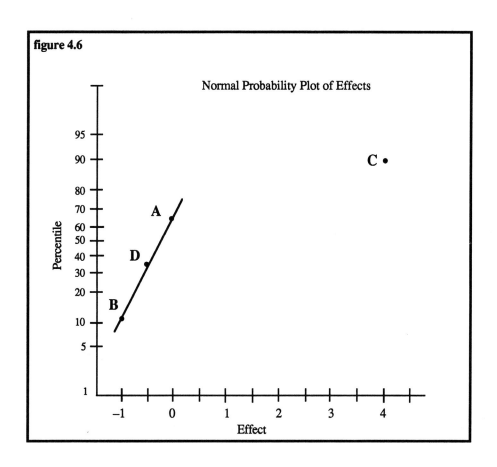

4.8 Robust Design Analysis

Robust designs are a type of designed experiment which identify input factor settings resulting in products and/or processes robust (insensitive) to noise factors. Taguchi refers to this type of experimentation as parameter design (see chapter 6 for more details). To provide the reader with an introduction to robust design and how the analysis is conducted, consider the following example from an injection molding process.

Statement of the Problem: An injection molding process engineer has been experiencing problems associated with part shrinkage which occurs after curing. This shrinkage problem has contributed to increased part variability and customer dissatisfaction.

Objective: Determine factors which contribute to variability in part shrinkage and find the best settings to minimize shrinkage.

Response: Part shrinkage is the amount of measured deviation from the desired part size and that of the manufactured part.

Some of the brainstorming in the form of a cause and effect or fishbone diagram is shown in figure 4.7. The asterisked factors are those determined to be uncontrollable in production; however they can be varied in the experiment. These 3 factors are referred to as noise or uncontrolled factors and will eventually appear in a separate design array called the outer array. A list of controllable and uncontrollable factors is shown in table 4.18.

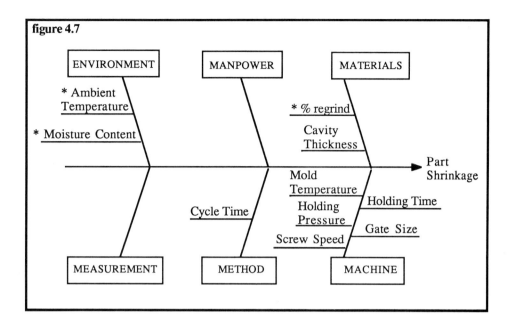

figure 4.7

table 4.18

Controllable Factors	Uncontrolled Factors
A: Cycle Time	H: Ambient Temperature
B: Mold Temperature	I: Moisture Content
C: Holding Pressure	J: Percent Regrind
D: Gate Size	
E: Cavity Thickness	
F: Holding Time	
G: Screw Speed	

Assuming the experimenter is satisfied with a 2 level design, the resulting robust design matrix will appear as shown in table 4.19.

table 4.19

	CONTROL FACTORS						J	1	2	2	1		
							I	1	2	1	2		
							H	1	1	2	2		
RUN	A	B	C	D	E	F	G					$\overline{Y}$	S
	1	1	1	1	1	1	1	2.6	2.7	2.8	2.8	2.725	.096
	1	1	1	2	2	2	2	0.8	3.0	0.8	3.2	1.950	1.330
	1	2	2	1	1	2	2	3.6	1.0	3.3	0.9	2.200	1.449
	1	2	2	2	2	1	1	2.5	2.4	2.5	2.3	2.425	.096
	2	1	2	1	2	1	2	3.5	3.6	3.5	3.5	3.525	.050
	2	1	2	2	1	2	1	2.6	4.7	3.6	1.5	3.100	1.369
	2	2	1	1	2	2	1	4.5	2.4	2.7	5.1	3.675	1.328
	2	2	1	2	1	1	2	2.4	2.5	2.3	2.4	2.400	.082

The marginal mean analysis of the signal ($\overline{Y}$) and the noise (ln S) appears in tables 4.20a and b.

table 4.20a

Analysis of $\overline{Y}$

Control Factors

	A	B	C	D	E	F	G
Avg 1	2.325	2.825	2.687	3.031	2.606	2.769	2.981
Avg 2	3.175	2.675	2.812	2.469	2.894	2.731	2.519
Δ	.850	-.150	.125	-.562	.288	-.038	-.462

table 4.20b	Analysis of ln S						
	\multicolumn{7}{c}{Control Factors}						
	A	B	C	D	E	F	G
ln S Avg 1	-1.01	-1.19	-1.07	-1.17	-1.04	-2.55	-1.02
ln S Avg 2	-1.23	-1.05	-1.16	-1.06	-1.19	.313	-1.21
Δ	-.22	.14	-.09	.11	-.15	2.863	-.19

Using the results of tables 4.20a and b, it is clear that factors A, D, and G shift the average, i.e. these are the location effects and factor F shifts the variability i.e. it is a dispersion factor. Therefore, to minimize shrinkage and the variability of shrinkage, the best settings are $A_1B_?C_?D_2E_?F_1G_2$, where factors with a question mark are set at levels based on economics, status quo and/or convenience. If the assumption of no interactions is correct, these settings will minimize shrinkage and its variability regardless of the percent regrind, ambient temperature and moisture content.

As a final note on the comparison of dispersion factor identification methods, this data was originally analyzed using signal-to-noise "smaller is better", S/N_S. The results shown in table 4.21 again failed to identify factor F as an important factor. Had we used signal-to-noise "nominal is best", the results shown in table 4.22 do point out the importance of factor F, however more cumbersome analysis is required to determine if F is a location or dispersion factor. In conclusion, the recommended procedure of looking at analysis of the signal ($\overline{Y}$) and the noise (ln S) separately was simpler and more straight forward.

table 4.21	Analysis Using S/N_S						
	\multicolumn{7}{c}{Factor}						
	A	B	C	D	E	F	G
Avg 1	-7.90	-9.29	-8.78	-9.90	-8.70	-8.74	-9.60
Avg 2	-10.20	-8.77	-9.28	-8.21	-9.36	-9.33	-8.43
Δ	-2.3	.52	-.50	1.69	-.66	-.59	1.17

table 4.22			Analysis Using S/N_N				
			Factor				
	A	B	C	D	E	F	G
Avg 1	16.03	19.12	17.65	19.63	17.29	30.87	18.28
Avg 2	20.57	17.48	18.94	16.96	19.30	5.72	18.32
Δ	4.54	−1.64	1.29	−2.67	2.01	−25.15	.04

4.9 Three Level Design Analysis and Comparison

The designs discussed in the previous section were all of 2 levels thus assuming a linear model is appropriate. If this assumption is incorrect, how does one proceed ? This case study will use a simulated process to present 2nd order modeling from 3 level designs.

"Plate 2" is a simulated process used for training engineers in Design of Experiments. In this problem, students are asked to optimize an "auto bumper" plating process using thickness as a quantitative response with time, temperature, percent nickel, ph, and percent phosphorous as the independent variables. From background information, it is believed that all of the above independent variables are important in impacting thickness. Additionally, there is concern about strong interactive as well as non-linear factors in the process. Crudely following the sequence of steps suggested by the "Experimental Design Information sheet" from chapter 1, we will address this problem.

Statement of the Problem

We are chartered with characterizing a new process (the plating process). This will be accomplished by developing a mathematical model for predicting plating thickness which will be used for optimizing various products requiring different plating thicknesses. An empirical model will provide the information required for adjusting the settings of the important independent variables in order to obtain the necessary thickness for the various products.

Objectives

(1). Determine the effect of time, temperature, percent nickel, ph, and phosphorous on plating thickness.

(2). Develop a "good" mathematical model.

QUALITY CHARACTERISTICS

Response	Type	Anticipated Range	How to measure?
Thickness	Q	0-2000	profilometer

FACTORS

Factor	Type	Control or Noise	Range	Interactions?
Time	Q	Control	4-12	possible
Temperature	Q	Control	16-32	possible
Nickel	Q	Control	10-18	possible
ph	Q	Control	2-10	possible
phosphorous	Q	Control	1.40-3.88	possible

Appropriate Experimental Design

To resolve the stated problem, different strategies will be demonstrated as described in the following cases.

Case I:

A two level fractional factorial of Resolution V is used to estimate all main effects and two-factor interactions. The design matrix with the response values is shown in table 4.23.

table 4.23

Run #	Time	Temp	Nickel	ph	Phos	Thickness
1	4.00	16.00	10.00	2.00	3.88	113
2	12.00	16.00	10.00	2.00	1.40	756
3	4.00	32.00	10.00	2.00	1.40	78
4	12.00	32.00	10.00	2.00	3.88	686
5	4.00	16.00	18.00	2.00	1.40	87
6	12.00	16.00	18.00	2.00	3.88	788
7	4.00	32.00	18.00	2.00	3.88	115
8	12.00	32.00	18.00	2.00	1.40	696
9	4.00	16.00	10.00	10.00	1.40	99
10	12.00	16.00	10.00	10.00	3.88	739
11	4.00	32.00	10.00	10.00	3.88	10
12	12.00	32.00	10.00	10.00	1.40	712
13	4.00	16.00	18.00	10.00	3.88	159
14	12.00	16.00	18.00	10.00	1.40	776
15	4.00	32.00	18.00	10.00	1.40	162
16	12.00	32.00	18.00	10.00	3.88	759

The computer output for the mathematical model is shown below.

table 4.24

	Term	Coeff.	Std. Error	T-Value	Signif.
1	1	420.937500	•	•	•
2	~ T	318.062500	•	•	•
3	~ TE	−18.687500	•	•	•
4	~ N	21.812500	•	•	•
5	~ P	6.062500	•	•	•
6	~ PHO	0.187500	•	•	•
7	~ T * TE	−7.062500	•	•	•
8	~ T * N	−6.062500	•	•	•
9	~ T * P	1.437500	•	•	•
10	~ T * PHO	3.812500	•	•	•
11	~ TE * N	8.937500	•	•	•
12	~ TE * P	2.437500	•	•	•
13	~ TE * PHO	−9.937500	•	•	•
14	~ N * P	15.187500	•	•	•
15	~ N * PHO	12.312500	•	•	•
16	~ P * PHO	−10.437500	•	•	•

No. cases = 16   Resid. df = 0   Cond. No. = 1

~ indicates factors are transformed.

table 4.25

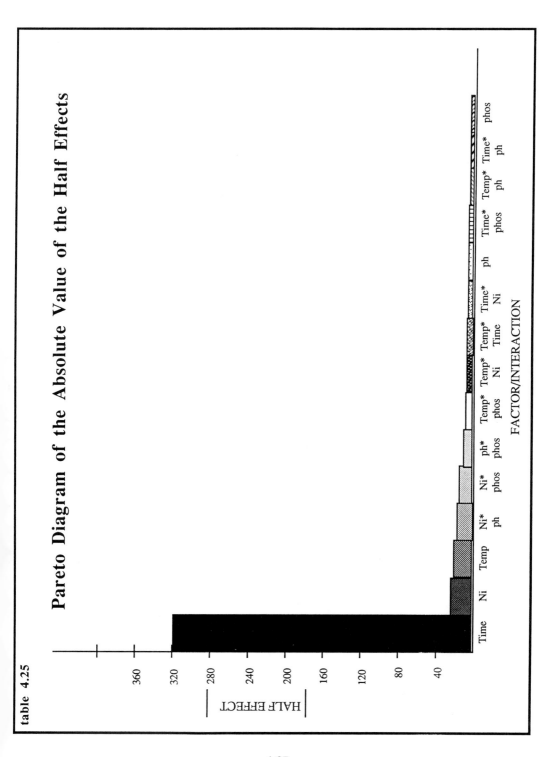

Pareto Diagram of the Absolute Value of the Half Effects

The lack of standard error estimates, t-values and significance is a result of using all the degrees of freedom for effects and thus none remain for estimating the experimental error. As previously demonstrated, a Pareto diagram for the effects can be used to separate the important (few) effects from the trivial (many) effects. The resulting Pareto diagram for the half-effects appears as indicated in table 4.25.

Based on the previous Pareto diagram of half effects it is obvious that time is an important factor. Nickel and temperature are ranked second and third, followed by everything else. At the outset, we discussed the possibility of non-linear effects. We will now tackle the question of non-linearity through the use of replicated centerpoints. This will be accomplished by complimenting our 16 original data points with 4 center-points. For a non-simulated problem this could be dangerous to do in a sequential fashion especially if there is a possibility that the process has shifted significantly. Therefore, the center-points are sometimes added during the first experiment phase. The centerpoint data is as follows:

	Time	Temperature	Nickel	ph	Phos	Thickness
1	8	24	14	6	2.64	351
2	8	24	14	6	2.64	373
3	8	24	14	6	2.64	353
4	8	24	14	6	2.64	321

Analysis of the combined 20 data points is shown in table 4.26.

```
table 4.26
                        Least Squares Summary ANOVA

  Source           df      Sum Sq.    Mean Sq.    F–Ratio     Signif.
1 Total (Corr)     19      1662583
2 Regression       15      1644873    109658       24.77      0.0035
3 Linear            5      1632409    326482       73.74      0.0005
4 Non–Linear       10        12464      1246        0.28      0.9527
5 Residual          4        17710      4427
6 Lack of fit       1        16331     16331       35.53      0.0094
7 Pure error        3         1379       460

                        R–sq. = 0.9893
                        R–sq–adj. = 0.9494

Model obeys hierarchy.  The sum of squares for linear terms is computed assuming
nonlinear terms are first removed.

F(1, 3) as large as 35.53 is a rare event => unlikely that model is correct.
```

Since the, "lack-of-fit" is significant, we need to determine which of our factors are non-linear and develop a mathematical model. As discussed in Chapter 3, an efficient way to accomplish a second order model is through the use of a Box-Wilson (central composite) Design. Making use of the applicable formulas, we will set the value of $\alpha = (n_F)^{1/4}$. Since $n_F = 16$, then $\alpha = 2$. The lows and highs from the fractional factorial portion of the experiment are, however, already at the extreme conditions. Because of this $\alpha = 1.0$ is used to construct a Central Composite Faced (CCF) design.

Since ph and phosphorous were screened out during the factorial portion of the experiment, we will not run α-points for these factors. Based on the above considerations, the resultant supplemental runs with associated thickness data are :

Run #	Time	Temp	Nickel	Thickness
1	12	24	14	736.1
2	4	24	14	96.0
3	8	32	14	328.9
4	8	16	14	303.5
5	8	24	18	358.2
6	8	24	10	347.7

By combining the 16 runs from the factorial portion of the design with the four centerpoints and the 6 runs from the α-portion of our design we obtain the following regression table:

table 4.27

	Term	Coeff.	Std. Error	T–Value	Signif.
1	1	342.7	11.747		
2	T	318.3	7.092		
3	TE	−15.2	7.092		
4	N	19.9	7.092		
5	T * TE	−7.1	7.522	−0.94	0.36
6	T * N	−6.1	7.522	−0.81	0.43
7	TE * N	8.9	7.522	1.19	0.25
8	T ** 2	80.1	18.047	4.44	0.00
9	TE ** 2	−19.7	18.047	−1.09	0.29
10	N ** 2	17.0	18.047	0.94	0.36

No. cases = 26   R–sq. = 0.9923   RMS Error = 30.1

Resid. df = 16   R–sq–adj. = 0.9880   Cond. No. = 5.68

Using backward elimination to remove unimportant terms leaves us with the final regression table shown below.

```
┌─────────────────────────────────────────────────────────────────┐
│  table  4.28                                                      │
│                                                                   │
│    Term         Coeff.      Std. Error     T–Value     Signif.    │
│  ─────────────────────────────────────────────────────────────── │
│   1  1          342.0        10.498                               │
│   2  T          318.3         6.998                               │
│   3  TE         −15.2         6.998        −2.17        0.04       │
│   4  N           19.9         6.998         2.85        0.00       │
│   5  T ** 2      78.4        12.618         6.21        0.00       │
│                                                                   │
│      No. cases = 26       R–sq. = 0.9902     RMS Error = 29.69     │
│      Resid. df = 21     R–sq–adj. = 0.9883    Cond. No. = 3.30     │
└─────────────────────────────────────────────────────────────────┘
```

The prediction equation generated from the orthogonally coded CCF design is:

Predicted thickness = 342.0 + 318.3(time) - 15.2(temp) + 19.9(nickel) + 78.4(time)^2

Step 12 from the Experimental Design Information Sheet in chapter 1 is "make some confirmation runs" to verify predicted results. Utilizing RS Discover software, a 95% confidence interval for the mean response when time = 4, temp = 16, and Nickel = 14 is calculated to be [72.9, 161.7]. This interval provides us something to judge our confirmation runs against. Ten confirmation runs at the above settings resulted in a mean value of 115.5 and a standard deviation of 20.4. Since the mean of our confirmation runs falls within our confidence interval, the model appears to be adequate.

Let's stop briefly and contrast our derived mathematical model with the equation in our simulation package. The actual uncoded equation in our simulation package is:

$$\text{Thickness} = 5(\text{time})^2 - 2.3(\text{temp}) + 4.0(\text{nickel})$$

with a standard deviation = 40

When temperature is 16, time is 4 and nickel is 14, the actual model would generate a mean value of 99.2 with a standard deviation of 40. The graphic on the next page will contrast the true population distribution with the modeled distribution for the 10 confirmation runs at temp = 16, time = 4, and nickel = 14.

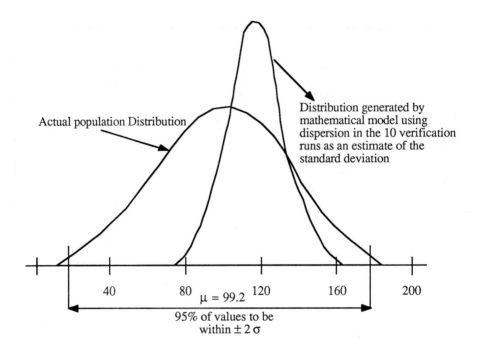

Actual population Distribution

Distribution generated by mathematical model using dispersion in the 10 verification runs as an estimate of the standard deviation

40   80 $\mu = 99.2$ 120   160   200

95% of values to be within $\pm 2\sigma$

As we can see from the above graphic, our model is off by approximately 20 units on the mean and we are nearly a factor of 2 off on the standard deviation. How can we improve on the mathematical model we determine? Large samples are required. A general rule of thumb is: For every term (excluding the constant) in our final prediction equation we need roughly 10 or more samples. Using this guideline, with the 4 non-constant terms in our model, 40 or more samples in the original experiment would be desired to obtain a "good" model. We used only 26. To estimate standard deviation the confirmation runs should have 15 or more data points. We used only 10. It is recognized however, that often times sample size is restricted due to time, money and resource constraints. In these cases we must simply recognize the limitations of our results.

Case II:

Instead of using a Central Composite Design to provide us with an empirical model, we could have used other approaches as well. For example, after the screening design was completed we could have made use of a D-optimal design, a Box-Behnken design, or a Taguchi design.

First, consider a computer generated D-Optimal design capable of estimating all linear, quadratic and 2 way interaction terms for the three most important terms. The design generated with the applicable thickness data is shown in table 4.29:

table 4.29				
RUN	TIME	TEMP	NICKEL	THICKNESS
1	12	32	10	717
2	4	16	10	136
3	12	16	18	787
4	4	32	18	78
5	4	16	18	87
6	4	32	10	80
7	12	16	10	760
8	8	24	10	282
9	8	32	14	318
10	4	24	14	96
11	12	32	18	682
12	12	24	14	747
13	12	24	14	764
14	12	24	14	742

The D-optimal design generated is only 12 runs, but we run replicates (2) on the last run. We are now ready to analyze the data. Results are as shown in table 4.30:

table 4.30

Term	Coeff.	Std. Error	T–Value	Signif.
1 1	314.786500	17.517042		
2 ~T	322.500000	6.491690		
3 ~TE	−26.625000	7.612182		
4 ~N	−7.375000	7.612182		
5 ~T ** 2	111.312500	17.517042	6.35	0.0031
6 ~T * TE	−10.375000	7.612182	−1.36	0.2446
7 ~T * N	5.375000	7.612182	0.71	0.5191
8 ~TE ** 2	29.937500	17.517042	1.71	0.1626
9 ~TE * N	−1.875000	7.612182	−0.25	0.8176
10 ~N ** 2	−40.062500	17.517042	−2.29	0.0841

No. cases = 14   R–sq = 0.9986   RMS Error = 21.53

Resid. df = 4   R–sq–adj. = 0.9953   Cond. No. = 7.392

~ indicates factors are transformed

Summary ANOVA

Source	df	Sum Sq.	Mean Sq.	F–Ratio	Signif.
1 Total (Corr.)	13	1283423			
2 Regression	9	1281569	142397	307.20	0.0000
3 Linear	3	1259538	419846	905.70	0.0000
4 Non–Linear	6	22030	3672	7.92	0.0325
5 Residual	4	1854	464		
6 Lack of fit	2	1588	794	5.97	0.1435
7 Pure error	2	266	133		

R–sq. = 0.9986

R–sq.–adj. = 0.9953

Model obeys hierarchy. The sum of squares for linear terms is computed assuming nonlinear terms are first removed. $F(2, 2)$ as large as 5.971 is not a rare event => no evidence of lack of fit.

From our summary ANOVA table it appears we have an acceptable fit with the model, but the "T-values" for numerous terms in our coefficient table indicate we can get rid of some terms with little loss in R^2. Our final coefficient table using backward elimination is shown in table 4.31:

table 4.31

Term	Coeff.	Std. Error	T–Value	Signif.
1 1	311.470588	17.084202		
2 ~T	323.657143	6.874237		
3 ~TE	–22.941176	8.053570	–2.85	0.0173
4 ~T ** 2	107.586555	18.415349	5.84	0.0002

No. cases = 14   R–sq. = 0.9957   RMS Error = 23.48

Resid. df = 10   R–sq.–adj. = 0.9944   Cond. No. = 5.328

~ indicates factors are transformed.

The prediction equation from the orthogonally coded D-Optimal design is

predicted thickness = 311.47 + 323.66(time) - 22.94(temp) + 107.59(time)^2

Had we chosen a Box-Behnken design for 3 factors it would appear as shown in table 4.32.

table 4.32

Run	TIME	TEMP	NICKEL	THICKNESS
1	4	16	14	129
2	12	16	14	792
3	4	32	14	94
4	12	32	14	702
5	4	24	10	37
6	12	24	10	738
7	4	24	18	133
8	12	24	18	714
9	8	16	10	339
10	8	32	10	302
11	8	16	18	319
12	8	32	18	344
13	8	24	14	364
14	8	24	14	342
15	8	24	14	404

Fitting the coded data with a regression model provided the statistics shown in table 4.33.

table 4.33

Term		Coeff.	Std. Error	T–Value	Signif.
1	1	370.000000	16.123999		
2	~T	319.125000	9.873892		
3	~TE	–17.125000	9.873892		
4	~N	11.750000	9.873892		
5	~T ** 2	69.375000	14.533976	4.77	0.0050
6	~T * TE	–13.750000	13.963792	–0.98	0.3700
7	~T * N	–30.000000	13.963792	–2.15	0.0844
8	~TE ** 2	-10.125000	14.533976	–0.70	0.5171

No. cases = 15   R–sq = 0.9954   RMS Error = 27.93

  Resid. df = 5   R–sq–adj. = 0.9872   Cond. No. = 4.522

~ indicates factors are transformed

Summary ANOVA

Source		df	Sum Sq.	Mean Sq.	F–Ratio	Signif.
1	Total (Corr.)	14	851473.7			
2	Regression	9	847574.0	94174.9	120.70	0.0000
3	Linear	3	818176.8	272725.6	349.70	0.0000
4	Non–Linear	6	29397.2	4899.5	6.28	0.0311
5	Residual	5	3899.8	780.0		
6	Lack of fit	3	1923.7	641.2	0.65	0.6535
7	Pure error	2	1976.0	988.0		

R–sq. = 0.9954

R–sq.–adj. = 0.9872

Model obeys hierarchy. The sum of squares for linear terms is computed assuming nonlinear terms are first removed. F(3, 2) as large as 0.649 is not a rare event => no evidence of lack of fit.

From the summary ANOVA, it appears the model is adequate. From the coefficient table, non-important terms can now be consolidated. The result is the following coefficients table:

table 4.34				
Term	**Coeff.**	**Std. Error**	**T–Value**	**Signif.**
1 1	363.769231	14.624921		
2 ~ T	319.125000	10.763650		
3 ~ N	11.750000	10.763650		
4 ~ T ** 2	70.153846	15.796720	4.44	0.0016
5 ~ T * N	-30.000000	15.222100	−1.97	0.0802
6 ~ N ** 2	–33.096154	15.796720	−2.10	0.0656

No. cases = 15   R–sq = 0.9902   RMS Error = 30.44

Resid. df = 9   R–sq–adj. = 0.9848   Cond. No. = 3.513

~ indicates factors are transformed

Thus, the prediction equation generated from the orthogonally coded Box-Behnken design is:

$$\text{predicted thickness} = 363.77 + 319(\text{time}) + 11.75(\text{nickel}) + 70.15(\text{time})^2 - 30(\text{time} * \text{nickel}) - 33.09(\text{nickel})^2$$

To model the desired linear, quadratic, and 2-factor interactions, the appropriate 3-level Taguchi design is the L27. In this instance, it is equivalent to the 3^3 full factorial.

The design and the generated thickness data are shown in table 4.35.

table 4.35

Run #	Time	Temp	Nickel	Thickness
1	4.00	16.00	10.00	113
2	4.00	16.00	14.00	152
3	4.00	16.00	18.00	147
4	4.00	24.00	10.00	65
5	4.00	24.00	14.00	53
6	4.00	24.00	18.00	130
7	4.00	32.00	10.00	83
8	4.00	32.00	14.00	40
9	4.00	32.00	18.00	94
10	8.00	16.00	10.00	339
11	8.00	16.00	14.00	303
12	8.00	16.00	18.00	381
13	8.00	24.00	10.00	348
14	8.00	24.00	14.00	342
15	8.00	24.00	18.00	420
16	8.00	32.00	10.00	327
17	8.00	32.00	14.00	255
18	8.00	32.00	18.00	322
19	12.00	16.00	10.00	745
20	12.00	16.00	14.00	780
21	12.00	16.00	18.00	740
22	12.00	24.00	10.00	772
23	12.00	24.00	14.00	769
24	12.00	24.00	18.00	755
25	12.00	32.00	10.00	735
26	12.00	32.00	14.00	726
27	12.00	32.00	18.00	757

After fitting the coded data with the appropriate model and eliminating unimportant terms, our final coefficients table is shown in table 4.36.

table 4.36

	Term	Coeff.	Std. Error	T–Value	Signif.
1	1	321.407407	13.192185		
2	~T	327.888889	7.225657		
3	~TE	−20.055556	7.225657	−2.78	0.0016
4	~N	12.166667	7.225657		
5	~T ** 2	87.888889	12.515205	7.02	0.0802
6	~N ** 2	24.055556	12.515205	1.92	0.0656

No. cases = 27   R–sq = 0.9902   RMS Error = 30.66

Resid. df = 21   R–sq–adj. = 0.9879   Cond. No. = 4.391

~ indicates factors are transformed

The prediction equation generated from the orthogonally coded L_{27} design is:

$$\text{predicted thickness} = 321.4 + 327.9(\text{time}) - 20.06(\text{temp}) + 12.2(\text{nickel})$$
$$+87.89(\text{time})^2 + 24.05(\text{nickel})^2$$

Lets now summarize the models we obtained with the various designs:

Design Type	R^2	Model # of Runs	Total # of Runs	Model (factors are orthogonally coded)
D-optimal	.9957	14	34	$311.47 + 323.65(\text{time}) + 22.94(\text{temp})$ $+ 107.59(\text{time})^2$
Box-Behnken	.9902	15	35	$363.77 + 319.13(\text{time}) + 11.75(\text{nickel}$ $+ 70.15(\text{time})^2 - 30.0(\text{time} * \text{nickel})$ $- 33.09(\text{nickel})^2$
CCF	.9902	26	26	$342.0 + 318.3(\text{time}) - 15.2(\text{temp})$ $+ 19.9(\text{nickel}) + 78.4(\text{time})^2$
L27	.9902	27	47	$321.41 + 327.89(\text{time}) - 20.06(\text{temp})$ $+ 12.17(\text{nickel}) + 87.89(\text{time})^2$ $+ 24.06(\text{nickel})^2$

The difference in predicted models is primarily due to large variation in the data coupled with the small sample size.

The actual simulation equation for the uncoded factors is:

$$Thickness = 5.0(time)^2 - 2.3(temp) + 4.0(nickel)$$

The program works as follows: You choose a specific uncoded value for time, temp and nickel and enter it into the program. The simulation will generate a response based on the equation above plus a normal error distribution with a $\sigma = 40$.

One way to test the effectiveness for each of the various designs is to obtain cross-validated R^2 values. Using the simulation equation without the added error you can obtain the expected value for all 27 combinations. (See table 4.37) Those 27 values will serve as the true response value in the crossvalidation. The cross-validated R^2 is obtained by correlating these true response values with the values obtained from the prediction equation for each design.

Design Type	R^2
D-Optimal	.996
Box-Behnken	.998
CCF	.999
L27	.998

From the above table we see that the cross validated R^2 for all of the models is quite good, with the best R^2 in this situation going with the CCF Design. A major reason for the large R^2 terms for each model is that the thickness values correlated against are actual values (there is no noise) calculated directly from the base equation. This was done to provide a fair comparison of the prediction ability of each model.

table 4.37

	Actual	Model			Predicted Values from Models			
Run #	time	temp	Ni	Thickness	D-opt	B/B	CCF	L27
1	4	16	10	83.2	118.3	39.9	97.3	113.4
2	4	16	14	99.2	118.3	114.8	117.3	101.5
3	4	16	18	115.2	118.3	123.5	137.3	137.7
4	4	24	10	64.8	95.4	39.9	82.1	93.2
5	4	24	14	80.8	95.4	114.8	102.1	81.4
6	4	24	18	96.8	95.4	123.5	122.1	117.6
7	4	32	10	46.4	72.5	39.9	66.9	73.2
8	4	32	14	62.4	72.5	114.8	86.9	61.4
9	4	32	18	78.4	72.5	123.5	106.9	97.6
10	8	16	10	323.2	334.4	318.9	337.3	353.4
11	8	16	14	339.2	334.4	363.8	357.2	341.5
12	8	16	18	355.2	334.4	342.4	377.2	377.7
13	8	24	10	304.8	311.5	318.9	322.1	333.3
14	8	24	14	320.8	311.5	363.8	342.0	321.4
15	8	24	18	336.8	311.5	342.4	362.0	357.6
16	8	32	10	286.4	288.5	318.9	306.9	313.2
17	8	32	14	302.4	288.5	363.8	323.8	301.4
18	8	32	18	318.4	288.5	342.4	346.9	337.6
19	12	16	10	723.2	765.7	738.2	733.9	769.1
20	12	16	14	739.2	765.7	753.0	753.9	757.2
21	12	16	18	755.2	765.7	701.7	773.8	793.5
22	12	24	10	704.8	742.7	738.2	718.7	749.1
23	12	24	14	720.8	742.7	753.0	738.7	737.2
24	12	24	18	736.8	742.7	701.7	758.6	773.4
25	12	32	10	686.4	719.8	738.2	703.5	729.0
26	12	32	14	702.4	719.8	753.0	723.5	717.1
27	12	32	18	718.4	719.8	701.7	743.4	753.4

Appendix 4.A   A Monte Carlo Simulation for Comparing Dispersion Factor Identification Methods [8]

To identify dispersion factors, the three most commonly used strategies are signal to noise ratios, ln S, and analysis of dispersion effects from residuals. Several papers have been written discussing the appropriateness of these strategies from a theoretical standpoint. This section is not intended to add to the collection of theoretical arguments; rather, it is our purpose to put these strategies to the test using simulated data. For purposes of this section, four modeling scenarios are presented and the competing strategies are compared based on K = 6 input factors and 32 available experiments. The designs chosen to satisfy the number of factors and limitations of resources are (1) a Taguchi L_8 (2^{6-3} fractonal factorial) with a sample size of four per run, (2) a Taguchi L_{16} (2^{6-2} fractional factorial) with a sample of two per run, and (3) a Taguchi L_{32} (2^{6-1} fractional factorial) with one response per run.

As you can see the resolution of the designs will increase as you decrease the sample size per run. The L_8 design is of resolution III indicating that the 6 main effects will be aliased with 2-way interactions. The resolution of the L_{16} design is IV indicating that the mains are free and clear of 2-way interactions but each 2-way interaction is aliased with another 2-way interaction. The last design, L_{32}, has resolution V which means the 2-way interactions are no longer aliased with other 2-way interactions. Ideally, the researcher prefers the highest design resolution possible but he/she is constrained by limited resources and number of factors plus the assumed need to replicate each run to estimate dispersion effects.

DESCRIPTION OF MODELS TESTED

It is impossible to model every different relationship of the response with the input factors while also incorporating several different types of error. However, since the problems encountered in industry are not always associated with additive error relationships having normally distributed error, it is important to include in this section some of the more common models.

ADDITIVE ERROR MODEL: The model is defined as

$$y = f\left(x_L, x_B\right) + \varepsilon\left(x_D, x_B\right)$$

where x_L is the set of factors which produce only location effects

x_D is the set of factors which produce only dispersion effects

x_B is the set of factors which produce both a location and a dispersion effect.

For this model, the strategies will be tested using normally and exponentially distributed error.

The actual model for the simulator is $y = w_0 + w_1 x_1 + w_2 x_2 + w_3 x_1 \bullet x_2 + \varepsilon$ where the w_i are generated randomly and each $w_i \geq 3\,\sigma$. In addition,

$$\varepsilon = \begin{cases} zM\sigma & \text{if } x_3 = +1 \\ z\sigma & \text{if } x_3 = -1 \end{cases}$$

where z has a standardized normal distribution in one and an exponential distribution ($\lambda = 0.5$) in the other. Therefore, x_3 is clearly the only dispersion effect and the amount of dispersion is controlled by the multiplier $M = 2, 3, 4,$ and 5.

MULTIPLICATIVE ERROR MODEL: This model is defined as

$y = f\left(x_L, x_B\right) \bullet \varepsilon\left(x_D, x_B\right)$. In this case the actual simulator model is

$$y = (w_0 + w_1 x_1 + w_2 x_2 + w_3 x_1 \bullet x_2) \bullet \varepsilon$$

$$\text{where } \varepsilon = \begin{cases} (4+z)\sigma M & \text{for } x_3 = +1 \\ (4+z)\sigma & \text{for } x_3 = -1 \end{cases}$$

and z has a standardized normal distribution. The values of M tested were 3 and 5.

EXPONENTIAL MODEL: The exponential model appears as

$$y = \exp(w_0 + w_1 x_1 + w_3 x_1 \cdot x_2 + \varepsilon)$$

$$\varepsilon = \begin{cases} z M \sigma \text{ for } x_3 = +1 \\ z \sigma \quad \text{ for } x_3 = -1 \end{cases}$$

and z has a standard normal distribution. The values of M tested were 3 and 5. A summary of the different tests is displayed in table 4.38.

---

table 4.38
A Summary of the Different Tests is shown below

STRATEGIES	DESIGNS	MODELING SCENARIOS
(1) Signal to Noise Ratios	(1) 4 replications of L_8	(1) $y = f(\underline{x}) + N(0, 1)$
- Smaller is better	(2) 2 replications of L_{16}	(2) $y = f(\underline{x}) * N(4, 1)$
- Larger is better	(3) 1 replication of L_{32}	(3) $y = \exp\lfloor f(\underline{x}) + N(3, .01) \rfloor$
- Nominal is better		(4) $y = f(\underline{x}) + \exp(\lambda = .5)$
(2) $\bar{y}$ plus ln S		
(3) Modified Box-Meyer		

RESULTS

The simulations were run 50 times for each strategy, model and design type. The difference in average marginal S/N_L, S/N_S, and S/N_N, and ln S for the replicated designs was used to identify the dispersion factor. The difference in residual standard deviations was used to identify dispersion factors for the unreplicated designs. The results for each model follow:

ADDITIVE ERROR MODEL: (Error term is normally distributed)

In the September 1988 IIE transactions, Pignatiello [7] discusses this model and concludes that maximizing S/N_N is accomplished by maximizing $\bar{y}$ and minimizing S^2. This is verified in that the results for S/N_N are very similar to ln S.

As stated in Box [3] S/N_L and S/N_S are based on ideas about location only and this is verified in that S/N_S and S/N_L never identified the dispersion effect. The results for this model are summarized in figure 4.8. Starting with the L_8 designs, you can see that S/N_N and ln S are about equal but they dropped substantially in percent correctly detecting x_3 as a dispersion effect. This result is discussed by Gunter [5] where he indicates that more information is required to identify dispersion effects than location effects. The L_{32} unreplicated method had very impressive results while providing much more information on interactions. Out of curiousity an unreplicated L_{16} was run which again verified Gunter's statement above. As previously stated, S/N_L and S/N_S were ineffective methods for identifying the dispersion effect.

ADDITIVE ERROR MODEL: (Error term exponentially distributed)

This test produced the following results:

table 4.39

Percent Correctly Detected

		Strategy				
		S/N_S	S/N_L	S/N_N	LnS	Mod B-M
Design Type	L_8 M = 3	0%	0%	30%	72%	N/A
	L_8 M = 5	0%	0%	75%	90%	N/A
	L_{32} M = 3	N/A	N/A	N/A	N/A	46%
	L_{32} M = 5	N/A	N/A	N/A	N/A	50%

For this method ln S was slightly better than S/N_N and the modified Box-Meyer approach was not as effective as before. The ineffectiveness of S/N_S and S/N_L is still confirmed in this type model.

MULTIPLICATE ERROR MODEL:

Pignatiello [7] discussed this model and concluded that S/N_N and ln S would produce similar results. The simulation analysis provided a confirmation to his conclusion.

The results are shown below.

table 4.40			Percent Correctly Detected				
					Strategy		
			S/N_S	S/N_L	S/N_N	LnS	Mod B-M
	L_8	M = 3	100%	100%	50%	45%	N/A
Design	L_8	M = 5	100%	100%	90%	90%	N/A
Type	L_{32}	M = 3	N/A	N/A	N/A	N/A	0%
	L_{32}	M = 5	N/A	N/A	N/A	N/A	0%

The success of S/N_L and S/N_S was anticipated because for the multiplicative error model, as the error increases so does the response. Thus, any dispersion effect is also a location effect. Therefore, the success of S/N_L and S/N_S can be attributed to identifying location effect and not the dispersion effect.

EXPONENTIAL MODEL: This test produced the results shown on the next page.

table 4.41							
			Percent Correctly Detected				
					Strategy		
			S/N_S	S/N_L	S/N_N	LnS	Mod B-M
Design Type	L_8	M = 3	0%	0%	96%	36%	N/A
	L_8	M = 5	96%	96%	98%	98%	N/A
	L_{32}	M = 3	N/A	N/A	N/A	N/A	0%
	L_{32}	M = 5	N/A	N/A	N/A	N/A	22%

Note the difficulty of all strategies except S/N_N in detecting a modest (M = 3) dispersion effect. However, S/N_S, S/N_L and ln S performed comparably to S/N_N in detecting large dispersion effects. As with the multiplicative error model, it was apparent that the exponential model had transformed non-homogeneous variance into location effects, enabling S/N_S and S/N_L to better detect the dispersion factors.

Of course, if the exponential model, $y = \exp(w_0 + w_1x_1 + w_2x_2 + w_3x_1x_2 + \varepsilon)$ is anticipated, then a transformation using the natural logarithm would produce $\ln(y) = w_0 + w_1x_1 + w_2x_2 + w_3x_1x_2 + \varepsilon$ which would yield the same results as the additive model. The modified Box-Meyer approach again had only marginal success with the nonlinear model; however, a log transformation as described above would compensate for this ineffectiveness.

CONCLUSIONS

The analysis of dispersion effects due to non-homogeneity of variance has become increasingly important as experimenters strive to optimize a response while minimizing its variance. Taguchi, Box, and others have suggested several techniques for analyzing dispersion effects within an experimental design, either simultaneous with or in addition to location effect analysis. This section compared the dispersion effect detection capabilities of five techniques (3 Taguchi signal-to-noise ratios, a response standard deviation analysis, and a modified Box-Meyer technique) via Monte Carlo simulation.

Simulation results clearly indicated that the standard deviation of responses for different factor combinations must typically change by at least a factor of 3 ($M \geq 3$) to be detected. This is consistent with the non-homogeneity threshold described by Keppel [6] for classical ANOVA analysis. Even with such a large dispersion effect, only S/N_N and ln S proved to be consistent detectors over the range of simulated models. Although comparable performers, the authors prefer using ln S over S/N_N due to its ease of interpretation. Because S/N_N simultaneously detects location and dispersion effects, it may be difficult to attribute the cause (either dispersion or location) to a particular factor. However, a traditional analysis of response means for location effects in conjunction with the ln S analysis for dispersion effects forgoes any question as to cause. The modified Box-Meyer method is as good as S/N_N and ln S for comparable number of runs if the underlying process model is linear, and it has the obvious advantages and efficiencies of an unreplicated design. However, its use should be questioned for process models which are truly unknown and possibly nonlinear unless the appropriate transformation is made. Taguchi's S/N_S and S/N_L are not recommended due to their inability to identify dispersion factors.

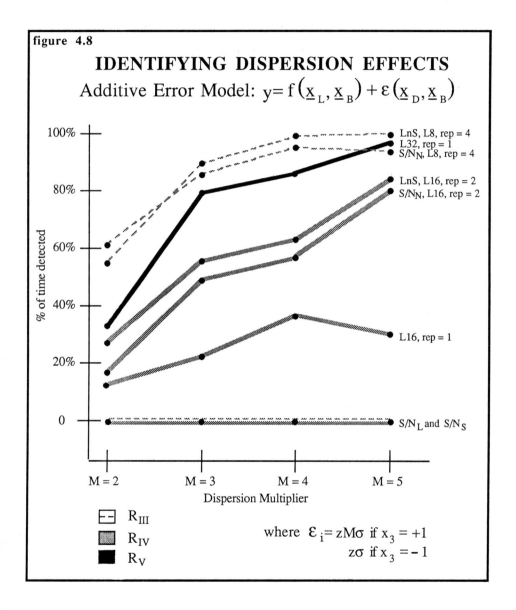

figure 4.8

IDENTIFYING DISPERSION EFFECTS

Additive Error Model: $y = f\left(\underline{x}_L, \underline{x}_B\right) + \varepsilon\left(\underline{x}_D, \underline{x}_B\right)$

where $\varepsilon_i = zM\sigma$ if $x_3 = +1$
$z\sigma$ if $x_3 = -1$

4-59

Chapter 4 Problem Set

1. Given the following settings and results:

A	B	C	D = ABC	Y_1	Y_2	$\bar{Y}$	S	ln S
−	−	−		50	70			
−	−	+		47	54			
−	+	−		44	35			
−	+	+		54	71			
+	−	−		57	53			
+	−	+		43	48			
+	+	−		42	39			
+	+	+		60	57			

a) Complete the rest of the table.

b) Find the prediction equation from these results. Only include significant main and 2-way interactions.

c) What settings will minimize the response and the variance of the response ?

2. An engineer needs to improve turbine blade quality by reducing thickness variability around a target value of 3mm. The brainstorming session identified 4 variables that are likely to affect the thickness. Those variables and their range of values are:

	−1	1
Metal Temperature	20 − 22° C	
Mold Temperature	3 − 5° C	
Pour Time	1 − 3 Sec	
Vendor	A or B	

The engineer has been limited to 20 experimental runs, so he decided to run a full factorial of 16 runs, saving 4 runs to confirm his results.

Run #	Metal Temp	Mold Temp	Pour time	Vendor	Thickness
1	20	3	1	A	2.314
2	20	3	1	B	2.307
3	20	3	3	A	2.321
4	20	3	3	B	2.233
5	20	5	1	A	4.204
6	20	5	1	B	4.260
7	20	5	3	A	4.221
8	20	5	3	B	4.198
9	22	3	1	A	2.687
10	22	3	1	B	2.748
11	22	3	3	A	2.697
12	22	3	3	B	2.728
13	22	5	1	A	4.769
14	22	5	1	B	4.868
15	22	5	3	A	4.787
16	22	5	3	B	4.849

a) Do the full analysis on this data so the engineer has a prediction equation. Look for main effects, interactions and dispersion effects.

b) What settings should the engineer use to meet the goal of 3mm thickness with minimum variability ?

3. A statistical process control analyst is trying to determine which factors have a major affect in a circuit board etching process. She wants to pare down the seven factors brought out in the brainstorming session to maybe two or three. The seven factors are:

	LOW	HIGH
A: Resist Thickness	.1mm	.5mm
B: Develop Time	80 sec	90 sec
C: Develop Concentration	3.1 : 1	2.7 : 1
D: Exposure	200	240
E: Develop Temperature	19° C	23° C
F: Circuit Line Thickness	1mm	3mm
G: Rinse Time	5 sec	10 sec

She ran the following L_8 design to find the significant effects.

Run	A	B	C	D	E	F	G	Response
1	−	−	−	−	−	−	−	74.48
2	−	−	−	+	+	+	+	70.07
3	−	+	+	−	−	+	+	75.71
4	−	+	+	+	+	−	−	68.79
5	+	−	+	−	+	−	+	84.98
6	+	−	+	+	−	+	−	81.57
7	+	+	−	−	+	+	−	84.03
8	+	+	−	+	−	−	+	80.97

a) Find out which factors effect the response variable.

b) If the goal is to maximize the response, what settings should be used ? Predict what the response should be.

c) The analyst ran confirmation runs to see if she had all the effects under control. If the mean of those runs is 86, is there any problem ? How about 90 ? 72 ?

4. Reference the Case Study on pages 7-25 through 7-30. In an attempt to get the width of the part to 9.380, the experimenter decided to look for interaction between the top four main effects on width:

D: Mold Temperature
A: Injection Velocity
E: Hold Pressure
B: Cooling Time

Run	D	A	E	B = DAE	Y_1	Y_2
1	−	−	−		9.3415	9.3416
2	−	−	+		9.36914	9.36916
3	−	+	−		9.34666	9.34664
4	−	+	+		9.36801	9.36809
5	+	−	−		9.3679	9.3680
6	+	−	+		9.34933	9.34937
7	+	+	−		9.3668	9.3669
8	+	+	+		9.35444	9.35446

a) Analyze his results and create a predictive equation for width. What are the best settings to reach the 9.38 goal for width ?

5. As a reliability engineer, you have been asked to weed out infancy failures in component-populated printed circuit boards. The four factors of interest are:

		LOW	HIGH
A:	Stress Temperature	80° C	125° C
B:	Thermo Cycle Rate	5°C/min	20° C/min
C:	Humidity	15%	95%
D:	g level for a 10 min sinusoid random vibration	3	6

The response is the number of electrical defects per board which has 1000 bonds.

a) Given the following design matrix and response data, determine the optimal screening method. (The more failures found, the better)

	A	B	–AB	C	–AC	–BC	D	Y_1	Y_2	Y_3
1	–	–	–	–	–	–	–	9	17	12
2	–	–	–	+	+	+	+	21	37	42
3	–	+	+	–	–	+	+	29	35	38
4	–	+	+	+	+	–	–	17	10	15
5	+	–	+	–	+	–	+	32	41	33
6	+	–	+	+	–	+	–	21	17	19
7	+	+	–	–	+	+	–	12	14	18
8	+	+	–	+	–	–	+	33	27	47

(Response header spanning Y_1, Y_2, Y_3)

6. A metal casting process for manufacturing turbine blades has four controllable factors:

Metal Temperature
Mold Temperature
Pour Speed
Raw Material

The blades must be 3mm thick; however, the ambient temperature causes the blades to expand and contract. The following experiment was run with ambient temperature as an outer array:

A Metal Temp	B Mold Temp	C Pour Speed	D Raw Material	TempL	TempM	TempH
1	1	1	1	3.06	3.14	3.06
1	1	−1	−1	3.01	3.00	3.05
1	−1	1	−1	2.81	2.81	2.80
1	−1	−1	1	2.80	2.88	3.01
−1	1	1	−1	2.62	2.61	2.62
−1	1	−1	1	2.61	2.66	2.72
−1	−1	1	1	2.42	2.43	2.52
−1	−1	−1	−1	2.42	2.43	2.41

a) Find a combination of settings that will produce the required thickness and is also robust to ambient temperature.

b) Plot the interactions of:

        A with Ambient Temperature

        B with Ambient Temperature

        C with Ambient Temperature

        D with Ambient Temperature

c) Use the plots from b) to verify the optimal conditions found in a).

Chapter 4 Bibliography

1. Barker, Thomas B. (1985) *Quality by Experimental Design*, Marcel Dekker.

2. Box, George E. P.; Hunter, William G., and Hunter, J. Stuart (1978) *Statistics for Experimenters*, John Wiley and Sons, Inc.

3. Box, G. E. P. (1988) *Signal-to-Noise Ratios, Performance Criteria, and Transformation*, **Technometrics**, 30, No. 1.

4. Box, G. E. P. and Meyer, R.D., *Report No. 1, Studies in Quality Improvement: Dispersion Effects From Fractional Designs*, Center for Quality and Productivity Improvement, University of Wisconsin, Madison.

5. Gunter, B. (1988) *Discussion; Signal-to-Noise Ratios, Performance Criteria, and Transformations*, **Techometrics**, 30, No. 1.

6. Keppel, G. **Design & Analysis, A Researcher's Handbook** (2nd ed.), Englewood Cliffs, NJ: Prentice-Hall, 1982.

7. Pignatiello, J.J. (1988) *An Overview of the Strategy and Tactics of Taguchi*, **IIE Transactions**, 20, 247-254.

8. Schmidt, S.R. and Boudot, J.R. *A Monte Carlo Simulation Study Comparing Effectiveness of Signal to Noise Ratio's and Other Methods for Identifying Dispersion Effects*, 1989 Rocky Mountain Quality Conference and 1989 ORSA/TIMS Annual Conference.

9. Taguchi, G. **System of Experimental Design**, Kraus International Publications, White Plains, New York.

CHAPTER 5 : OPTIMIZATION

5.1   Introduction and Overview

If you have used the techniques discussed in Chapter 4 and found factor settings that provide a satisfactory response, then there may be no need for further investigation. Unfortunately, using the simplest optimization techniques may not always lead to desirable results. When your best response does not prove satisfactory, what can you do? There are 4 things to consider in selecting the next plan of action:

(1)   Were all potential factors and associated 2-way interactions included in the design?
(2)   Is there too much noise in the data?
(3)   Did you select the appropriate number of levels for each factor?
(4)   Did you select the appropriate range for each factor?

If your problem is related to (1) or (2) then you may need to brainstorm again and make sure all relevant effects are measured. If interactions are the only problem, a foldover of the first design may provide the answers you need. If factors were left out of the first design you'll need to build a new design, which will consume lots of resources. To counter excessive noise, you can add more important factors to explain the noise or replicate each run to reduce the impact of noise. Once you've encountered these problems due to an inadequate brainstorming process, the consequences involved in correcting the problem should convince you to invest more time in future brainstorming.

If (3) and (4) are contributors to your problem then you probably need to select a better method to optimize the response. This chapter is designed to compare 3 methods of finding the optimal input factor settings: (1) one-at-a-time designs, (2) orthogonal arrays (Taguchi methods) and (3) response surface methodology (RSM). Although the first two methods are simple and easy to implement, they typically will produce results that are 60-80% of the total improvement to be made. If you need 80-100% of the total improvement, you need another method. To illustrate this point, consider the example that follows.

Given a hypothetical process with 2 input factors, f_1 = temperature (°F) where $100 \le f_1 \le 400$ and f_2 = pressure (psi) where $100 \le f_2 \le 200$, the objective is to optimize some measure of percent yield, y. Figures 5.1a, 5.1b, 5.2a and 5.2b present the one–at–a–time approach. In figure 5.1a, the researcher uses brainstorming information to select $f_2 = 150$ as a place to start his investigation. He sets the pressure knob at 150 psi and varies the temperature over its range to find the optimal temperature setting, which appears to occur at 200°F with a corresponding yield of 85.3%.

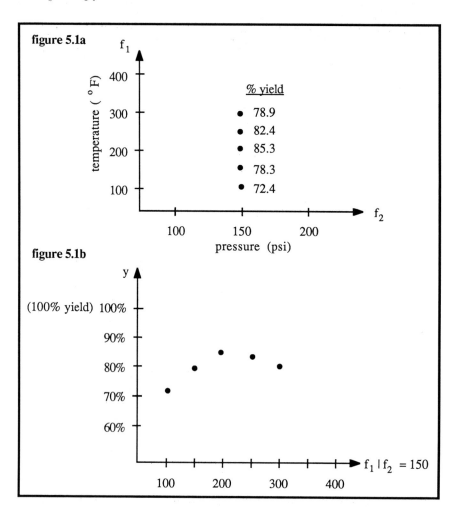

figure 5.1a

% yield
• 78.9
• 82.4
• 85.3
• 78.3
• 72.4

figure 5.1b

The temperature knob is now fixed at 200°F and pressure is varied over its range as shown in figures 5.2a and 5.2b. These figures reveal that the best X_2 setting (given $f_1 = 200$) is $f_2 = 150$. At this point, the researcher would conclude his best results occur at (f_1=200, f_2=150) and he should make several confirmation runs to ensure the results produce a desired average yield and a tolerable variance.

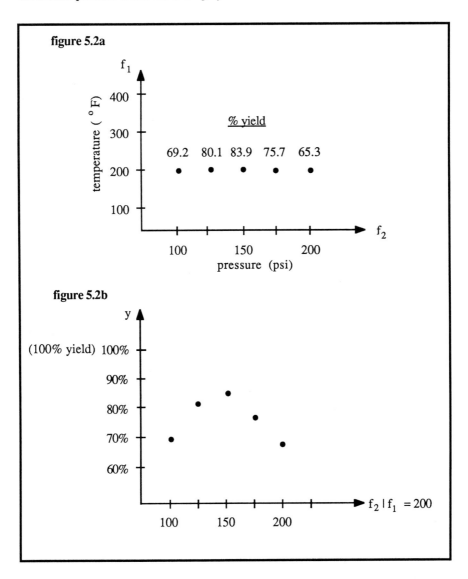

figure 5.2a

figure 5.2b

Unfortunately, if you chose the one–at–a–time approach and your results are less than desirable, it is confusing as to how you might improve the process. In this example, the response contours are displayed in figure 5.3. The one–at–a–time approach will use up large amounts of resources and if results are unsatisfactory, there is no efficient way to seek improvements.

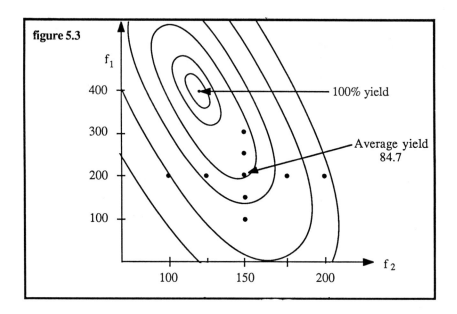

figure 5.3

Using the Taguchi approach, or any other orthogonal array approach, without RSM would consist of the following:

(1) The researcher would select an experimental region over which he chooses to investigate the response.

(2) Since the input factors in this example are both continuous, either a 2 or 3 level design would typically be used to evaluate the response over the experimental region.

(3) Using analysis techniques as described in Chapter 4, a best setting for each of the factors is obtained.

Referring back to the example, assume that the experimental region is decided to be $\{100 \le f_1 \le 200$ and $125 \le f_2 \le 175\}$. The results for a 3 level design appear in figure 5.4. Based on average marginals, the best settings would be $f_1 = 200$ and $f_2 = 150$ which produces an estimated average response of $\hat{\mu} = 82.82$.*

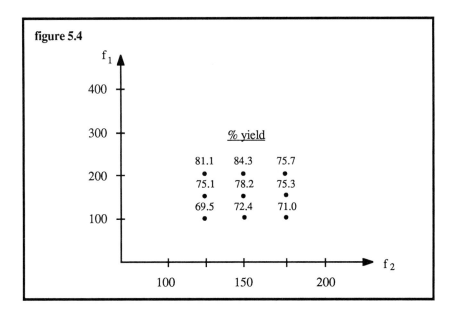

figure 5.4

Obviously, the failures of the one–at–a–time and the Taguchi approach associated with this example are due to a poor choice of factor values over which to investigate. As contrived as this data is, the point to be made is that this unfortunate choice of factor ranges can occur with old and new products even from the most experienced researchers, resulting in less than desired results. When this situation arises, a

---

$$* \; \hat{\mu} = \overline{y} + (\overline{f}_{1(200)} - \overline{y}) + (\overline{f}_{2(150)} - \overline{y})$$
$$= 75.844 + (80.366 - 75.844) + (78.3 - 75.844)$$
$$= 82.82$$

sequential approach is needed to efficiently lead you in the direction of the optimal. The previous two methods do not have this flexibility and, if modified to produce an optimum search, they will typically be inefficient. What is recommended is an approach based on RSM.

5.2 Response Surface Methodology

The basic idea associated with response surface methodology and experimental design is to build an efficient design to determine if the experimental region contains the optimum. If the optimum lies in the experimental region, build a mathematical model to locate it, else continue experimenting along a gradient of steepest ascent (direction of greatest improvement) to find the center of a new experimental region to be tested. At this new point, build another efficient design in the new experimental region and test whether it contains the optimum. You continue the process until the optimum is located. In order to conduct the process just described, we must be able to represent y as a function of the input factors. If the true statistical relationship is represented by $y = h(x_1, x_2, \cdots x_k) + C$, our objective is to estimate $h(x_1, x_2, \cdots x_k)$ with as few experimental runs as possible. To accomplish this task, a Taylor Series approximation of $h(x_1, x_2, \cdots x_k)$ is recommended. In most cases, a second order Taylor Series is a sufficient approximation. For k input factors, the 2nd order Taylor Series approximated at $\underline{c} = (c_1, c_2, \ldots c_k)$ is

$$h(\underline{x}) \approx h(\underline{c}) + \sum_1^k (x_i - c_i) \frac{\partial h}{\partial x_i}\bigg|_{\underline{x}=\underline{c}} + \sum_1^k \frac{(x_i - c_i)^2}{2!} \frac{\partial^2 h}{\partial x_i^2}\bigg|_{\underline{x}=\underline{c}}$$

$$+ \sum\sum_{i>j} (x_i - c_i)(x_j - c_j) \frac{\partial^2 h}{\partial x_i \partial x_j}\bigg|_{\underline{x}=\underline{c}} + R_c.$$

For coded data, i.e. $x_i = 2 \dfrac{f_i - \bar{f}_i}{d_i}$, you can estimate the surface at the center of the experimental region by setting $\underline{c} = \underline{0}$. Thus,

$$h(x) = h(0) + \sum_{1}^{k}(x_i)\frac{\partial h}{\partial x_i}\bigg|_{x=0} + \sum_{1}^{k}(x_i^2)\frac{\partial^2 h}{\partial x_i^2}\bigg|_{x=0} + \sum\sum_{i>j}(x_i)(x_j)\frac{\partial^2 h}{\partial x_i \partial x_j}\bigg|_{x=c} + R_0$$

If the 2nd order model is a good fit, then R_0 should be relatively small.

The Taylor Series can be arranged such that it resembles the 2nd order polynomial model below. The weights, b_0, b_i, b_{ij} and b_{ii}, are easily obtained from a multivariable regression software package.

$$y = b_o + \sum_{1}^{k}b_i x_i + \sum_{1}^{k}b_{ii}x_i^2 + \sum\sum_{i>j}b_{ij}x_i x_j + R_o$$

where

$$b_0 = h(0)$$

$$b_i = \frac{\partial h}{\partial x_i}\bigg|_{x=0}$$

$$b_{ii} = \frac{\partial^2 h}{\partial x_i^2}\bigg|_{x=0}$$

$$b_{ij} = \frac{\partial^2 h}{\partial x_i \partial x_j}\bigg|_{x=0}$$

R_0 is the error term.

The error term, $\underline{e} = R_0$, can be broken into two parts:

(1) pure error - estimated through replication
(2) lack of fit - estimated by the difference in the model error and pure error
   - this error indicates whether the model type (linear, quadratic, etc) is appropriate.

Using the central composite design to accomplish RSM, the replications required to estimate pure error will occur at the center point. Consider the previous example where x_1(high) = 200, x_1(low) = 100, x_2(high) = 175 and x_2 (low) = 125. The first order design with replicated center points would appear as shown below, where x_i is the transformed value of f_i.

f_1	f_2	x_1	x_2	$x_1 \cdot x_2$	y
200	175	1	1	1	75.9
200	125	1	-1	-1	82.1
100	175	-1	1	-1	70.1
100	125	-1	-1	1	69.7
150	150	0	0	0	75.6
150	150	0	0	0	76.2

The next step is to test the model adequacy. Using regression and a 0.10 significance level, the first order model which fits the data is

$$\hat{y} = 75.9 + 4.55x_1 - 1.45x_2 - 1.65x_1x_2.$$

The computer output shown in table 5.1 indicates non-significant lack of fit for a quadratic term.

---

table 5.1

DEP VAR:　Y　　N:　6　　MULTIPLE R:　.999　　SQUARED MULTIPLE R:　.998

ADJUSTED SQUARED MULTIPLE R:　.991　　STANDARD ERROR OF ESTIMATE: 0.424

VARIABLE	COEFFICIENT	STD ERROR	STD COEF	TOLERANCE	T	P(2 TAIL)
CONSTANT	75.900	0.300	0.000	1.0000000	253.000	0.003
x1	4.550	0.212	0.888	1.0000000	21.449	0.030
x2	-1.450	0.212	-0.283	1.0000000	-6.835	0.092
x1 • x2	-1.650	0.212	-0.322	1.0000000	-7.778	0.081
quadratic term	-1.450	0.367	-0.163	1.0000000	-3.946	0.158

ANALYSIS OF VARIANCE

SOURCE	SUM-OF-SQUARES	DF	MEAN-SQUARE	F-RATIO	P
REGRESSION	104.913	4	26.228	145.713	0.062
RESIDUAL	0.180	1	0.180		

The insignificant quadratic lack of fit indicates that the optimum lies outside of the initial experimental region. Therefore, the next step is to find a gradient vector which will point us toward the optimum response. This is obtained by taking the derivative of the 1st order model with respect to each of the factors. The gradient vector, $\underline{g}$, is

$$\underline{g} = \left(\frac{\partial \hat{y}}{\partial x_1}, \frac{\partial \hat{y}}{\partial x_2}\right)\Bigg|_{\substack{x_1=0 \\ x_2=0}} = (4.55, -1.45)$$

To experiment along the gradient in reasonable increments, you divide $\underline{g}$ by the smallest absolute value of the components of $\underline{g}$. In this case, we divide by 1.45 indicating that the first new experiment, R_1, should be conducted at $(3.14, -1)$. Subsequent experiments, R_i will take place at coordinates found by adding 3.14 to the x_1 component and subtracting 1 from the x_2 component. Experimentation along the gradient direction should continue until curvature is detected or until you reach the factor limitations. Then determine the estimated coordinates of the minimum or maximum along the gradient, make these coordinates the center of a new experimental region, conduct a new designed experiment and test for quadratic lack of fit. If the quadratic term is significant, add axial points to estimate the 2nd order model and locate the stationary point.

For our example, the experimental results for R_1, R_2 and R_3 are 91.0, 89.4 and 77.0, respectively. The best response along the gradient appears to occur at R_1, therefore let the new experimental center point be $(3.14, -1)$. (See figure 5.5). The new design matrix and corresponding response values are shown below.

f_1	f_2	x_1	x_2	y
357	100	4.14	-2.0	87.0
357	150	4.14	0	83.1
257	100	2.14	-2.0	75.7
257	150	2.14	0	86.9
307	125	3.14	-1.0	91.0
307	125	3.14	-1.0	90.1

To convert x_i back to the original values, use the formula

$$x_i = 2\left(\frac{f_i - \bar{f}_i}{f_{max} - f_{min}}\right)$$

For example, $x_1 = 4.14$ results in $4.14 = \dfrac{2(f_1 - 150)}{100}$ which simplifies to $f_1 = 357$.

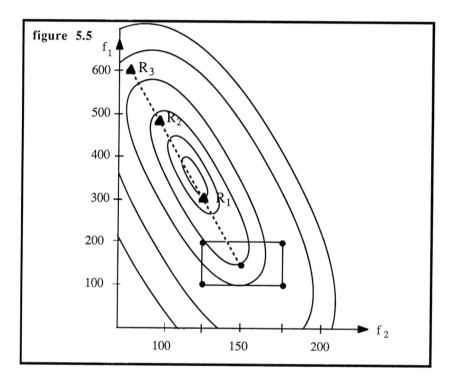

figure 5.5

To simplify the model building and analysis for this new experimental region, the design matrix will be altered to reflect R_1 as the center of the experimental region. This is accomplished by the following data table.

w_1	w_2	y
1	1	87.0
1	-1	83.1
-1	1	75.7
-1	-1	86.9
0	0	91.0
0	0	90.1

where $\qquad w_i = 2\left(\dfrac{f_i - \overline{f}_i}{f_{i_{max}} - f_{i_{min}}}\right).$

$\overline{f}_i$ = mean for f_i over the new experimental region

$f_{i_{max}}$ = maximum value of f_i over the new experimental region

$f_{i_{min}}$ = minimum value of f_i over the new experimental region.

i.e. $\qquad \overline{f}_1 = 307,\ f_{1_{max}} = 357,\ f_{1_{min}} = 257.$

$\qquad \overline{f}_2 = 125,\ f_{2_{max}} = 150,\ f_{2_{min}} = 100.$

Table 5.2 contains the least squares regression analysis of the data.

table 5.2

DEP VAR: Y N: 6 MULTIPLE R: .999 SQUARED MULTIPLE R: .997

ADJUSTED SQUARED MULTIPLE R: .987 STANDARD ERROR OF ESTIMATE: 0.636

VARIABLE	COEFFICIENT	STD ERROR	STD COEF	TOLERANCE	T	P(2 TAIL)
CONSTANT	90.550	0.450	0.000	1.0000000	201.222	0.003
w1	1.875	0.318	0.299	1.0000000	5.893	0.107
w2	-1.825	0.318	-0.291	1.0000000	-5.735	0.110
w1 • w2	3.775	0.318	0.602	1.0000000	11.864	0.054
quadratic	-7.375	0.551	-0.679	1.0000000	-13.381	0.047

ANALYSIS OF VARIANCE

SOURCE	SUM-OF-SQUARES	DF	MEAN-SQUARE	F-RATIO	P
REGRESSION	156.908	4	39.227	96.857	0.076
RESIDUAL	0.405	1	0.405		

Since the quadratic term is significant at the .1 level, it is decided to add axial points to build the full 2nd order model. The value for α (the axial point distance from the center) is

$$\alpha = (n_F)^{\frac{1}{4}} = (4)^{\frac{1}{4}} = 1.414,$$

where n_F = # of runs in the factorial portion of the design. Therefore, the completed data set is

w_1	w_2	y
1	1	87.0
1	-1	83.1
-1	1	75.7
-1	-1	86.9
0	0	91.0
0	0	90.1
1.414	0	85.4
-1.414	0	77.6
0	1.414	80.5
0	-1.414	77.4

A graphical display is shown in figure 5.6. The computer output for the fitted model is shown in table 5.3.

table 5.3

DEP VAR: Y   N: 10   MULTIPLE R: .930   SQUARED MULTIPLE R: .865

ADJUSTED SQUARED MULTIPLE R: .695   STANDARD ERROR OF ESTIMATE: 3.019

VARIABLE	COEFFICIENT	STD ERROR	STD COEF	TOLERANCE	T	P(2 TAIL)
CONSTANT	90.549	2.135	0.000	1.0000000	42.412	0.000
w1	2.316	1.068	0.399	1.0000000	2.170	0.096
w2	-0.365	1.068	-0.063	1.0000000	-0.342	0.750
w1 · w2	3.775	1.510	0.460	1.0000000	2.501	0.067
$(w1)^2$	-3.787	1.412	-0.546	.8164322	-2.681	0.055
$(w2)^2$	-5.063	1.412	-0.730	.8164322	-3.584	0.023

ANALYSIS OF VARIANCE

SOURCE	SUM-OF-SQUARES	DF	MEAN-SQUARE	F-RATIO	P
REGRESSION	232.776	5	46.555	5.107	0.070
RESIDUAL	36.465	4	9.116		

5-12

Using a significance level of .10, the parsimonious model is

$$\hat{y} = 90.549 + 2.316W1 + 3.775W1 \cdot W2 - 3.787(W1)^2 - 5.063(W2)^2.$$

The stationary point is found by differentiating $\hat{y}$ with respect to both factors and solving simultaneously for w_1 and w_2.

$$\left\{ \begin{array}{ll} 1 & \dfrac{\partial \hat{y}}{\partial w_1} = 2.316 + 3.775\, w_2 - 7.574\, w_1 = 0 \\[2ex] 2 & \dfrac{\partial \hat{y}}{\partial w_2} = -10.126 w_2 + 3.775\, w_1 = 0 \end{array} \right.$$

Solving 2 for w_1 results in $w_1 = 2.682 w_2$.

Substituting this for w_1 in 1 results in

$$3.775 w_2 - 7.574(2.682) w_2 = -2.316 \quad \text{or} \quad w_2 = 0.140.$$

Now, substituting this value in 2 results in $w_1 = .375$. Thus, our stationary point is estimated to be at $(w_1, w_2) = (.375, .140)$.

Define B as the second order matrix shown below:

$$B = \begin{bmatrix} b_{11} & \dfrac{b_{12}}{2} \\[2ex] \dfrac{b_{21}}{2} & b_{22} \end{bmatrix} \qquad \text{where}$$

$$\left\{ \begin{array}{l} b_{11} \text{ is the coefficient of } (w_1)^2 \\ b_{22} \text{ is the coefficient of } (w_2)^2 \\ b_{12} = b_{21} \text{ is the coefficient of } (w_1 \cdot w_2) \end{array} \right.$$

The eigenvalues of B are -6.419 and -2.433. Since both eigenvalues are negative, the stationary point $(w_1, w_2) = (.375, .140)$ is a maximum. The untransformed coordinates of the stationary point are (325.75, 128.5). See figure 5.6 for a summary of the experimentation used to find the optimum.

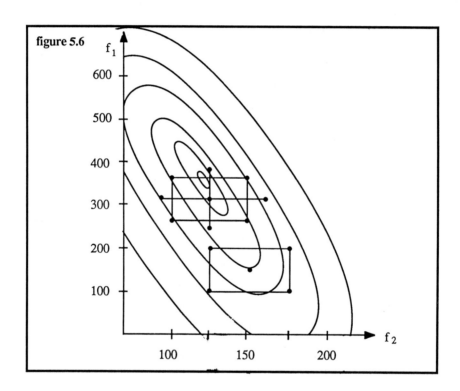

figure 5.6

The following table of possible eigenvalues summarizes the characteristics of stationary points.

eigenvalue sign		stationary point
(1) all positive	→	minimum
(2) all negative	→	maximum
(3) some negative some positive	→	saddle point
(4) all one sign with one or more close to zero	→	rising or falling ridge

For a more indepth study of RSM see Box and Draper [1], Montgomery [2] or Myers [3].

Appendix 5.A Example demonstrating the need to used coded data when examining Second Order models.

Run	A	B	C	x_1	x_2	x_3	Y_1	Y_2	Y_3	Y_4	Y_5	$\bar{Y}$	S
1	75	60	71	0	−	+	1310	1330	1340	1329	1350	1332	14.873
2	70	60	68	−	−	0	1361	1369	1368	1373	1379	1370	6.633
3	75	60	65	0	−	−	1382	1386	1390	1383	1390	1386	3.768
4	80	60	68	+	−	0	1368	1381	1367	1386	1385	1377	9.236
5	75	65	68	0	0	0	1376	1393	1379	1385	1401	1387	10.257
6	80	65	71	+	0	+	1348	1365	1365	1366	1382	1365	12.029
7	80	65	65	+	0	−	1394	1376	−	1386	1413	1392	15.671
8	75	65	68	0	0	0	1407	1416	1404	1412	1414	1411	4.980
9	70	65	71	−	0	+	1355	1369	1379	1379	1379	1372	10.545
10	70	65	65	−	0	−	1411	1417	1421	1418	1419	1417	3.768
11	75	65	68	0	0	0	1374	1400	1403	1403	1404	1397	12.833
12	75	70	71	0	+	+	1387	1393	1396	1400	−	1394	5.477
13	80	70	68	+	+	0	1394	1402	1405	1417	1405	1405	8.264
14	75	70	65	0	+	−	1403	1413	1407	1415	1414	1410	5.179
15	70	70	68	−	+	0	1398	1400	1413	1392	1393	1399	8.408

Column group headers: factors (A B C), Coded factors (x_1 x_2 x_3), Replicated Response (Y_1 Y_2 Y_3 Y_4 Y_5).

Using the average response in a 2nd order model results in the following computer output.

DEP VAR: $\bar{Y}$ N: 15 MULTIPLE R: .951 SQUARED MULTIPLE R: .904
ADJ SQUARED MULTIPLE R: .731 STD ERROR OF ESTIMATE: 11.397 F = 5.232 P = .042

VARIABLE	COEFFICIENT	STD ERROR	STD COEF	TOLERANCE	T	P(2 TAIL)
CONSTANT	-934.713	4599.512	0.000	1.0000000	-0.203	0.847
A	-6.475	46.409	-1.113	0.0003	-0.140	0.894
B	4.375	43.719	0.752	0.0003	0.100	0.924
C	74.602	97.241	7.696	0.0002	0.767	0.478
A^2	-0.092	0.237	-2.365	0.0051	-0.386	0.715
B^2	-0.332	0.237	-7.416	0.0007	-1.398	0.221
C^2	-1.060	0.659	-14.876	0.0002	-1.609	0.169
A*B	-0.010	0.228	-0.171	0.0013	-0.044	0.967
A*C	0.300	0.380	4.207	0.0007	0.790	0.465
B*C	0.633	0.380	8.539	0.0007	1.667	0.156

The low tolerance values indicate high multicollinearity which is due to the use of the raw data. The P values are large due to inflated standard errors. The following simple example illustrates this point.

A	A^2	vs	x	x^2
1	1		−1	1
1	1		−1	1
2	4		0	0
2	4		0	0
3	9		1	1
3	9		1	1

$$R^2 = 0.979 \qquad R^2 = 0$$

Using the coded data in a 2nd order model

$$x_i = 2\left(\frac{f_i - \overline{f}}{f_{max} - f_{min}} \right)$$

results in the following computer output.

DEP VAR: $\overline{Y}$ N: 15   MULTIPLE R: .949   SQUARED MULTIPLE R: .901
ADJ SQUARED MULTIPLE R: .723   STD ERROR OF ESTIMATE: 11.618 F = 5.052 P = .045

VARIABLE	COEFFICIENT	STD ERROR	STD COEF	TOLERANCE	T	P(2 TAIL)
CONSTANT	1398.333	6.708	0.000	1.0000	208.464	0.000
x_1	-2.375	4.108	-0.081	1.0000	-0.578	0.588
x_2	18.250	4.108	0.625	1.0000	4.443	0.007
x_3	-17.375	4.108	-0.595	0.9890	-4.230	0.008
x_1^2	-2.667	6.046	-0.062	0.9890	-0.441	0.678
x_2^2	-7.917	6.046	-0.185	0.9890	-1.309	0.247
x_3^2	-9.167	6.046	-0.215	1.0000	-1.516	0.190
$x_1 \cdot x_2$	-0.250	5.809	-0.006	1.0000	-0.043	0.967
$x_1 \cdot x_3$	4.500	5.809	0.109	1.0000	0.775	0.474
$x_2 \cdot x_3$	10.250	5.809	0.248	1.0000	1.764	0.138

The high tolerance values indicate that there is little to no multicollinearity. Notice that P values here are useful for finding the significant effects.

The parsimonious model is: $\hat{y} = f(x_2, x_3)$ which results in

$$\hat{y} = 1387.600 + 17.875(x_2) - 17.75(x_3).$$

To predict $\bar{y}$ for any (x_2, x_3)

$$\hat{\bar{y}} = 1387.600 + 17.875(x_2) - 17.75(x_3)$$

i.e. for $(x_2 = 0, x_3 = 0)$, $\hat{\bar{y}} = 1387.600$.

The prediction interval is

$$\hat{\bar{y}} \pm t(1 - \tfrac{\alpha}{2}, n - p)\hat{\sigma}\sqrt{1 + \underline{x}'_h (X'X)^{-1} \underline{x}_h}$$

where $\underline{x}_h$ = coordinates of desired point $(1, x_{2_h}, x_{3_h})$

and $X = \begin{bmatrix} 1 & x_{21} & x_{31} \\ 1 & x_{22} & x_{32} \\ \vdots & \vdots & \vdots \\ 1 & x_{2n} & x_{3n} \end{bmatrix}$ $X'X = \begin{bmatrix} 15 & 0 & 0 \\ 0 & 8 & 0 \\ 0 & 0 & 8 \end{bmatrix}$ $(X'X)^{-1} = \begin{bmatrix} \frac{1}{15} & 0 & 0 \\ 0 & \frac{1}{8} & 0 \\ 0 & 0 & \frac{1}{8} \end{bmatrix}$

A 95% prediction interval at (0,0) is

$$1387.60 \pm 2.179(11.618)\sqrt{1 + (1,0,0)\begin{bmatrix} \frac{1}{15} & 0 & 0 \\ 0 & \frac{1}{8} & 0 \\ 0 & 0 & \frac{1}{8} \end{bmatrix}\begin{bmatrix} 1 \\ 0 \\ 0 \end{bmatrix}}$$

$$= 1387.60 \pm 2.179(11.618)\sqrt{\frac{16}{15}}$$

$$= 1387.60 \pm 26.146$$

$$= [1361.454, 1413.146]$$

Chapter 5 Problem Set

1. When should you use Response Surface Methodology ? What indicators in ANOVA is (are) there that you should use RSM ?

2. What is the gradient vector and what does it have to do with RSM ?

3. Given you have searched along the gradient and have selected a new experimental region, how do you know this area contains a stationary point ?

   How can you tell if that stationary point is a:

   > maximum ?
   > minimum ?
   > saddle point ?
   > a ridge ?

4. State the main reason for using coded variables in regression analysis.

5. If the gradient vector from a 2 term model is:

$$\underline{g} = (-2.67, 1.33)$$

Where should you conduct the first experiment for RSM ? Find the second test values as well. If the second test value yields the maximum, what values should you use in a L_4 design ?

Chapter 5 Bibliography

1. Box, G.E.P. and Draper, N.R. (1987) *Empirical Model Building and Response Surfaces,* John Wiley and Sons, Inc. New York.

2. Montgomery, Douglas C. (1984) *Design and Analysis of Experiments*, John Wiley and Sons, 2nd Edition.

3. Myers, Raymond H. (1976) *Response Surface Methodology*, Virginia Polytechnic Institute.

CHAPTER 6: TAGUCHI PHILOSOPHY, DESIGN AND ANALYSIS *

6.1 Introduction

An analysis of Japanese success in the American marketplace and their ability to produce quality products is a matter of current interest. The Japanese did in fact turn around a reputation of manufacturing junk to one of manufacturing quality products that can and still do obtain higher market value than similar U.S. products. The tools used by the Japanese to accomplish this remarkable turnaround are focused at the product design phase versus quality inspection during manufacturing. By designing quality into the product from its conception, fewer resources are required to produce a quality product. The use of Taguchi Methods in parameter design, known as Quality Engineering, typically results in increased productivity, improved quality, and increased efficiency. The designs used by Taguchi were developed primarily by British and American statisticians as fractional-factorials, Plackett-Burman, Latin Square and mixed designs. The idea of orthogonality has been around a long time; however, Taguchi's approach is different in that he (1) stresses only a few basic designs most commonly used in industry, (2) provides a table of these designs, and (3) avoids statistical rigor by providing a cookbook approach to the analysis.

The Taguchi approach to quality control emanates from the following definition of quality: "The **quality** of a product is the (minimum) loss imparted by the product to society from the time the product is shipped" [11]. From this definition, a loss function is developed that translates any deviation of a product from its target parameter into a financial measure. The parameter(s) utilized in this measure are those which have an impact on quality. Any deviation from the target of the parameter(s), either through variability (noise) or by a deviation of the average from target, translates into increased loss. The process of reducing variability and moving the mean to the target is based on designed experiments and emphasizes a phase of product and process design that Taguchi calls **parameter design.**

---

* This chapter contains much of the information found in Warsavage and Schmidt [12].

Despite much criticism of the Taguchi approach to designed experiments by traditional statisticians, the approach continues to expand. The expansion may be due to the simplicity associated with the presentation of the material by the American Supplier Institute and through the use of success stories such as those from Ford Motor Co., ITT, Digital Equipment Corporation, Xerox, Nippon Telephone and Telegraph Co., Toyota and Fuji Film. Historically, engineers would vary only one factor at a time or intuitively select the best combination of factor settings based on engineering judgement. This can be extremely inefficient and will rarely lead to an optimal design. Traditional statistics were certainly an alternative; however, the presentation was not always simple and due to its perceived complexity, non-statisticians were less inclined to use it. One positive result of the discussions surrounding the use of Taguchi methods has been a great awakening of interest in industry to numerous experimental design approaches. Many engineers (as compared to just a couple of years ago) are now actively learning how to apply various types of orthogonal arrays and analysis approaches. Yet with all the discussions and learning that is taking place, probably fewer than 10% of the experiments done in industry today utilize a "designed experiment" approach. Much remains to be done.

The remainder of this chapter is divided into three parts. Under the topic of Taguchi philosopy and methodology, the loss function and Taguchi's approach to experimental design are discussed. The Taguchi approach to experimental design is presented in the second section. In the last section, the Taguchi approach to experimental design is illustrated.

6.2   Taguchi Philosophy and Methodology

6.2.1 Loss Function

Specification limits are the means by which manufacturers determine the details of a manufacturing product and process. These limits are frequently used in such a way as to imply that if the parameter of a part falls within these limits, the product is satisfactory; if it falls outside of these limits, it is unsuitable. A target value may be given, but hitting the target has not attained an importance beyond just being within spec. Taguchi implements his definition of quality so that any deviation from target results in incremental loss. He accomplishes this through a loss function which Chernoff and Moses [1] address as a part of statistical decision theory. Taguchi's contribution is the application of the loss function to the measurement of quality.

He most frequently uses a quadratic loss function (based on a Taylor series approximation) where the function increases as the parameter deviates either side of the target (see figure 6.1). The quadratic loss function is symmetric around the target and loss is minimized by producing all items as close as possible to the target. Since the target cannot always be met, the strict use of spec limits implies that no loss is incurred if all items are within spec. However, this does not hold in practice because a product just within specification is more like an out of spec product than a product right on target. As an example, consider a door built within spec but at the upper limit, and the door frame within spec but at the lower limit. This obviously presents a problem which will lead to reduced quality, inefficiency and increased costs to the manufacturer. The loss function therefore reveals the impact of deviation from target.

Taguchi employs the use of a quadratic function to model loss because of its simplicity in the development of a signal to noise ratio used to optimize the process. Other functions may be argued more accurate in that a deviation to one side of the target may be more of a problem than a deviation to the other side. The important point is that the development of a loss function results in evaluating costs associated with a deviation from the target and simplicity of the quadratic loss function results in ease of implementation. Based on this approach, a more analytical method of assigning spec limits can be utilized instead of subjectivity alone. Improper specs result in unnecessary increased costs if they are too tight, or they can pass on problems to internal or external customers if they are too lenient.

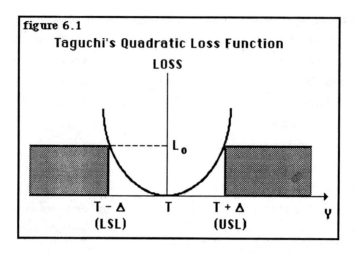

figure 6.1

Taguchi's Quadratic Loss Function

Another important aspect of the loss function is that it maps deviations from the target into a financial measure. Everyone understands money, and since it is a common measure, comparisons can be made between products and processes. Operationally, the average loss would be computed using a sample large enough to characterize the process measured on a critical parameter. Using a quadratic loss function, the average loss associated with each product would be computed as follows [4]:

$$L_i = k(y_i - T)^2$$

where

y_i is the quality characteristic of interest for product i

T is the quality charactistic target

k is a constant that converts deviation to a monetary value.

To determine the value of k for a specific target, you need to estimate the loss at some measured value of the characteristic in question. Referring to figure 6.1, assume you estimate the loss, L_0, at a point on the horizontal axis, y_0 (in this case $y_0 = LSL$). Then using $L_0 = k(y_0 - T)^2$ to represent the quadratic loss function and substituting known values for L_0, y_0 and T, you simply solve for k.

For a specified target T (nominal is best) the total loss, L, for n products is

$$L = k \sum_i \left(\frac{1}{n}\right) (y_i - T)^2$$

where   n = sample size

y$_i$ = values of the critical parameter

T = target value.

The decomposition of the loss function for all n products is developed as follows:

$$L = k \sum \frac{1}{n} (y_i - T)^2$$

$$= \frac{k}{n} \sum (y_i - \bar{y} + \bar{y} - T)^2$$

$$= \frac{k}{n} \sum [(y_i - \bar{y})^2 + (\bar{y} - T)^2 + 2(y_i - \bar{y})(\bar{y} - T)]$$

$$= \frac{k}{n} \sum (y_i - \bar{y})^2 + k(\bar{y} - T)^2 + (\bar{y} - T)\frac{k}{n} \sum (y_i - \bar{y})$$

Since the last term results in a value of 0, $L = k \left[\sigma_y^2 + (\bar{y} - T)^2 \right]$. Thus, loss is partitioned into variance of the response and deviation of the average response from the target. Therefore, to minimize loss, it is clear that not only do you minimize the average response from the target, but you also must minimize response variability.

One illustration of these points is cited by L. Sullivan [10] on the color intensity of television sets. Figure 6.2 illustrates the distribution of the product on this parameter for Sony Japan and Sony U.S.A. Notice that Sony Japan was producing some products outside the spec limits, but Sony U.S.A. was not. Under

6-5

conventional methods, Sony U.S.A. would be preferred. Yet, if given a choice, customers chose Sony Japan. A loss function can model this preference, whereas strict adherance to 'spec' limits cannot.

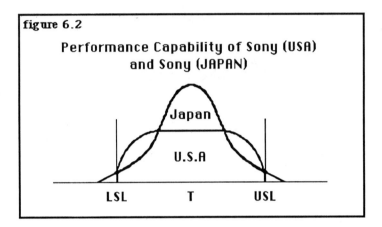

figure 6.2

Performance Capability of Sony (USA) and Sony (JAPAN)

Japan

U.S.A

LSL   T   USL

Ideally, you would want the variability of the product to appear as shown in figure 6.3. Since the majority of the response measurements occur near the target, the

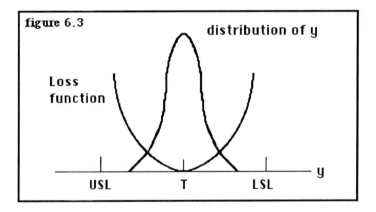

figure 6.3   distribution of y

Loss
function

USL   T   LSL   y

loss is reduced. The use of traditionally designed experiments to evaluate the optimal parameter settings for each factor will minimize the $(\bar{y} - T)^2$ part of the loss function. In order to minimize the σ_y^2 portion, Taguchi has developed an outer array (or noise array) in conjunction with a transformation of the response called signal to noise. This measure is discussed further in section 6.3.3.

Two quantitative measures which relate σ_y^2 to the specification limits are C_p and C_{pk}.

$$C_p = \frac{USL - LSL}{6\sigma}$$ where $C_p < 1$ is unsatisfactory, $1.0 \le C_p \le 1.33$ is marginal and $C_p > 1.33$ is desired. Since C_p assumes $\bar{y} = T$, which is not always the case, a new measure incorporating the product mean is C_{pk}. To calculate C_{pk},

$$C_{pk} = \frac{\min(USL - \bar{y}, \bar{y} - LSL)}{3\sigma}$$ where $C_{pk} < 1$ is unsatisfactory, $1.0 \le C_{pk} \le 1.33$ is marginal and $C_{pk} > 1.33$ is desired. The higher the C_{pk} value, the more the σ^2 portion of the loss function is minimized and the closer $\bar{y}$ is to T. Note that σ is typically estimated with S, the sample data standard deviation.

6.2.2 Quality Engineering

Currently there is a lot of emphasis placed on techniques to control the building of a product once it has entered the manufacturing phase. These techniques are referred to as statistical quality or process control. They were discovered and used by Shewhart in the 1920's and recently made popular by Deming's success with the Japanese. Taguchi refers to these techniques, as well as some innovations of his own, as **On-line Quality Control.**

One of Taguchi's major contributions is to direct attention to the contributions that statistics and quality engineering can make in the design phase of a product. These techniques are thus referred to as **Off-line Quality Control.** Statistics and a quality oriented approach taken in the design phase can save large amounts of resources in terms of time and effort to manufacture a product. This approach also leads to a reduction of costly scrapping and reworking of the product.

6.2.3 Quality by Design

As previously discussed, the Taguchi approach emphasizes and elaborates on the design phase of a product's life. He subdivides the design phase into three steps or subphases: system design, parameter design, and tolerance design.

System Design. This is the inventive phase where engineers make the prototype. Historically, this is an area in which the U.S. has excelled to the point of leading the world in patents, whereas the Japanese have contributed little. Initially, the Japanese borrowed western product design and then manufactured it better; however, they are currently placing more emphasis on system design. In certain areas, the Japanese are surpassing the U.S.

6.2.4 Parameter and Tolerance Design.

Taguchi argues that the Japanese have added or at least emphasized two additional steps between the prototype and the hand-off to manufacturing. The most important is the parameter design step. In this step, Taguchi analyzes the effect of each of the controllable factors and determines the best factor settings in relation to the noise factors. Factors that most affect performance are determined as well as factor settings that are optimal in terms of performance, cost, ease of manufacture, and reliability. At this stage, one should avoid factor settings difficult to maintain and control. If those settings cannot be avoided, then tight tolerances must be set. The setting of those tolerances is addressed in the **tolerance design subphase.** The identification of critical factors and their optimal values are found by experimentation using designed experiment techniques. If tolerances must be determined, these are also determined experimentally.

6.2.5 Controllables and Noise

The factors that affect the performance of a product by increasing variability are referred to as noise or noise factors. Product reliability and ease of manufacture are the result of specifically designing the product to be resistant to noise. In contrast, those factors that are relatively easy to control are called controllables. Some factors can actually fall into both categories. For example, the temperature of a wave solder process can be set closer to 150 degrees than 200 degrees; however, even if 150 degrees is the target, the measured temperature may vary a few degrees within a run or between runs. In the first case, the temperature is considered a controllable and in the second it is noise. The Taguchi strategy investigates the effect of large and small changes. Another noise factor might be ambient humidity which may be systematically varied during experimentation in the parameter design phase, but remains uncontrollable in the operational phase of the process.

Taguchi's consideration of noise results in some changes in the way that an experimental design is set up and analyzed. Noise, a pseudonym for experimental error, is usually considered in statistical thinking as the element of unpredictability of the outcome. It is assumed to be independent of the factor settings. If there is an indication that this element of variability changes with the factor settings then a transformation of the data (frequently logarithmic) is made before the analysis. In traditional experimental design, you would assume that there are no errors in the factor settings. Any small errors, such as they are, are absorbed by randomization. Taguchi recommends that these errors or noise can be directly analyzed using experimental design functions.

An example from Taguchi [11] illustrating the application of this thinking concerns the characteristic function of a transistor (see figure 6.4). Consider the following scenario:

(1) Target output voltage is 100.

(2) Gain has a large effect on output voltage and is non-linear.

(3) The circuit engineer has many options in terms of choosing the gain setting.

One approach would be for the circuit engineer to set the transistor gain at 18 so as to obtain the desired output voltage of 100. Please note, however, that small deviations from the target of 18 will result in relatively large variations in our output voltage. A better approach would be for the circuit engineer to select a transistor gain at a much flatter portion of the curve (for example, 30), thus minimizing the potential resultant variation in output voltage. By using another component in the circuit (which has a linear effect on output voltage), the engineer can adjust the mean value from 120 to 100. Provided the reconfiguration did not increase the variability in the output voltage, the result would be more satisfactory.

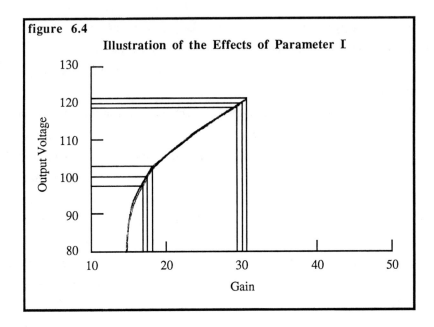

figure 6.4

Illustration of the Effects of Parameter I

In general, running transistors near the point where the slope of the output voltage is small may result in a higher quality product. This of course oversimplifies the problem but it does illustrate a valuable point in that some factor settings can be used to reduce variability and others to obtain the response closest to the target.

6.3 Taguchi Strategy of Experimentation

Taguchi's strategy is characterized by three elements: (1) orthogonal arrays based on fractional factorials, Plackett-Burman or Latin Square experimental designs, (2) design modification that incorporates information about potential noise factors, and (3) analysis of a transformation that he calls the signal to noise ratio along with the analysis of the means.

6.3.1 Saturated Fractional Factorial Experimental Designs

A fractional factorial experiment is saturated when the design only allows for the estimation of main effects. For example, instead of estimating 4 main effects with 16 runs, it is possible to estimate 15 main effects independently of each other with those 16 runs. In this case, all the interactions are aliased with these main effects.

Fractionated designs, when used as part of a screening process, are effective when there are large numbers of factors. Since large numbers of factors usually include some which do not affect the response measure, the screening process provides a way to remove from consideration (screen out) those unimportant factors as efficiently as possible. There are potentially hundreds of factors in most industrial models and very little is known about their effects. The screening process is therefore a way to start reducing the number of factors in a systematic fashion.

6.3.2 Designs that Assess Information About Noise

Randomization is the key to reaching conclusive results in a time and resource efficient manner. Randomization in factorial designs implies that runs of all different factor combinations will be completed in a random order. There are two reasons for wanting to randomize: the critical assumption of independence is usually ensured through random sampling and randomization, and long–term trends and/or systematic noise factors not included in the design matrix will become part of the error by being spread in an equally likely fashion over the experimental conditions [6]. Since randomization in industrial experiments is frequently time consuming and often impossible, Taguchi does not emphasize the use of randomization but attempts to compensate for it by incorporating noise in the design. He therefore assumes all potential noise factors can be measured and will be included in the design. If this is not the case, there exists a strong possibility that the results will be ambiguous.

Taguchi's robust designs use two sets of factors: an inner array of controllables and an outer array of noise factors. At each combination of the controllables, an outer array of noise factors is run. So at each combination of the controllables, a separate estimate of the effect of noise factors can be made. This is similar to a sensitivity analysis by J. S. Hunter, [3]. The primary objective of robust designs is to determine optimal settings for the controllable factors which will result in a desired response which is invariant to noise.

6.3.3 Signal to Noise Ratio

Sensitivity to noise is analyzed through a signal to noise statistic which involves both location and dispersion effects. Many different estimates of signal to noise have been tested, but typically one of three are used. When a specified target is the criterion, a statistic related to the coefficient of variation is:

$$(S/N) = 10 \log_{10} \frac{\bar{y}^2}{S^2}$$

$$\begin{cases} \bar{y} = \sum \frac{y_i}{n} \\ S = \sqrt{\frac{\sum (y_i - \bar{y}_i)^2}{n-1}} \end{cases}$$

Taking the logarithm and multiplying by ten converts y to decibel units. This statistic is closely related to the conventional signal to noise ratio which is $\bar{y}/s$. The optimum factor levels for any process is found by maximizing (S/N). In the case of a zero or infinite target, the above formula is inappropriate because (S/N) is maximized by closeness to the target values. These statistics can be derived from the formulae for average loss, resulting in:

$$(S/N) = -10 \log_{10} MSD$$

where $\quad MSD = \frac{1}{n}\sum y_i^2$, for a zero target (smaller is better),

or $\quad MSD = \frac{1}{n}\sum \frac{1}{y_i^2}$, for an infinite target (larger is better).

A typical Taguchi stragegy involves analyzing one of these statistics with or without a graphical analysis of means.

6.4   Example of a Taguchi Experimental Design

In this section, the Taguchi strategy is illustrated with an example of a wave solder process optimization. In a printed circuit board assembly plant, parts are inserted either manually or automatically into a bare board with a circuit printed on it. After most of the parts are inserted, the board is put through a wave solder machine. This process is a means of connecting, both electrically and mechanically, all the parts into the circuit. Boards are placed on a conveyor, and taken through a series of steps. They are bathed in a flux mixture to remove oxide. To minimize warpage, they are preheated before the solder is applied. The soldering process takes place as the boards move across the wave of solder. After that, the board usually goes through a device similar to a dishwasher to remove any debris. Because of the many factors used in controlling the process, it is an excellent candidate for experimentation. For wave soldering processes which have not been studied statistically, it is not uncommon for 10-20% of the assembly work force to be involved in solder touchup.

The data used in this example are simulated to illustrate the Taguchi strategy. The objective of the experiment is to minimize the number of solder defects per million joints. Table 6.1 lists the factors and their settings decided upon for this example. The choice of parameters to be used must be carefully considered in the brainstorming phase. Experts at other plants with different types of wave solder machines might choose a different set of factors. The approach is to begin exploring systematically the factors affecting the wave solder process. For the inner array of controllables, five factors are considered. After brainstorming, it was decided to test each factor at two levels and include an interaction between solder pot temperature and conveyor speed.

The factors in the noise array are selected as well. Because several different types of assemblies are run through this wave solder process, two different types of assemblies were used. The objective is to find one setting for the wave solder process that is suitable for both types of assemblies. The design will also indicate if assembly type interacts with any of the controllables. In addition to product noise, both the conveyor speed and solder pot temperature will be moved around the initial setting given by the controllables array. This is because it is difficult to set the conveyor speed with any degree of accuracy and it is also difficult to maintain solder pot temperature. So the team of engineers chose to include these variables in the noise array variables to determine how much noise affects the process.

Eight runs will be used to test the effects of the five controllables in a Taguchi L8 design (see table 6.2a). Notice that for each factor, there are four runs with the factor set at the low setting and four runs with the factor set at the high setting. This balancing is a property of the orthogonality of the set of runs. Table 6.2b lists the array of noise factors to be run at each of the eight settings of the controllables. This is a Taguchi L4 design. The combination of the inner and outer arrays results in each run of the controllables being repeated over the 4 combinations of the noise factor.

table 6.1

A Designed Experiment for Wave Solder
An Example of the Use of Orthogonal Arrays

Controllable

Factors	Levels	
	Low	High
(1) Solder Pot Temperature (S)	480° F	510° F
(2) Conveyor Speed (C)	7.2 ft/m	10 ft/m
(3) Flux Density (F)	.9 °	1.0 °
(4) Preheat Temperature (P)	150 F	200 F
(5) Wave Height	0.5"	0.6"

Noise

Factors

(1) Product Noise	Assembly #1, Assembly #2
(2) Conveyor Speed Tolerance	-0.2, +0.2 ft/m
(3) Solder Pot Tolerance	-5° F, +5° F

table 6.2a

A Designed Experiment for Wave Solder
An Example of the Use of Orthogonal Arrays

Controllables Design
Inner Array

Run	Solder Pot Temperature	Conveyor Speed	Flux Density	Preheat Temperature	Wave Height
1	510	10.0	1.0	150	0.5
2	510	10.0	0.9	200	0.6
3	510	7.2	1.0	150	0.6
4	510	7.2	0.9	200	0.5
5	480	10.0	1.0	200	0.5
6	480	10.0	0.9	150	0.6
7	480	7.2	1.0	200	0.6
8	480	7.2	0.9	150	0.5

Note:

For 5 variables with 2 levels, a full factorial would need
$2**5 = 32$ runs. By choosing runs cleverly and ignoring most
interactions,the space was covered with 8 combinations (L8).

table 6.2b

A Designed Experiment for Wave Solder
An Example of the Use of Orthogonal Arrays

Outer Array

At each combination of the inner array,
an outer array of noise factors is run.

RUN				
1	2	3	4	Parameter
Assm#1	Assm#1	Assm#2	Assm#2	Product Noise
- 0.2	+ 0.2	- 0.2	+ 0.2	Conveyor Tolerance
- 5	+ 5	+ 5	- 5	Solder Tolerance

Table 6.3 presents the results of the experiment. The rightmost 2 columns give the averages and signal to noise ratio for each of the eight settings of the controllables or inner array. Because the objective is to minimize the outcome variable, solder defects per million, the formula for signal to noise is

$$-10 \log \left(\tfrac{1}{n} \sum y_i^2 \right).$$

An inspection of the results in table 6.3 reveals that run number 4 maximized (S/N) and therefore, our "experimental champion" is S = 510, C = 7.2, F = .9, P = 200 and W = .5. It is quite possible that some of the factors are more influential than others. To determine which factors are most important, marginal averages are computed by averaging the four results at a particular level of a factor. For example, the first four runs of the inner array are with the solder pot temperature set at the higher level, the second four are at low solder pot temperature. So these sixteen results (4 runs of the inner array each with 4 runs of the outer array) are averaged together to obtain 170 average solder defects per million. The compilation of these results are presented in table 6.4 and displayed graphically in figure 6.5. Analysis of the table and graph indicates that low flux density, high solder pot temperature and high wave height reduced the number of solder defects. It was this combination that influenced run number four. The effect for conveyor speed is marginal; however, economic considerations, such as throughput, might influence the decision to run the conveyor at a higher speed. Preheat temperature also appears insignificant with economic considerations favoring the lower temperature. Notice from figure 6.6 that the anticipated interaction between conveyor speed and solder pot temperature does not appear important. The marginals for these lines are found by computing the four combinations of solder pot temperature and conveyor speed. The points are plotted against the levels of solder pot temperature and the lines for constant conveyor speed are drawn. Notice that the lines are nearly parallel, indicating that the effect for conveyor speed is the same at each level of solder pot temperature. From the S/N column of table 6.4, a "paper champion" would be

table 6.3

A Designed Experiment for Wave Solder

Combined Inner and Outer Arrays
Results

RUN

Noise Factors	1	2	3	4
Product Noise	#1	#1	#2	#2
Conveyor Tolerance	+0.2	+0.2	-0.2	+0.2
Solder Tolerance	- 5	+ 5	+ 5	- 5

Controllable Factors

Run	Solder	Conveyor	Flux	Preheat	Wave						Mean S/N*
1	510	10.0	1.0	150	0.5	194	197	193	275	215	-46.75
2	510	10.0	0.9	200	0.6	136	136	132	136	135	-42.61
3	510	7.2	1.0	150	0.6	185	261	264	264	244	-47.81
4	510	7.2	0.9	200	0.5	47	125	127	42	85	-39.51 *
5	480	10.0	1.0	200	0.5	295	216	204	293	252	-48.15
6	480	10.0	0.9	150	0.6	234	159	231	157	195	-45.97
7	480	7.2	1.0	200	0.6	328	326	247	322	305	-49.76
8	480	7.2	0.9	150	0.5	186	187	105	104	145	-43.59

* EXPERIMENTAL CHAMPION

table 6.4

A Designed Experiment for Wave Solder

Analysis

Parameter	Level	Mean	S/N
Solder Pot Temperature	480	225	-46.87
	510	170	-44.17 *
Conveyor Speed	7.2	195	-45.17
	10.0	200	-45.87
Flux Density	0.9	140	-42.91 *
	1.0	255	-48.11
Preheat Temperature	150	200	-46.03
	200	194	-45.01
Wave Height	0.5"	174	-44.50
	0.6"	220	-46.54
Interaction		200	-45.68
		194	-45.36

* These are the optimum level settings for each factor based on S/N. Factors without an asterisk are not significant and their levels can be based on other considerations.

figure 6.5

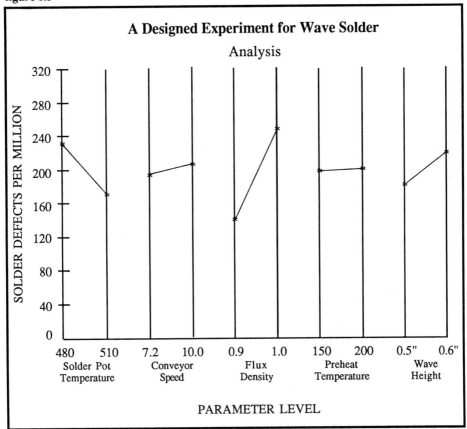

S = 510, C = 10.0, F = .9, P = 150 and W = 0.5.   Where C, P and W were set
based on economic considerations. These combinations were never tested in the
experiment, thus the term "paper champion". To predict the S/N at these
settings, regression analysis is used to generate a prediction equation for S/N
using only the significant effects of S and F.

$$S/N = -45.52 + 1.35(S) - 2.60(F)$$

where coded S and F are (+) for high and (–) for low. Using this equation, the
predicted value for our "paper champion" is -41.57. Since there were no
replicated runs for S/N, if significance tests and/or prediction intervals are of
interest, the df for terms not in the math model are pooled as an estimate of
experimental error.

Although this S/N value for the "paper champion" is a little smaller than that
of the "experimental champion", the "paper champion" warrants further
investigation because it could possibly cost less to produce. Confirmatory runs
should be made for both champions to check for a C_{pk} value at these factor
settings and to make the final selection of factor settings.

The use of S/N to simultaneously analyze the mean response and variability
provides a simple approach to the analysis. The reader is referred, however, to
Appendix 4.A where S/N analysis did not appear sensitive to information on
variability. The combining of $\bar{y}$ and s in a S/N fashion may not always be the
ideal approach. One should consider separate analysis of $\bar{y}$ and s (or ln s) to
supplement S/N analysis.

figure 6.6

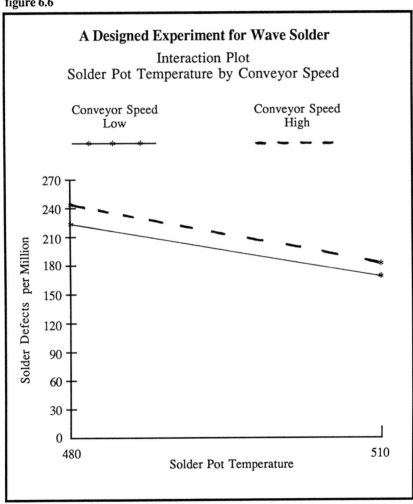

6.5   A Second Taguchi Example

Example 1. Given 5 factors A, B, C, D and E, at 2 levels with suspected A x B and A x C interaction, the objective is to <u>maximize Y</u>. Use an L8 with 2 observations per run and factors assigned to columns using the following linear graph.

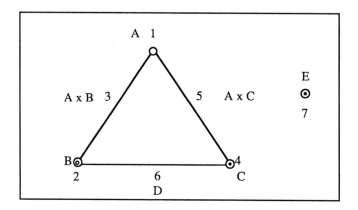

The design appears as

Run	A 1	B 2	AxB 3	C 4	AxC 5	D 6	E 7	Response Y_1	 Y_2	Mean $\overline{Y}$	S/N	R
1	1	1	1	1	1	1	1	62	66	64	36.11	4
2	1	1	1	2	2	2	2	63	68	65.5	36.30	5
3	1	2	2	1	1	2	2	69	70	69.5	36.84	1
4	1	2	2	2	2	1	1	64	65	64.5	36.19	1
5	2	1	2	1	2	1	2	73	77	75	37.49	4
6	2	1	2	2	1	2	1	47	42	44.5	32.90	5
7	2	2	1	1	2	2	1	38	39	38.5	31.71	1
8	2	2	1	2	1	1	2	57	62	59.5	35.47	5

The graphical analysis of means appears as shown on the next page.

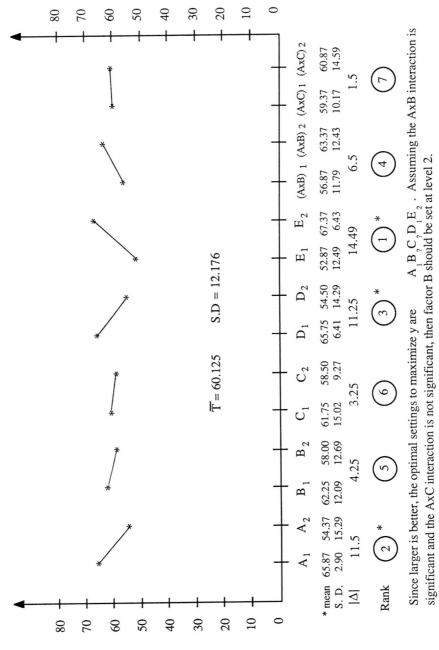

	A_1	A_2	B_1	B_2	C_1	C_2	D_1	D_2	E_1	E_2	$(AxB)_1$	$(AxB)_2$	$(AxC)_1$	$(AxC)_2$
*mean	65.87	54.37	62.25	58.00	61.75	58.50	65.75	54.50	52.87	67.37	56.87	63.37	59.37	60.87
S. D.	2.90	15.29	12.09	12.69	15.02	9.27	6.41	14.29	12.49	6.43	11.79	12.43	10.17	14.59
$|\Delta|$	11.5		4.25		3.25		11.25		14.49		6.5		1.5	
Rank	②	*	⑤		⑥		③	*	①	*	④		⑦	

$\overline{T} = 60.125$ S.D = 12.176

Since larger is better, the optimal settings to maximize y are $A_1 B_? C_? D_1 E_2$. Assuming the AxB interaction is significant and the AxC interaction is not significant, then factor B should be set at level 2.

These graphs indicate that the A x B interaction is probably insignificant. However, to demonstrate the incorporation of interaction terms, this example will be analyzed with and without A x B significant.

From the data table, the "experimental champion" is run #5 where the parameter settings are $A_2B_1C_1D_1E_2$.

A graphical analysis of means where A x B is insignificant results in a "paper champion" of $A_1B_?C_?D_1E_2$ where B and C are set at the most economically suitable levels. Assume we choose B_1 and C_1; then our "paper champion" is $A_1B_1C_1D_1E_2$. This run was not made, but can be estimated by the following:

$$\hat{\mu} = \overline{T} + (\overline{A}_1 - \overline{T}) + (\overline{B}_1 - \overline{T}) + (\overline{C}_1 - \overline{T}) + (\overline{D}_1 - \overline{T}) + (\overline{E}_2 - \overline{T})$$

where I will set $(\overline{B}_1 - \overline{T}) + (\overline{C}_1 - \overline{T}) = 0$ since these effects are insignificant. Thus, $\hat{\mu} = \overline{A}_1 + \overline{D}_1 + \overline{E}_2 - 2(\overline{T}) = 65.87 + 65.75 + 67.37 - 2(60.125) = 78.74$. In terms of $\overline{Y}$, this setting appears to be better than the "experimental champion" and should be included in the confirmatory runs.

An analysis of means based on a significant A x B interaction would first have to consider an A x B interaction graph as shown below.

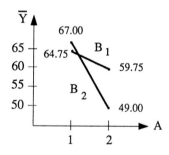

This graphic indicates that A_1B_2 is the better setting and since interactions will

take precedence on parameter settings, the new "paper champion" is $A_1B_2C_?D_1E_2$.*
Again, for whatever reason, set C at C_1. The new estimate is

$$\hat{\mu} = \overline{T} + (\overline{A}_1 - \overline{T}) + (\overline{B}_2 - \overline{T}) + (\overline{C}_1 - \overline{T}) + (\overline{D}_1 - \overline{T}) + (\overline{E}_2 - \overline{T})$$
$$+ \left[(\overline{A_1B_2} - \overline{T}) - (\overline{A}_1 - \overline{T}) - (\overline{B}_2 - \overline{T}) \right]$$

and again set $(\overline{C}_1 - \overline{T}) = 0$. Thus,

$$\hat{\mu} = \overline{A_1B_2} + \overline{D}_1 + \overline{E}_2 - 2(\overline{T})$$
$$= 67.00 + 65.75 + 67.37 - 2(60.12)$$
$$= 79.88$$

This is a better prediction than any of the previous ones. The bottom line is that you might want to have a confirmatory experiment to test which of these 3 is the best.

(1) "Experimental champion" $A_2B_1C_2D_1E_2$, $\hat{\mu} = 75.00$

(2) "Paper champion" without A x B interaction $A_1B_1C_1D_1E_2$, $\hat{\mu} = 78.74$

(3) "Paper champion" with A x B interaction $A_1B_2C_1D_1E_2$, $\hat{\mu} = 79.88$

This can be accomplished by setting $D = D_1$, $E = E_2$ and C at the most economic level. Then run an L4 to test A, B, and A x B.

The above analysis was based only on signal ($\overline{y}$) information. Since each run was repeated, you can also analyze the data using variability per run or S/N. First, consider variability per run where the range (R) is a measure of variability. Regression analysis output is shown below.

---

* If factor A is a continuous one and also one that is hard to hold at a fixed value, you can increase response variability by setting B at level 1 because the slope of B_1 over the range of A is much less than that of B_2. If this is done, you would have to accept the decrease in $\overline{y}$ associated with a variance improvement.

```
DEP VAR:    R      N:   8    MULTIPLE R:   1.000    SQUARED MULTIPLE R:   1.000

ADJUSTED SQUARED MULTIPLE R:      1.000    STANDARD ERROR OF ESTIMATE: 0.000


VARIABLE  COEFFICIENT  STD ERROR  STD COEF   TOLERANCE      T     P(2 TAIL)

CONSTANT     3.250      0.000        .       1.0000000      .        .
   A         0.500      0.000        .       1.0000000      .        .
   B        -1.250 *    0.000        .       1.0000000      .        .
   AB       -0.500      0.000        .       1.0000000      .        .
   C         0.750      0.000        .       1.0000000      .        .

   AC       -0.500      0.000        .       1.0000000      .        .
   D        -0.250      0.000        .       1.0000000      .        .
   E         0.500      0.000        .       1.0000000      .        .
```

This analysis indicates that factor B has a larger dispersion effect than the others. The negative coefficient indicates that B_2 has less variability than B_1. This can also be seen by looking at the data.

Using S/N as the measure of interest, the experimental champion is run #5, $A_2B_1C_1D_1E_2$; however, the regression analysis shown below differs somewhat. Remember from Chapter 4 that marginal effects are related to the regression coefficients. For example, the regression coefficients below are

$$\left(\overline{S/N}_{(2)} - \overline{S/N}_{(1)}\right) \div 2$$ for each effect (assuming level 2 is (+) and level 1 is (−) for the design matrix).

```
DEP VAR:    S/N     N:   8    MULTIPLE R:   1.000    SQUARED MULTIPLE R:   1.000

ADJUSTED SQUARED MULTIPLE R:      1.000    STANDARD ERROR OF ESTIMATE: 0.000


VARIABLE  COEFFICIENT  STD ERROR  STD COEF   TOLERANCE      T     P(2 TAIL)

CONSTANT    35.376      0.000        .       1.0000000      .        .
   A        -0.984 *    0.000        .       1.0000000      .        .
   B        -0.324      0.000        .       1.0000000      .        .
   AB        0.479      0.000        .       1.0000000      .        .
   C        -0.161      0.000        .       1.0000000      .        .
   AC        0.046      0.000        .       1.0000000      .        .
   D        -0.939*     0.000        .       1.0000000      .        .
   E         1.149      0.000        .       1.0000000      .        .
```

Based on these findings, the paper champion is $A_1 B_1 C_1 D_1 E_2$ which produces a predicted S/N of $\hat{\mu}_{S/N}$ =35.376 - .984 (-1) - .939 (-1) + 1.149 = 38.448.

Combining all three analyses ($\bar{y}$, R and S/N), I would be inclined to set A_1 to maximize $\bar{y}$ and S/N, B_2 to minimize R, C based on economical considerations, D_1 to maximize $\bar{y}$ and S/N, and E_2 to maximize $\bar{y}$ and S/N.

As a side light, the following table provides information as to significance of effects using S/N as the dependent variable while pooling SS for C and AC into an error term.

DEP VAR: SN   N: 8   MULTIPLE R: .996   SQUARED MULTIPLE R: .992

ADJUSTED SQUARED MULTIPLE R: .972   STANDARD ERROR OF ESTIMATE: 0.336

VARIABLE	COEFFICIENT	STD ERROR	STD COEF	TOLERANCE	T	P(2 TAIL)
CONSTANT	35.376	0.119	0.000	1.0000000	298.236	0.000
A	-0.984	0.119	-0.524	1.0000000	-8.293	0.014
B	-0.324	0.119	-0.172	1.0000000	-2.729	0.112
AB	0.479	0.119	0.255	1.0000000	4.036	0.056
D	-0.939	0.119	-0.500	1.0000000	-7.914	0.016
E	1.149	0.119	0.611	1.0000000	9.684	0.010

ANALYSIS OF VARIANCE

SOURCE	SUM-OF-SQUARES	DF	MEAN-SQUARE	F-RATIO	P
REGRESSION	28.021	5	5.604	49.788	0.020
RESIDUAL	0.225	2	0.113		

This example highlights the shortcomings of the S/N ratio for identifying dispersion effects. As discussed in Chapter 4, the authors recommend that you also analyze $\bar{y}$ and ln(s) separately for determining location and dispersion effects.

6.6 Comparison of Traditional and Taguchi Methods

The traditional designed experimental flowchart might appear as follows:

1. Brainstorm (as discussed in Chapter 1).

2. Effects to be tested typically include main and 2-way interaction effects.

3. Determine desired levels of α and β in order to estimate the appropriate sample size to measure effects.

4. Build an orthogonal design (usually R_{IV}) to satisfy (2) and (3) above.

5. Randomize the order of the runs to spread all unmeasured noise factors evenly across the other effects. This step ensures internal validity of the experiment and allows for causality analysis.

The Taguchi approach flowchart would appear as follows:

1. Brainstorm (as discussed in Chapter 1).

2. Effects to be tested usually consist of only main effects. Taguchi methods can incorporate 2-way interactions but priority is given to main effects for simplicity.

3. Discussions about α and β normally are not addressed. Sample size, n, is found based upon the number of effects to be tested from (2) and the number of levels of each factor.

4. Orthogonal designs are tabled and and because of de-emphasis of interactions are usually of R_{III}.

5. Randomization is de-emphasized. Factors are placed in up to 4 groups based on their difficulty level to be reset during experimentation. The hardest factors to reset typically are assigned to columns with only one change of levels (for 2-level designs) and the easiest to reset are assigned to alternating 1, 2 values. Thus, randomization is not enforced, making the experiment vulnerable to confounding of unsuspecting noise factors. Advocates of Taguchi will argue that if all important controllable and noise factors are in the design then the

confounding problem is resolved. Traditionalists will still be concerned about factors which were not included due to incomplete brainstorming or inability to be measured.

6. Run the experiment and collect the data.

6. Run the experiment and collect the data.

7. Perform hypothesis tests on all desired effects and classify as significant or not significant using the t or F test. Use replication at some or all points to estimate error and/or pool higher order inter- action SS for the error. Recent emphasis is on the use of NPP plots for unreplicated designs.

7. Usually, hypothesis tests are not emphasized; instead, a graphical analysis is conducted or the S/N ratio is used. If ANOVA is conducted, the error estimate is based on the pooling of insignificant effect sums of squares. Rules for pooling of effect sums of squares are not rigorous. F values are tested for significance per classical procedures; however, it is recom- mended that final unpooled F ratios greater than 2 not be ignored [4].

8. Build a parsimonious mathematical model to (i) form prediction inter- vals and (ii) estimate the optimal response through the use of Response Surface Methodology.

8. Important effects are determined graphically to select a "paper champion" or the "experimental champion" based on the best $\bar{y}$ or largest S/N. Prediction equations are generated but prediction intervals are not calculated. The mathematical model is used for prediction but not to locate the optimal as in Response Surface Methodology.

9. If the optimal response lies outside the sample region, conduct another experiment in the direction of the optimal.

9. No iterative experimentation is used.

10. Find the optimal through an iterative experimental procedure.

10. The true optimal is not really found. Rather, settings for the best response over the experimental region are based on experimental and/or paper champions.

11. Set the process knobs at the optimal settings and go on-line. Recent emphasis is also on confirmation runs.

11. The importance of confirmation runs is stressed. Confirmatory runs are made prior to going on-line.

12. Assumptions include normality, independence and equal variability. The F test is robust to minor violations of normality and homogeneity especially for large samples balanced for each experimental condition. If you are concerned about large violations, you can use a Pareto Principle approach, i.e., rank order effects based on F values and select the larger ones as being important.

12. Assumptions are similar to the traditional method. Independence may be a problem due to lack of randomization. Unequal variance is accounted for by the transformation used in S/N calculation.

In addition to the 12 points previously discussed, the Taguchi method stresses simplicity. The idea of using tabled orthogonal arrays and marginal mean analysis is very appealing to engineers. It is obviously an easy way to get started in designing good experiments. However, engineers should not be content to limit themselves to the basics. Eventually they will want to progress to the more sophisticated designs and analysis techniques discussed in Chapters 3, 4 and 5.

Probably the most important contribution of the Taguchi method is in the area of robust designs where parameter settings are invariant to noise. However, the signal to noise statistic does not appear to be the best metric for finding dispersion effects as pointed out in Chapter 4 and section 6.5. In additon, the Taguchi method will infrequently find the optimal; it is designed to be 60-80% effective. The Taguchi advocates will maintain that further improvements are so costly that more sophisticated efforts are not cost efficient. One could argue that this is true in many applications; however, there are applications that require more fine tuning in finding the optimal. For 80-100% effectiveness, the traditional method of response surface methodology (RSM) techniques should be used.

Chapter 6 Problem Set

1-6. Redo the problems from chapter 4 using the appropriate S/N formula for that question. Compare the results between using S/N and the results you originally found.

7. If using signal-to-noise, how do you find the optimal levels for a process ?

8. Name 3 advantages of Taguchi's methodology. Name 3 short-comings.

9. Why does Taguchi use a quadratic for his loss function ? What two components does the function partition into ?

10. When does a factor belong in an outer array ?

Chapter 6 Bibliography

1. Chernoff, H. and Moses, L. E. (1959). *Elementary Decision Theory*. John Wiley and Sons, Inc.

2. Fisher, R. A. (1966). *Design of Experiments*, 8th ed. Hafner (MacMillan).

3. Hunter, J. S. (1985). "Statistical Design Applied to Product Design". *Quality Progress*, (Vol. 17, No. 4).

4. *Introduction to Quality Engineering, 5-Day Seminar Course Manual*, American Supplier Institute, Inc. 1987.

5. Kackar, Raghu N. (1986). "Taguchi's Quality Philosophy: Analysis and Commentary", *Quality Progress*, Dec.

6. Kirk, Roger E. (1982). *Experimental Design*, Brooks/Cole Publishing Co.

7. Lindquist, E. F. (1953). *Design and Analysis of Experiments in Psychology and Education,* Boston, Haughton Mifflin.

8. Norton, D. W. (1952). "An empirical investigation of some effects of non-normality and heterogeneity of the F-distribution" unpublished doctoral dissertation, State University of Iowa.

9. Rogan, J. C. and Kesselman, H. J. (1977). "Is the ANOVA F-test robust to variance heterogeneity when sample sizes are equal? An investigation via a coefficient of variation." American Educational Research Journal, 1977, Vol 14, 493-498.

10. Sullivan, L. P. (1984). "Reducing Variability: A New Approach to Quality," *Quality Progress,* Vol. 15, No. 4.

11. Taguchi, G. (1986). *Introduction to Quality Engineering*. Asian Productivity Organization.

12. Warsavage, B. and Schmidt, S. (1987). "Taguchi Philosophy and Methodology" presented at an IBM Technical Symposium.

Chapter 7: Case Studies

7.1 Introduction

This chapter contains a collection of real industrial case studies, simulated studies, and training type case studies. The complexity of these studies ranges from the fairly simple to the state of the art. The studies have not been altered by the authors of this text and thus it is very possible that many of us would have conducted a similar study using a different approach. The purpose of this chapter is not to critique each case study, but rather document the types of applications that exist. If the reader has serious concerns about any particular study, please share these concerns with the authors by writing or calling:

STEPHEN R. SCHMIDT

14220 Gleneagle Dr

Colorado Springs, Co 80921

(719) 488 - 3554

The long range plan is to add more case studies to this text and develop a second volume dedicated to applications. If you would like to publish your case study in the next printing of this text please contact us.

Submitted by: J. Merritt
                Honeywell
                Colorado Springs, CO

**"Results of a First Run Fractional Factorial for the TRE 10X Photo Process
on CMOS III Gate Policy"**

Background

At the time of this experiment, the author's process was generating an
unacceptable rate of rework on the CMOS III 10X stepper process because of poor
resolution.   In an effort to evaluate this process and determine which process
parameters have the most influence on poor resolution, a 2^{8-4} fractional factorial
was devised.

Experimental Objective

To detect the relative magnitudes of 1st order effects and to screen out less
significant factors.

Controlled Factors	"low" (-1)	"center" (0)	"high" (+1)
(A) Resist thickness	12,400 Å	13,700 Å	15,000 Å
(B) Softbake time	20 min	30 min	40 min
(E) Develop temperature	19°C	21°C	23°C
(G) Focus	430	480	530
(C) Softbake temperature	95°C	100°C	105°C
(D) Develop concentration	3.1:1	3.0:1	2.7:1
(F) Develop time	80 sec	90 sec	100 sec
(H) Exposure	200	220	240

Response

A quantitative Critical Dimension (CD).   A CD was measured at five points
across the wafer for each experimental combination.   For this data, the sample mean
and sample standard deviation were calculated.

Experimental Array

Design #	A	B	C	D	E	F	G	H
1	−	−	−	−	−	−	−	−
2	+	−	−	−	+	+	+	−
3	−	+	−	−	+	+	−	+
4	+	+	−	−	−	−	+	+
5	−	−	+	−	+	−	+	+
6	+	−	+	−	−	+	−	+
7	−	+	+	−	−	+	+	−
8	+	+	+	−	+	−	−	−
9	−	−	−	+	−	+	+	+
10	+	−	−	+	+	−	−	+
11	−	+	−	+	+	−	+	−
12	+	+	−	+	−	+	−	−
13	−	−	+	+	+	+	−	−
14	+	−	+	+	−	−	+	−
15	−	+	+	+	−	−	−	+
16	+	+	+	+	+	+	+	+
* 17	0	0	0	0	0	0	0	0

NOTE: Multiple centerpoints were run so as to obtain an experimental error estimate. This estimate was used to sort out important from unimportant effects for both the mean and standard deviation.

Experimental Conduct

The experiments were all conducted during a 10 hour period. Randomization was conducted where it was possible.

Calculation of Effects

Table I
Mean CD

Factor	Description	Effect (in order of absolute magnitude)
A	Resist Thickness	.504
H	Exposure	− .401
B	Softbake Time	.389
F	Develop Time	− .386
A∗C	Resist Thickness ∗ Softbake Temp	− .369
A∗B	Resist Thickness ∗ Softbake Time	− .359
C	Softbake Temp	.309
E	Develop Temp	− .301
B∗C	Softbake Time ∗ Softbake Temp	− .129
G	Focus	− .121
D∗E	Develop Concentration ∗ Develop Temp	.111
C∗D	Softbake Temp ∗ Develop Concentration	.069
A∗D	Resist Thickness ∗ Develop Concentration	−.059
D	Develop Concentration	.039
B∗D	Softbake Temp ∗ Develop Concentration	.024

Calculation of Effects

Table II
Standard Deviation of CD

Factor	Description	Effect (in order of absolute magnitude)
C	Softbake Temp	−.117
A	Resist Thickness	.109
B*C	Softbake Time * Softbake Temp	−.090
H	Exposure	−.084
A*B	Resist Thickness * Softbake Time	−.070
D	Develop Concentration	.066
C*D	Softbake Temp * Develop Concentration	−.063
F	Develop Time	−.052
E	Develop Temp	−.037
A*D	Resist Thickness * Develop Concentration	−.036
G	Focus	.032
B	Softbake Time	.024
D*E	Develop Concentration * Develop Temp	.021
A*C	Resist Thickness * Softbake Temp	−.016
B*D	Softbake Temp * Develop Concentration	.014

Analysis (Mean CD)

From the variation observed in the centerpoints, it was determined that any effect greater than 0.30 would be considered significant. The major surprise for this characteristic was that the Developer Concentration (Factor D) appeared to have little or no effect on the mean CD.

Analysis (standard deviation of CD)

The data in Table II indicates that Softbake Temperature (Factor C) may be a dispersion effect. The sign of the effect indicates smaller variability may result at the higher softbake temperature. Technically, this appears readily explainable since a lower softbake temperature is less efficient in removing solvents from the photoresist that may interfere with the uniformity of exposure.

Conclusions

The experiment was successful in screening out major mean effect factors and interactions as well as identifying a potential dispersion reduction factor.

Submitted by: Dale Owens
Kurt Manufacturing
Minneapolis, Minnesota

Reduction of Measurement Device Variability
Using Experimental Design Techniques

Background

Kurt Manufacturing, headquartered in Minneapolis, Minnesota, is a precision machine shop of considerable size and expertise. Kurt's clients include the nation's elite from aerospace, communications, and computers. Starting in the mid-1980s, Kurt embarked upon an aggressive program to implement gage studies and Process Control techniques across their entire operation. Findings from initial gage studies indicated that in certain situations, the inherent variability of some measurement devices was "large" in relation to the specified engineering tolerances. Kurt's own interval criteria is to use only measurement systems which have an inherent variability of less than 10% of the engineering tolerance.

In what follows, we will learn how Kurt made use of a simple orthogonal array so as to come up with a gaging process with vastly improved variability.

Case Study

When setting up a new process with dedicated gaging, it is not only important to have a capable machine, but a capable gaging device as well. At Kurt, our goal is to have a gage R&R (repeatability and reproducability) of less than 10% of specification and a process C_{pk} greater than 1.33.

An initial gage R&R study on our device in question yielded a gage error of 80.5% of the characteristic tolerance. Obviously, this was found to be far from acceptable. Accordingly, a team composed of the operator, supervisor, manufacturing engineer, quality engineer, and statistical facilitator met together to determine the variables and levels for a designed experiment.

Extensive brainstorming resulted in the following variables and levels:

(1) **Type of indicator**
> level 1 = current situation = .0001" increment indicator
> level 2 = .0005" indicator

(2) **Indicator support**
> level 1 = existing special stand
> level 2 = height stand

(3) **Inspection movement**
> level 1 = keep part stationary and move the indicator
> level 2 = keep the indicator stationary and move the part

(4) **Type of inspection master**
> level 1 = existing part master
> level 2 = use gage blocks

(5) **Indicating method**
> level 1 = bring the indicator straight into the part
> level 2 = sweep the indicator into the part from the side

Since this was strictly a screening design, it was decided to not test for two-factor or higher order interactions. After some discussion, the experimental goal was established as "to reduce the amount of gage error". Byproducts of meeting this overall goal would result in reduced cost through a reduction in inspection time and more accurate and precise measurement readings. The experimental objectives of the experiment included determining the effect of factors (1) through (5) on repeatability. Repeatability was quantified through the use of the sample standard deviation.

The team decided to make use of a 2^{5-2} fractional factorial (Taguchi L8). The test plan called for checking one part five times in succession per inspection setup. This approach was used for all eight experimental combinations. The experimental setup was as follows:

Exp #	Increment	Indicator Support	Movement	Master	Method	Result *
1	.0001	tree	part	part	straight	.00004
2	.0001	tree	part	gage	sweep	.00001
3	.0001	stand	indic	part	straight	.00001
4	.0001	stand	indic	gage	sweep	.00005
5	.00005	tree	indic	part	sweep	.00000
6	.00005	tree	indic	gage	straight	.00002
7	.00005	stand	part	part	sweep	.00003
8	.00005	stand	part	gage	straight	.00003

* Result - sample standard deviation of five readings.

By simply "eyeballing" the results, we find that combination #5 is the best. Verification runs by the company showed that this particular combination produced a gage R&R value of 15.2%, a big improvement over previous settings. Efforts were taken by the company to further decrease the gage error.

"Corrosion Study"

The following data was provided to us by an engineer wishing to study an inhibitor of corrosion on a metallic alloy. The questions to be answered in this study were: Is time an important factor? Do corrosion thicknesses change over time? To answer this question, samples were assessed at zero months, one month, 2 months, and 3 month intervals. Values shown are amount of material remaining.

Table I

Zero Months	One Month	Two Months	Three Months
99	104	108	96
114	77	107	68
85	80	69	98
*222	110	sample destroyed	105
127	84	89	104
124	90	124	90

*The value of 222 in column "Zero Months" was discarded for technical reasons. Investigation revealed it was not representative. Because of some additional confusion, only five sample values were available in the "Two Months" column.

Two approaches were taken to answer the question posed. The first approach was to provide a graphical representation of the data. It appears in Figure 1. From this chart we see no clear cut answer to the question. The second assessment approach, a one-way ANOVA resulted in the following statistics:

Source	df	SS	M S	F	"P Value"
Factor	3	1134.9	378.3	1.41	.274
error	18	4844.3	269.1		
total	21	5979.2			

As we can see from the table, at a significance level of .05, there is insufficient evidence to reject the hypothesis that the sample came from populations with equal means.

Based upon the two above analysis approaches, it was concluded there was no major change in corrosion thickness over time.

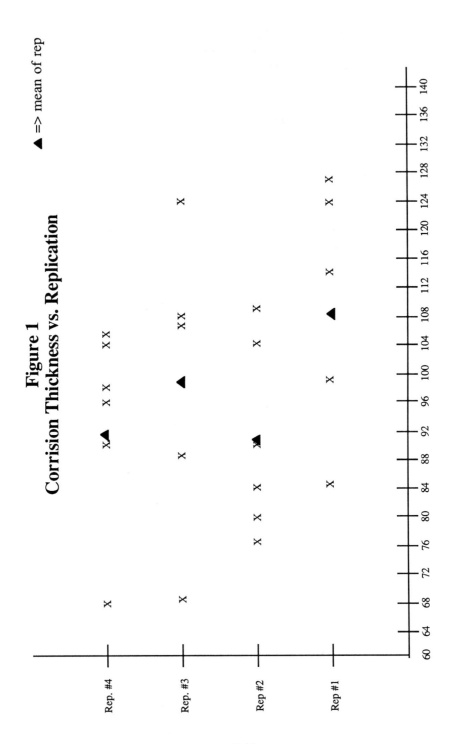

Figure 1
Corrision Thickness vs. Replication

▲ => mean of rep

.Submitted by: Karen Cornwell

Digital Equipment Corporation

Linear Motor Design
(Originally presented at the Rocky Mountain Quality Conference, June 1987)

Case Study: The force constant typically drops off approximately 10% on the ends of the motor. To reach optimum seek times, a more linear distribution is required.

Objective: Determine the effect various end cap and motor configurations have on the force constant and leakage flux values.

Analysis: The experiment was conducted using Taguchi's performance parameter (the signal to noise ratio). This made it possible to look at the force constant at both the mean level and variation around the mean. The analysis indicates that the 0.180 groove end cap and standard motor provide the best results.

Results: The improved configuration represents a major improvement over the standard configuration.

I.   Introduction:

In an attempt to improve the force constant linearity, the end cap was modified with a groove which created an air gap between the magnet and end cap. In the current configuration, the end cap contacts the magnets. With the addition of the air gap, the magnetic lines of flux will be channeled more toward the center pole rather than through the end cap. The same principle was also attempted with a groove cut at the front of the motor. There is a limited amount of space available for an air gap or additional steel.

The Taguchi analysis technique was employed in developing an optimum motor/end cap configuration which resulted in the maximum force constant with the least variability.

The force constant is measured using a load cell. Current is applied to the coil in the forward and reverse directions. The resulting force is measured on a strain gage. The force constant is simply the measured force divided by the applied current.

II. Experimental Objective

Determine the effect of end cap, front motor cut, and center cut on the dependent variables force constant and leakage flux.

III. Identification of Dependent Variables(s)

Force Constant

Leakage Flux

IV. Identification of Independent Variables

Noise	Control

Tester Repeatability and positional location	A. End Cap
	  1. .100 depth groove
	  2. .155 depth groove
	  3. .180 depth groove
	  4. .200 depth groove
	  5. .220 depth groove
	  6. 44D Stainless Steel
	  7. 303 Stainless Steel
	  8. Standard

	B. Front Cut Motor
	C. Center Cut Motor

V. Setup of O.A.

We progressed through several experimental stages in conducting the experiment. Briefly, the stages (with applicable results) were as follows:

Stage 1:

We believed that several of the end cap configurations would not produce acceptable results for leakage flux. Because of this, we first tested the leakage flux at six levels for end cap. Results of this experiment were as follows:

End Cap	Leakage Flux (must be < 25)
standard	16
.155	17
.180	18
.200	23
.225	31
standard (303 stainless)	58

From this series of experiments we found only the first four to have an acceptable leakage flux.

Stage 2:

Using the information from stage 1, we set up an L8 with three control factors. Factor settings for the experiment were:

	Variable	Level 1	Level 2
(A)	end cap	standard	.180
(B)	front	standard	.120
(C)	center	standard	.080

The experimental design with applicable results was:

RUN	A	B	A*B	C	A*C	B*C	A*B*C	S/N *(of force constant)
1	1	1	1	1	1	1	1	11.97
2	1	1	1	2	2	2	2	10.97
3	1	2	2	1	1	2	2	11.08
4	1	2	2	2	2	1	1	10.22
5	2	1	2	1	2	1	2	13.12
6	2	1	2	2	1	2	1	12.10
7	2	2	1	1	2	2	1	12.35
8	2	2	1	2	1	1	2	11.23

* Signal to noise $= 20 \log \left[\dfrac{\bar{y}^2}{s^2} \right]$, where $\bar{y}^2$ and s^2 were calculated from 8 data points.

MARGINAL MEAN RESULTS

FACTOR	LEVEL 1	LEVEL 2	$|\Delta|$
A	11.06	12.20	1.14
B	12.04	11.22	.82
A*B	11.63	11.63	0.00
C	12.13	11.13	1.00
A*C	11.60	11.67	.06
B*C	11.64	11.60	.01
A*B*C	11.66	11.60	.06

From the above, it appears that end cap at 0.180 cut is best, a standard front is best, and standard center is best. (Note: Leakage flux was acceptable for all combinations tested, i. the average was well below the upper specification of 25).

Stage 3:

In stage 3, we decided to panic a bit. Perhaps we had been premature in going to 2 levels instead of 4 levels on the end cap. Because of this, we set out to conduct a 4 x 3 full factorial using end cap and motor configuration as factors. The design, with resulting S/N ratios, was as follows:

		MOTOR		
		Center Cut	Front Cut	Standard
	standard	10.79	11.08	11.83
	.100	11.24	11.66	12.65
END CAP	.155	12.11	12.32	12.69
	.180	12.10	12.35	13.17
	.200	12.09	12.11	12.45

A graphical comparison of the results is provided in Figure 1.

Figure 1

S/N VS. END CAP CONFIGURATION

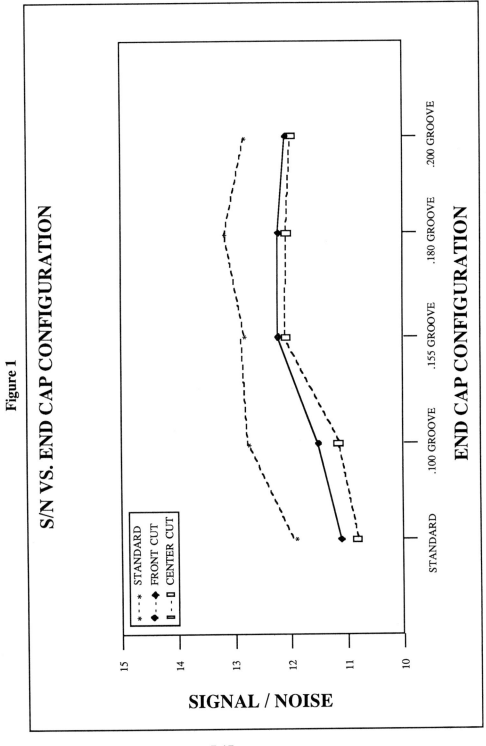

As is apparent from the graph, the standard motor with the 0.180 groove is the best performer.

Stage 3A:

After extensive discussion, we decided to re-analyze the raw data using the standard deviation. Our data table appeared as follows:

END CAP	CENTER CUT	FRONT CUT	STANDARD
standard	.041	.039	.030
.100	.037	.034	.025
.155	.031	.028	.025
.180	.031	.026	.022
.200	.031	.026	.026

Graphically, these results were as follows:

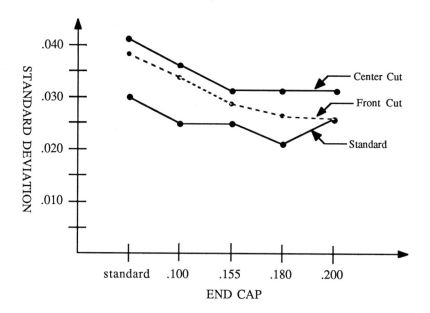

Analysis using the standard deviation indicates that we have the least dispersion with a 0.180 end cap and a standard cut motor.

Stage 4:

We conducted verification testing of the predicted best combinations. Results obtained were in agreement with our predicted best combination.

Conclusions:

Based upon the above findings, it is recommended to go with the standard motor and the 0.180 end cap.

Submitted by: Doug Sheldon
                 Ramtron Corporation

"Factorial Experiment for Bottom Electrode Stress"

Abstract:: Test Lot XXX was run using a 2-level, 3-factor orthogonally designed experiment to characterize Bottom Electrode film stress. The results show that current factor A thickness (200) is too thin for good, controllable adhesion and must be made thicker.

Factors and Levels:

Factor	"Low"	"High"
A	200	1500
B	500	4300
C	no	yes

Responses: Stress and resistance

Experimental Stages:

Stage (1). Stress and resistance measurements were taken with A and C at all possible combinations (providing us with a 2-level, 2-factor, full factorial design).

Stage (2). Adding Factor B (3rd process step) at two levels provided us with a 2^3 full factorial design.

Data and Analysis:

Stage (1). Table I lists the applicable results after deposition of factor A

TABLE I

Run	Factor A	Factor C	Stress xi	Stress $\bar{x}$	Resistance yi	Resistance $\bar{y}$
1	200	no	-10, -1.26	-5.63	40, 101	70.5
2	1500	no	1.31, 1.23	1.27	6, 9.5	7.75
3	200	yes	7.76, 14.4	11.08	111, 82.5	96.75
4	1500	yes	1.47, 2.43	1.95	7, 10.5	8.750

Parameter	Stress Effect	Resistance Effect
A	-1.115	-75.375
C	8.690	13.625
A * C	-8.015	-12.625

For "stress", note that the A * C interaction effect is the second largest of the three. This means the effect of factor A on stress cannot be considered independently of C. Figure #1 provides a graphical representation of the interaction. At thin values of factor A the stress varies greatly. At thick values for A there is less difference in stress regardless of whether C is in or out of the experiment.

For resistance, the most significant factor is A. As factor A thickness increases to 1500, the resistance drops an order of magnitude. The fact that the thin factor A values have a large mean with relatively high dispersion indicates a good factor A layer may not be forming. Resistance values with factor A set at 1500 seems to have much less dispersion. The effect of having factor C present is to increase resistance.

Stage (2). Once factor A was deposited and measured, factor B was deposited and stress was measured. The applicable data with resultant effects are shown in table II.

TABLE II

Run	A	B	C	Stress	Parameter	Effect
1	200	500	no	1.67	C	-.22
2	1500	500	no	2.3	B	-.06
3	200	4300	no	.85	C*B	.40
4	1500	4300	no	2.2	A	1.41
5	200	500	yes	.53	A*C	.42
6	1500	500	yes	2.2	A*B	.26
7	200	4300	yes	.72	A*B*C	-.10
8	1500	4300	yes	2.7		

From the above, only factor A appears to stand out. The interaction terms as well as factors A and C appear to be of no importance. Figure #2 provides a graphical representation of the effects.

Conclusions:

The current thickness of factor A (200) appears to be too stressful and prone to great dispersion. It is recommended the setting for factor A be increased to 1500. With a setting of 1500 for factor A the occurrence of factor C is not required. If, however, we go to 1500 for factor A, we will need to reduce factor B. The lower limit of factor B would be where possible factor A interdiffusion into factor B becomes a problem. Results from previous tests indicate it is possible to run with very thin layers of factor B.

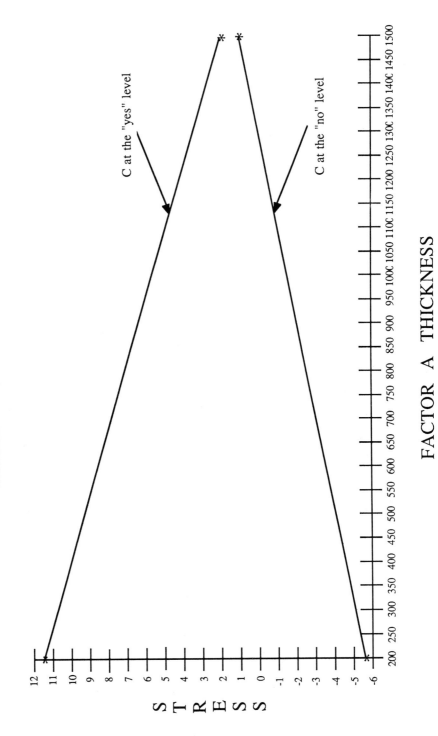

FIGURE 1
STRESS VS FACTOR A

C at the "yes" level

C at the "no" level

FACTOR A THICKNESS

STRESS

FIGURE 2
Effects of Factor B thickness on Stress

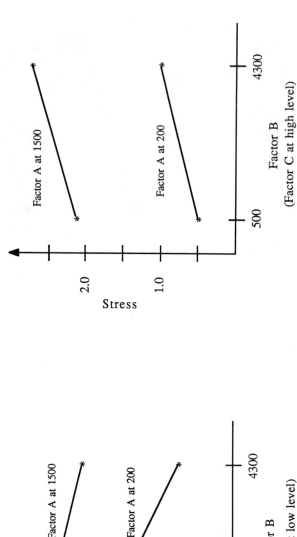

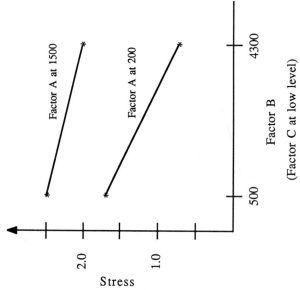

"Experimental Design on an Injection Molded Plastic Part "

Background: Shrinkage is a major concern with some injected molded parts. Typically, a molded die for a part will be built oversize to allow for part shrinkage after it has been produced. In the following situation, a new die had been produced for the production of a part. What are the proper process settings to produce a product with the proper dimensions?

Objective: Determine the effect of injection velocity, cooling time, barrel zone temperature, mold temperature, hold pressure, and back pressure on shrinkage of part XXXX (as defined by the 9.380 and 14.500 dimensions).

Brainstorming: The brainstorming process lasted approximately four hours. Attendees included a process engineer, process supervisor, quality control inspector, quality control supervisor, and a person with some rudimentary training in experimental design. Barrel zone temperatures was a tough one. Barrel zones actually had six different control points. Some discussion was conducted as to whether we should first run a series of experiments about zone temperatures. After a series of discussions, it was concluded to look at only a "low" and "high" group of settings for barrel zone temperatures in the screening design chosen.

Design Type: An eight run fractional factorial at 2-levels was chosen (Taguchi L8). The controlled factors with applicable levels chosen were:

Control Factor		Low	High
A	Injection Velocity	1.0	3.1
B	Cooling Time	40 sec	50 sec
C	Barrel Zones	"low temps"	"high temps"
D	Mold Temperature	100	150
E	Hold Pressure	200	1100
F	Back Pressure	50	150

We decided to look only for main effects in this particular design.

Test Plan: We chose to run all experiments during the same shift using raw materials from the same lot and containers. We also chose to use only "fresh" raw materials. No reground product was to be used. Additionally, our test procedure called for cycling in the settings, running until these settings were stable at the specified settings, running 10 shots, and then running the parts for the experiment. Five experimental parts were measured at each design setting with applicable dimensional measurements recorded. Since factor D was a factor which was very time consuming to change, we assigned it to the first column of our array, then randomized the run order within the two levels of factor D.

Design and Resultant Data

Design #	D	C	A	B	E	F	Length
1	1	1	1	1	1	1	14.5000, 14.5005, 14.5000, 14.5000, 14.5005
2	1	1	1	2	2	2	14.5075, 14.5090, 14.5070, 14.5065, 14.5065
3	1	2	2	1	1	2	14.5045, 14.5050, 14.5045, 14.5045, 14.5045
4	1	2	2	2	2	1	14.5100, 14.5105, 14.5105, 14.5110, 14.5105
5	2	1	2	1	2	1	14.5105, 14.5110, 14.5105, 14.5120, 14.5100
6	2	1	2	2	1	2	14.5045, 14.5055, 14.5065, 14.5050, 14.5050
7	2	2	1	1	2	2	14.5150, 14.5140, 14.5155, 14.5150, 145145
8	2	2	1	2	1	1	14.5055, 14.5065, 14.5055, 14.5055, 14.5060

Design #	D	C	A	B	E	F	Width
1	1	1	1	1	1	1	9.3575, 9.3560, 9.3570, 9.3585, 9.3590
2	1	1	1	2	2	2	9.3650, 9.3640, 9.3640, 9.3640, 9.3645
3	1	2	2	1	1	2	9.3545, 9.3545, 9.3545, 9.3550, 9.3540
4	1	2	2	2	2	1	9.3630, 9.3630, 9.3625, 9.3635, 9.3635
5	2	1	2	1	2	1	9.3555, 9.3560, 9.3560, 9.3555, 9.3560
6	2	1	2	2	1	2	9.3580, 9.3565, 9.3550, 9.3540, 9.3545
7	2	2	1	1	2	2	9.3600, 9.3580, 9.3585, 9.3590, 9.3585
8	2	2	1	2	1	1	9.3565, 9.3560, 9.3565, 9.3565, 9.3560

Analysis:

Table of Average Marginals

(14.5 - nomimal, length)

Level	D	C	A	B	E	F
1	14.5057	14.5059	14.5070	14.5076	14.5040	14.5068
2	14.5092	14.5089	14.5078	14.5072	14.5109	14.5080
|Effect|	.0035	.0030	.0008	.0004	.0069	.0012

Table of Average Marginals

(9.380 - nominal, width)

Level	D	C	A	B	E	F
1	9.3599	9.3583	9.3593	9.3567	9.3560	9.3582
2	9.3566	9.3582	9.3573	9.3598	9.3605	9.3583
|Effect|	.0033	.0001	.0020	.0031	.0045	.0001

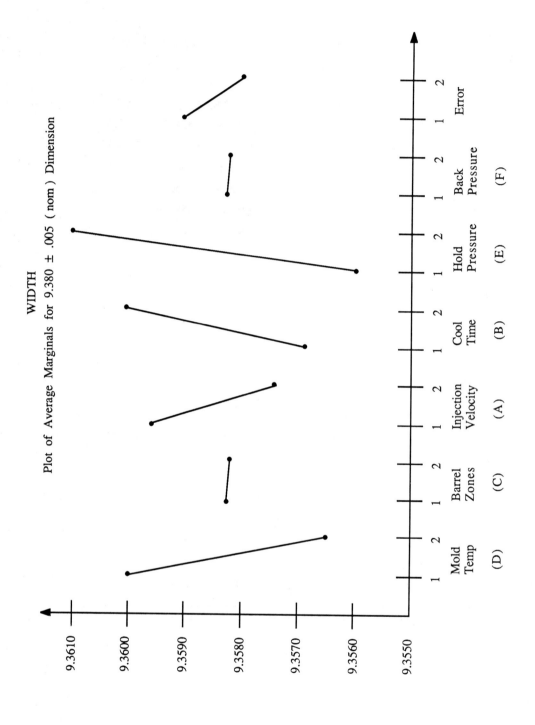

WIDTH

Plot of Average Marginals for 9.380 ± .005 (nom) Dimension

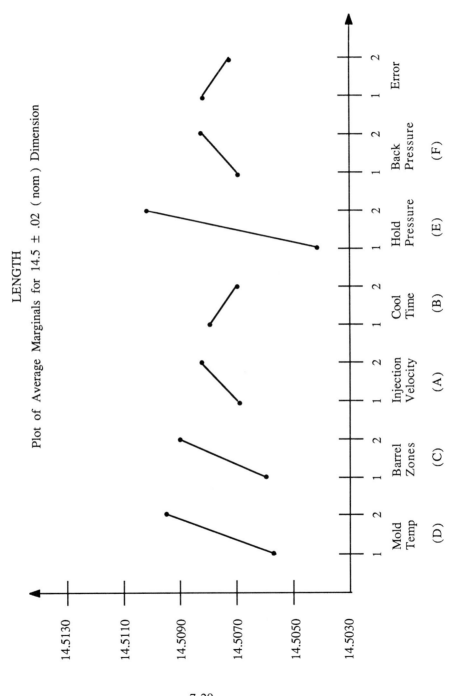

LENGTH

Plot of Average Marginals for 14.5 ± .02 (nom) Dimension

Interpretation of Results

LENGTH	WIDTH
Best settings for 14.5 (nominal)	Best settings for 9.380 (nominal)

Mold temp (1)	Mold temp (1)
Barrel zones (1)	Barrel zones (don't care)
Hold pressure (1)	Hold pressure (2)
Inj. velocity (don't care)	Inj. velocity (1)
Cool time (don't care)	Cool time (2)
Back pressure (don't care)	Back pressure (don't care)

The only major conflict exists on hold pressure. Because of problems with holding the 9.380 (nom) dimension we will select hold pressure as level 2. NOTE: *Under no operating conditions during the experiment were we able to meet the specification on the 9.380 dimension.*

Best conditions appear to happen at mold temperature (1), barrel zones (1), hold pressure (2), injection velocity (1) and cool time (2). Back pressure, with little impact, is best set at level 1.

Findings from Verification Run: Twenty-five verification runs were conducted at the "best conditions" as predicted by our experiment. Under no conditions were we able to meet the criteria established for the 9.380 (nominal).

Conclusions: Based upon the above, it was determined to conduct a redesign of the injection die.

CONSIDER THE METRIC
(Are you defining your inputs sensibly?)

Jack E. Reece, Ph.D.
3685 Hickory Hill Drive
Colorado Springs, CO 80906
719-540-8521

1. Introduction:

Engineers and scientists eager to try new tools to improve their abilities to solve processing problems readily adapt to the the matrix approach to experimentation required by statistical experimental design. Once a person understands the structures of the factorial, fractional-factorial, or various response surface design matrices, setting up an experiment involving common input factors such as time, temperature, power, catalyst type, etc., becomes almost second nature.

However, some particularly troublesome processes for experimental design are those involving combinations of reactive ingredients in a fixed or flowing environment. One example includes the mixtures of gases used to deposit a coating or to etch a given layer selectively in the semiconductor industry. Another might be a batch or flow process in a chemical industry involving a mixture of reactants in either a static or flow reactor.

Most texts on experimental design devote most of their attention to the justification of the technique of experimental design and to the statistics and analysis of results. Few devote other than passing reference to assisting the user in defining the metric for these cases. The experimental design literature has generally treated these mixture experiments as a special class and devised special treatments for them (see, for example, "Experiments with Mixtures" by John A. Cornell, New York, Wiley, 1981). The transformations of input metrics described below can accommodate a wide variety of experimental situations.

2. Dealing with Mixtures:

2.1 Experiments with Photographic Emulsions:

Conventional photographic emulsion formulations contain light sensitive elements, chemical and photo sensitizers, binders and other materials. Developing an image in the systems requires treating the exposed film with an external developing chemistry.

At one time the author conducted product development research on a family of photographic emulsions whose entire chemistry (including the developers) resided in the light sensitive coating. Developing an image in an exposed coating for these materials required only heat.

The developing chemistry required a hindered phenolic reducing agent (A) and another agent generally considered a "co-developer" or development "accelerator" B. The system worked best when the emulsion contained an excess of the phenolic compared to the accelerator. However, too much or too little of the phenolic in the emulsion gave unacceptable image properties; too much of the combination might cause the entire unexposed area to darken upon development; too little gave very slow development with poor image density. Proper balance of these ingredients allowed the "tuning" of the emulsion for specific photographic applications.

In attempting to optimize the performance of this emulsion for a particular application, the author decided to use the amount (or concentration) of the two materials in a fixed batch and their ratio as factors in the the design rather than to manipulate the amount of each in the emulsion separately. This approach produced two equations with two unknowns:

$$A + B = \text{Total weight in fixed batch (X1)}$$
$$A/B = \text{Ratio (X2)}; \quad A = B*X2$$

$$B*X2 + B = X1$$
$$B = X1/(1 + X2)$$
$$A = X1 - B$$

Where A and B are the amounts of developer and actually added to the emulsion, and X1 and X2 are the factors included in the design matrix. Using this approach obviously requires knowledge of the likely functional limits of each factor.

This relationship provides the following two-factor matrix in X1 and X2; solving the algebraic expression above provides the actual amounts of each reactant to add:

X1	X2
-1	-1
+1	-1
-1	+1
+1	+1

Thus the design is orthogonal in X1 and X2, but not with respect to the weights of each ingredient actually added to each batch.

The logic of this approach is that it considers the stoichiometry (chemistry) of the system and includes the concentrations of these two reactions and their "psuedo-molar" ratios. The practical significance of the approach is that it provided the author a convenient metric for managing these two variables and optimizing the emulsion characteristics.

2.2 Experiments Involving Gas Flows:

2.2.1 Two Reactive Gases:

An extension of the approach described above provides a means for controlling two reactive gases flowing into a reaction chamber to provide a product or to modify (etch, for example) a layer on a substrate. In this example G1 and G2 have flows measured in SCCM (Standard Cubic Centimeters per Minute) or equivalent units. A logical approach to characterizing this system would use a range of the actual flow of each gas in the design matrix.

Based on his experience in optimizing photographic emulsions, the author prefers to use a "chemical" approach to this problem. The usefulness of the method has been demonstrated in actual practice in that models prepared in this fashion have shown smaller residual sums of squares, less lack of fit sums of squares, and fewer outlier or maverick points in an investigation. Models prepared this way better explain the relationships between inputs and outputs in the system being investigated.

The equations are adaptations of those shown above:

$$G1 + G2 = \text{Total Flow (TF)}$$
$$G1/G2 = \text{Ratio (R)}$$

TF and R become the factors in the experiment, and one calculates the flow settings used to investigate the process as demonstrated above.

2.2.2 <u>More than Two Gases (All Reactive)</u>:

Additional gases necessarily complicate the algebra necessary to solve for the actual flows. The example below assumes we must explore three gas flows, all of which play a reactive part in the process.

$$G1 + G2 + G3 = \text{Total Flow (TF)}$$
$$G1/G2 = \text{Ratio1 (R1); } G1 = G2*R1$$
$$G3/G2 = \text{Ratio2 (R2); } G3 = G2*R2$$

$$G2*r1 + G2 + G2*R2 = TF$$
$$G2 = TF(1 + R1 + R2)$$

Choosing one of the gases as the denominator in both of the ratio expressions simplifies the algebra somewhat; one convenient approach uses the major component of the gas mixture as that denominator.

2.2.3 <u>Several Reactive Gases and a Carrier (or Diluent)</u>:

A further complication occurs when one of the gases in a mixture of three or more is a diluent or carrier. Here we can conveniently use the ratios of reactants as well as the <u>concentration</u> of reactants in the stream:

$$G1 + G2 + G3 + D = \text{Total flow (TF)}$$
$$G1/G2 = \text{Ratio1 (R1); } G1 = G2*R1$$
$$G3/G2 = \text{Ratio2 (R2); } G3 = G2*R2$$
$$(G1 + G2 + G3)/TF = \text{Concentration (R3)}$$

$$(G2*R1 + G2 + G2*R2)/TF = R3$$
$$G2 = R3*TF(1 + R1 + R2)$$

Using the expression for the concentration of the reactants both simplifies the algebra and probably better describes the relationship among inputs and outputs.

2.3 Chemical Reactions in Batches:

Consider the case in which a chemist wishes to design experiments to study the reaction of two species in a vessel to produce a particular product. In this case the vessel has a finite capacity, commonly expressed in volumetric or weight measure. Most chemists prefer to consider reaction components in terms of mole ratios of the reactants based on some presumed equation for the chemistry of the process. The moles of reactant X1 = k_1*wt X1; moles of X2 = k_2*wt X2, where k_1 and k_2 are constants whose value depends on the respective molecular weights of X1 and X2. The mole ratio X1/X2 = k_1*wt X1/k_2*wt X2 or k_3*(wt X1/wt X2). Therefore, one may use the raw weights of each reactant, understanding that multiplying that ratio of weights by k_3 would yield the actual mole ratio.

Quite commonly, chemical reactions in a batch reactor require an inert solvent or diluent -- note the similarity between this problem and the one described in the section above. So another concern is the concentration of the reactive species in the reactions mixture. Assume for this example that the maximum weight allowed in this reaction vessel is 1000 units. Then:

$$X1 + X2 + S = 1000$$
$$X1/X2 = Ratio1 \ (R1); \ X1 = X2*R1$$
$$(X1 + X2)/1000 = Concentration \ (R2)$$
$$(X2*R1 + X2)/1000 = R2$$
$$X2 = R2*1000/(1 + R1)$$
$$X1 = X2*R1$$
$$S = 1000 - (X1 + X2)$$

Thus the controlled factors in this experiment are the ratios of the reactants to each other and their concentration in the system.

3. Dealing with Temperature Gradients:

Figure 1 represents a tube furnace similar to those used in the semiconductor industry to prepare coatings on silicon wafers, for example. Among the variables one might manipulate in characterizing the performance of this device are the three temperature zones B1, B2, and B3. In this particular furnace configuration reactive gases enter through the ports J in the end cap C, and proceed down the furnace to the exhaust pump outlet H. Because the process depletes the reactive species in the gas flow, "tilting" the furnace temperature from low to high from the entry to the exit is a common practice.

One might decide to investigate this process using entry temperatures that average 20 degrees cooler than the center and exit temperatures 20 degrees warmer:

$$
\begin{aligned}
\text{Entry} \quad &-- \quad 790 \text{ to } 810 \\
\text{Center} \quad &-- \quad 810 \text{ to } 830 \\
\text{Exit} \quad &-- \quad 830 \text{ to } 850
\end{aligned}
$$

The two-level three-factor matrix that results is:

Entry	Center	Exit
790	810	830
810	810	830
790	830	830
810	830	830
790	810	850
810	810	850
790	830	850
810	830	850

While the general trend of the temperature settings is higher toward the exit in each run, the distribution of settings produced contains center temperatures equal to entry and center temperatures equal to exit. The author suggests an alternative metric: consider the center zone the key or master zone and offset the other zones with respect to it.

```
Entry    -- -30 to -10
Center   -- 810 to 830
Exit     --  10 to  30
```

Entry	Center	Exit
-30(780)	810	10(820)
-10(800)	810	10(820)
-30(780)	830	10(840)
-10(800)	830	10(840)
-30(780)	810	30(840)
-10(800)	810	30(840)
-30(780)	830	30(860)
-10(800)	830	30(860)

Now each run contains an upward temperature bias from entry to exit; the severity differs from run to run. The design is orthogonal in the offset settings, but not in the actual temperatures run.

4. Summary:

The metrics described in the preceding sections ultimately have provided useful regression models describing the behavior of one or more outputs as a function of these inputs. These metrics by no means represent all those an investigator might elect to use rather than the obvious one.

Experiments and the time necessary to conduct them are expensive. Experiments using the matrix approach of statistical experimental design such as factorials, etc., can make our investigations much more cost effective and provide more reliable information than investigations done by some other method. But the engineer or scientist can improve on the results obtainable from a statistical experimental design if she or he will take the time to step back from a problem and consider alternative, possibly novel, metrics for input variables before starting the experiments. Conversion of factors to a new metric after all runs are complete may give useful information, but the non-orthogonality of the matrix resulting in that case may cast doubt on the validity of any regression models derived.

Figure 1

LPCVD Tube Furnace Configuration

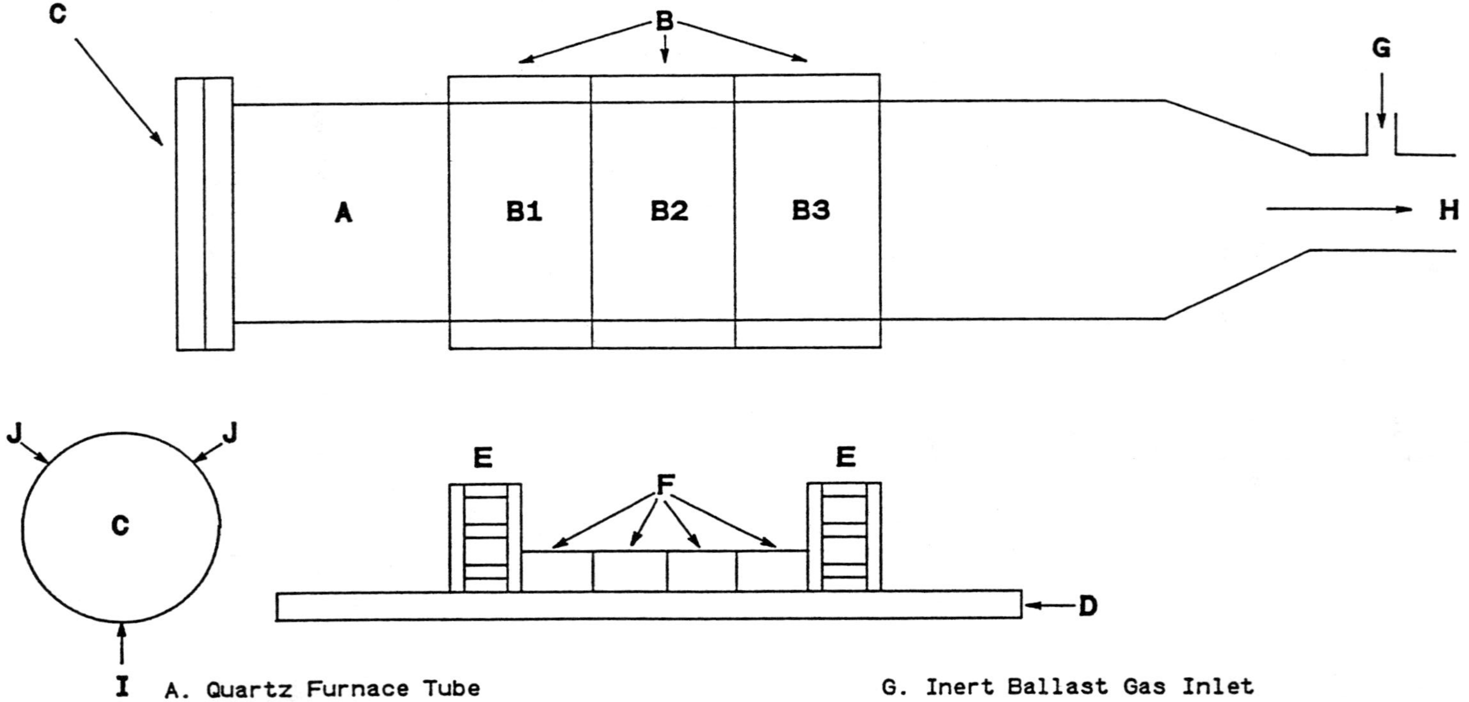

A. Quartz Furnace Tube
B. Heating Zones
C. Metal Sealing Flange
D. Quartz Carrier for Wafer Boats
E. Quartz Flow Baffles
F. Quartz "Boats" for Wafers

G. Inert Ballast Gas Inlet
H. Outlet to Vacuum System and Exhaust
I. Closed-loop Pressure Sensor
J. Reactive Gas Mixture Inlet (from Mixing Manifold)

MINIMIZING VARIABILITY

Jack E. Reece, Ph.D.
3685 Hickory Hill Drive
Colorado Springs, CO 80906

1. Introduction:

The techniques of statistical experimental design evolved from the work led by Sir Ronald A. Fisher during World War I at Rothamstad Agricultural Experiment Station in London as he and his colleagues strived to increase crop yields. Generally scientists and engineers using these methods concentrated most commonly on the "location" of a process, trying to increase yields, decrease byproducts, or understand a process enough to set and control inputs to maintain a desired result or combination of results (1,2).

Particularly in recent years the widespread teaching of so-called "Taguchi Methodology" has focused attention not only on the location of a process, but also on its dispersion or variability (3). Within that methodology are a number of metrics designed to describe variability and location in one parameter. Classical statisticians have attacked these metrics, creating considerable confusion among "lay practitioners" of the art of statistical experimental design (4).

Engineers tend to focus on the "bottom line". What interests them are methods they can exploit to find answers in terms they and their managers can understand. This paper discusses the analysis of the results of a designed experiment in the semiconductor industry in which the investigators sought to understand the factors controlling coating uniformity during a batch process (5).

The study was successful using one of the metrics described below. Having obtained access to an advanced experimental design package called RS/Discover Version 2.0 (marketed by BBN Software Products Corporation, Cambridge, MA.) which supports a variety of "performance statistics" and which contains a sophisticated multiple regression utility, the author elected to re-examine that data to compare predictions of optimal regions from several such statistics.

Figure 1

LPCVD Tube Furnace Configuration

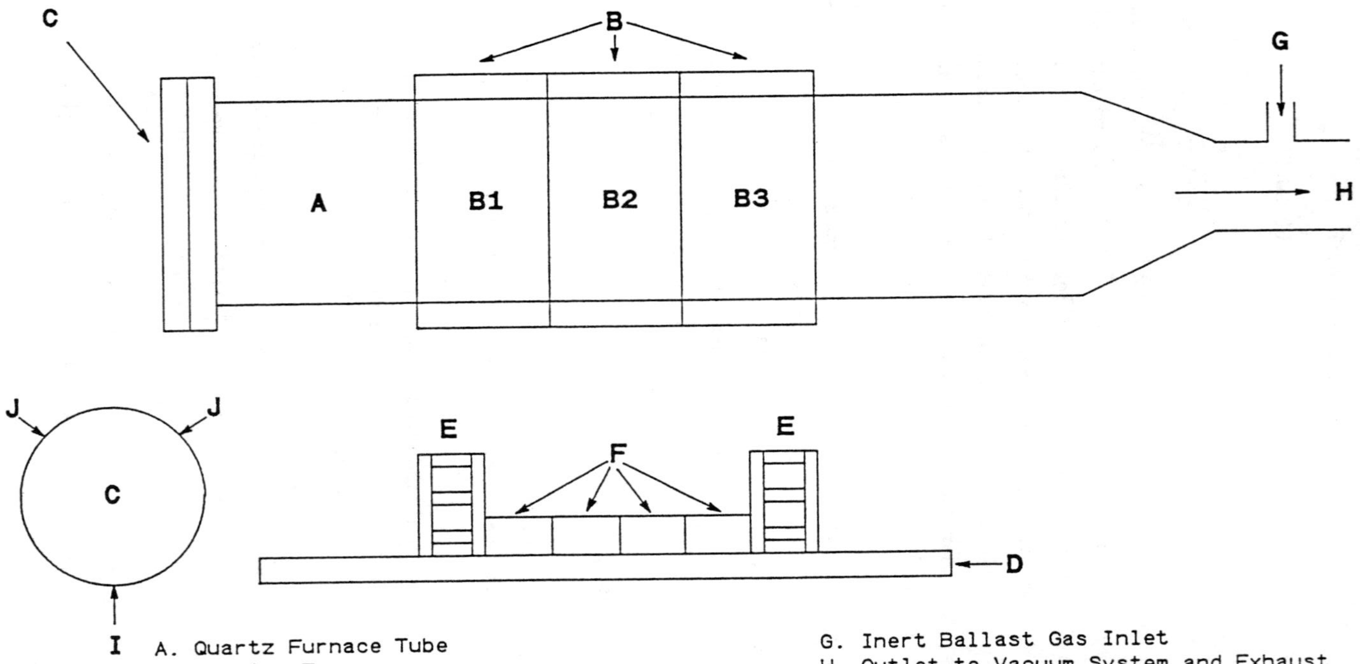

A. Quartz Furnace Tube
B. Heating Zones
C. Metal Sealing Flange
D. Quartz Carrier for Wafer Boats
E. Quartz Flow Baffles
F. Quartz "Boats" for Wafers

G. Inert Ballast Gas Inlet
H. Outlet to Vacuum System and Exhaust
I. Closed-loop Pressure Sensor
J. Reactive Gas Mixture Inlet (from Mixing Manifold)

2. The Process:

Figure 1 illustrates the furnace used to prepare the samples for the experimental study.

One inserts a batch of silicon wafers mounted on edge in the boats 'F' on the quartz carrier 'D' and places them in the furnace. The objects 'E' on the carrier are quartz baffles which provide turbulent flow through the furnace and which have thermal mass to preheat the reactive gases before they contact the wafers. From a mixing manifold (not shown) the gas mixture containing two reactive species enters at the positions 'J' in the end cap 'C' and passes through the furnace tube 'A' to the exhaust pump and scrubbing system at 'H'. Heaters around the tube at 'B1', 'B2', and 'B3' heat the gas stream to about 800 degrees centigrade. Metering an inert ballast gas at the port 'G' controls the pressure.

In characterizing the process the engineers manipulated six variables: CLP (Closed Loop Pressure); CTMP (temperature of zone B2); Z1OFF (offset of zone B1 with respect to zone B2; Z3OFF (offset of zone B3 with respect to zone B3); TF (total flow of reactive gases); and RAB (ratio of gas A to gas B). Because they expected complex, nonlinear relationships among the input and outputs of the process, they elected to run a Box-Wilson Central Composite design requiring 53 trials to allow estimation of curvilinear effects. Each trial contained 12 pilot wafers (among the rest used for ballast); and each boat contained 3 pilots. The engineers gathered 10 thickness measurements from each wafer in each run -- a total of 6360 thickness measurements.

The experimental objective was to identify a process window which provided reasonable deposition rates with minimum variation within or among wafers.

3. Performance Statistics:

3.1 Taguchi's "Target is Best":

As stated above, the experimental objective is to get the process on target with minimum variability. Taguchi suggests a parameter called "Target is Best" or SN_T for this purpose. In mathematical terms this parameter is:

$$10\log_{10}(\bar{y}^2/s^2)$$

where $\bar{y}$ is the average value observed in each run and
s is the observed standard deviation in each run.

Taguchi uses this parameter to define the optimum operating point for the process in that the combination of factor settings which maximizes this function defines the optimum. Classical statisticians object to the use of this statistic because it blends the mean with a measure of variability. The absence of close correlation between the variability and the mean causes this statistic to lose information compared to separate measurements of mean and variability in that case.

The form of the statistic reflects Taguchi's conversion of these numbers to "decibel" readings because of his engineering background. This statistic is actually a variant of the classical coefficient of variation statistic:

$$CV = s/\bar{y}$$
$$10*\log_{10}(CV) = 10*\log_{10}(s/\bar{y})$$
$$-20*\log_{10}(CV) = 20*\log_{10}(\bar{y}/s)$$
$$= 10*\log_{10}(\bar{y}^2/s^2)$$

Thus Taguchi's "Target is Best" statistic is the same as $-20*\log_{10}CV$.

3.2 Performance Statistic Based on Range:

A common metric used for expressing the "uniformity" of a variety of different kinds of measurements within a group of measurements in the semiconductor industry is:

$$UNIF = range/(2*\bar{y}) * 100$$

This is also a variant of the coefficient of variation CV in that for small samples one may estimate the standard deviation by dividing the range by 4:

$$UNIF = (2/2)*(range/2)*(1/\bar{y})*100$$
$$= 2*(range/4)*(1/\bar{y})*100$$
$$= 2*s/\bar{y}*100 = 2*CV*100$$

The author recommends against using the statistic based on the range because of the influence outlier-prone data can have -- the standard deviation is far more robust to outliers than the range. Because the natural limit of either statistic is 0, one should transform the response to preclude negative predictions before attempting to prepare a regression model; common transforms are the logarithm, square root, inverse, inverse square root, etc.

3.3 Performance Statistic Based on the Standard Deviation:

Dr. George Box of the University of Wisconsin, Madison, and his colleagues have published a series of papers dealing with modeling variability independent of the mean or location. They recommend:

$$-\log10(s)$$

At once you can see that this is a very simple metric, requiring the calculation of the standard deviation of each set of observations and finding the logarithm of each standard deviation. The negative sign causes this function to follow the "target is best" statistic in that one seeks to maximize its value. The results presented below show how well it performs for the data used here in which the standard deviation appears proportional to the mean.

Another form of this statistic is somewhat more difficult to generate:

1   Find the logarithm of each observation in each group.
2   Determine the standard deviation of each group of logarithms of
     the observations.
3   Find the logarithm of that standard deviation.

For comparison purposes, the author generated this metric and called it "LGSDLGTHK".

3.4 Other Performance Statistics:

All of the metrics described so far assume that the distribution of the observations for each run is normal, although the transforms in each minimize the impact of non-normality. In the particular case described in this paper, only 17 of the sets of observations of the 53 indicated a normal distribution using the Pearson's Chi Square test. Therefore metrics independent of the nature of the distribution of the observations can be quite useful in some cases.

One such statistic is the InterQuartile Range (IQR). If one orders observations from low to high and defines the end of the first quartile (25% of the observations) and the end of the third quartile (75% of the observations), the difference between the two numbers is the IQR. The range defined by the IQR represents 50% of the observations. Recall that +/- 1 standard deviation from the mean in a normal distribution represents about 2/3 of the area under the normal curve or 2/3 of the data. Thus one may model the IQR of a group of data or its $-\log_{10}$ and obtain results analogous to the metric recommended by George Box and associates.

Another statistic independent of the distribution is the Median. The median is the middle value in the ordered data when the count of the data is odd; or the average of the two middle values in the ordered data when the count is even. The ratio of the IQR to the Median provides a metric quite analogous to the coefficient of variation and is robust to large departures from normality. Another approach to the problem described here went as follows:

> Each run contained 4 wafer boats which held three wafers on which we made 10 measurements of thickness.
>
> Because the distribution of the measurements tended to be non-normal, calculating the mean of each boat helped normalize the distribution.
>
> If we then calculated the standard deviation among the boat means and created a form of the coefficient of variation using that standard deviation and the **grand mean** of each run, we had a metric describing the relative spread of the data in each boat in each run:
>
> CV' = STDEV(MEANS OF BOATS 1 TO 4 PER RUN)/GRAND MEAN PER RUN

To convert this metric to a measure relatable to the +/- percent variation around a central value, we used the following:

$$BTUNIF = 2*CV'*100$$

This contrived metric is the one the author used initially in helping engineers model the process discussed here. An important part of its usefulness was that it gave "uniformity" values engineers could understand. It and those based more directly on the coefficient of variation translate to a +/- variation about a central point.

4. About the Software:

BBN Software Products Corporation markets RS/Discover[TM] and its companion multiple regression analysis package MULREG[TM]. Together these packages provide a comprehensive environment for the design, analysis, and interpretation of statistical experimental designs. In its current form Version 2.0, RS/Discover supports experiments having inner and outer arrays and calculates a variety of performance statistics automatically when requested by the user.

The MULREG package associated with RS/Discover 2.0 is extremely useful in that it incorporates the Box-Cox system of response transformations and can advise the user on which is statistically appropriate. Furthermore it supports both least squares and robust (bisquare) regression and can transform inputs based on variance inflation factors associated with a particular factor's settings. To give each metric the best opportunity possible to describe the variability seen in this study the author used every utility the MULREG routines indicated were statistically appropriate.

The author completed the initial study of this process using RS/Discover 1.2 and its accompanying multiple regression utilities. These products were the state of the art for menu-driven user-friendly statistical packages in 1987, but did not support the sophistication and convenience available in 1989. Nor did they support experiments with outer arrays.

One could consider the 120 thickness measurements in each experiment part of an outer array. While they are not true 'experiments' or 'replicates', they provide an estimate of how the thickness of material deposited varied from one end of the furnace to another. The author created a suitable experiment within the software then inserted the data from the previous experiment into the tables created. The system calculated the performance statistics 'Target-is-Best', IQR, and LogS.

The software also allows the user to define custom performance statistics
whose metric is more familiar in his environment. Examples of these include
the 'BTUNIF', IQR/Mean, UNIF, and LGSDLGTHK discussed above. In these cases
the system provides space for them in the worksheet generated, but the user
must calculate or insert them.

5. Consider the Data:

Space considerations prohibit the display of the entire data base. Rather
Tables 1 and 1a show the conditions of each run, the mean thickness observed
in each wafer boat, the grand mean thickness for each run as well as the
median, stdev among boats, minimum and maximum thicknesses, standard deviation
of each run, IQR, and standard deviation of the logarithm base 10 of
thicknesses within each run.

Figure 2 plots the observed means of the boats versus the run number. The
error bars on this graph represent the minimum and maximum thickness observed.
From it one can see that normally the wafers decreased in thickness as boat
number increased (boat 1 was near the entrance; boat 4 near the exit); but not
always. Some of the runs had very tight distributions of thicknesses -- the
purpose of regression modeling is to find the simplest expression relating the
input factors to the observed distribution of thicknesses.

6. The Regression Models:

We intend to model variability independent of the mean or median of the
response. The small table at the bottom of the next page shows the linear
correlation coefficients between the mean thickness value observed and the
metric created to estimate variability. Note the $-\log10(S)$ statistic
correlates at about -0.45 with the mean -- this translates to an $R**2$ of about
.20. With this amount of data, the correlation is probably statistically
significant. Note that those metrics involving a ratio of the standard
deviation to the mean correlate less well with the mean. We can decide the
importance of this correlation when we compare how well each statistic
identifies the region of minimum variability in the process. (Incidentally,
the previous results using older software identified such a region, and
confirming runs proved it.)

Correlation Matrix

Location and Variability Metrics

Term	Mean Thickness
Mean Thickness	1.000
Median Thickness	0.996
-LOG10(S)	-0.449
IQR	0.432
2*SD/Mean*100	0.243
Target is Best	-0.196
LOG10(SD(LOG10(THICK)))	0.190

Table 1

Factor Settings

EXP NUM	CLP	TOT FLOW	RAB	CTMP	Z1 OFF	Z3 OFF	EXP NUM	CLP	TOT FLOW	RAB	CTMP	Z1 OFF	Z3 OFF
1	425	350	5	790	-20	20	28	425	350	5	790	-20	20
2	478	287	4	786	-24	24	29	478	287	6	786	-24	16
3	372	287	6	794	-16	24	30	478	287	6	786	-16	24
4	478	287	6	794	-24	24	31	478	413	4	794	-24	24
5	372	413	6	786	-24	16	32	478	287	4	794	-24	16
6	478	413	6	794	-16	24	33	425	350	3	790	-20	20
7	425	350	5	790	-20	20	34	372	413	4	794	-24	16
8	478	413	6	786	-24	24	35	425	350	5	790	-30	20
9	372	287	4	794	-24	24	36	372	287	4	786	-16	24
10	478	413	4	786	-16	24	37	372	413	4	786	-16	16
11	372	413	6	794	-24	24	38	425	350	5	790	-20	20
12	425	350	5	790	-20	30	39	478	287	6	794	-16	16
13	425	350	5	790	-20	20	40	372	287	4	794	-16	16
14	372	413	4	794	-16	24	41	372	413	4	786	-24	24
15	372	287	6	786	-16	16	42	478	413	4	786	-24	16
16	478	413	6	794	-24	16	43	372	413	6	786	-16	24
17	478	287	4	794	-16	24	44	425	350	5	790	-20	20
18	425	350	7	790	-20	20	45	425	500	5	790	-20	20
19	372	287	4	786	-24	16	46	478	413	6	786	-16	16
20	478	287	4	786	-16	16	47	372	413	6	794	-16	16
21	372	287	6	786	-24	24	48	425	350	5	790	-10	20
22	300	350	5	790	-20	20	49	425	350	5	780	-20	20
23	372	287	6	794	-24	16	50	425	350	5	790	-20	20
24	425	350	5	790	-20	20	51	425	200	5	790	-20	20
25	425	350	5	790	-20	10	52	425	350	5	790	-20	20
26	425	350	5	800	-20	20	53	550	350	5	790	-20	20
27	478	413	4	794	-16	16							

Table 1a
Results

EXP NUM	BOAT1 AVG	BOAT2 AVG	BOAT3 AVG	BOAT4 AVG	BOAT STDEV	GRNDMN	GRNDMED	MIN	MAX	GRAND STDEV	IQR	STDEV LOG10(THK)
1	995	942	901	895	46.71	933.2	923.5	851	1084	49.42	70.0	0.023
2	968	935	919	908	24.67	932.6	922.0	852	1090	44.16	51.5	0.020
3	893	809	749	732	72.28	795.7	781.5	690	936	68.49	115.0	0.037
4	933	872	797	811	68.27	853.2	859.0	705	1017	69.32	102.5	0.036
5	1003	945	884	868	59.37	925.0	911.0	813	1138	76.24	107.0	0.035
6	1183	1070	996	976	94.16	1056.1	1034.0	903	1401	109.81	156.5	0.044
7	972	925	884	880	43.98	915.4	911.0	837	1053	47.68	63.0	0.022
8	963	951	932	944	15.46	947.4	938.0	886	1052	38.18	55.5	0.017
9	950	920	892	903	29.27	916.2	911.0	853	1019	35.39	49.0	0.017
10	1152	1092	1072	1129	41.67	1111.5	1112.5	1009	1301	64.15	96.0	0.025
11	951	931	909	916	21.02	927.1	922.5	867	1004	31.84	47.5	0.015
12	951	919	905	936	23.43	927.8	922.0	872	1037	36.23	53.5	0.017
13	992	928	888	894	52.50	925.4	916.5	838	1177	58.77	70.5	0.027
14	1105	1056	1032	1077	37.42	1067.4	1063.0	980	1198	49.20	75.0	0.020
15	786	716	658	634	63.95	698.5	682.5	611	843	63.05	102.5	0.039
16	1096	1031	946	904	74.78	994.3	986.5	865	1199	85.66	140.0	0.037
17	1055	936	872	871	92.99	933.7	908.5	812	1312	92.84	118.0	0.041
18	876	824	759	730	58.88	797.0	796.0	694	921	63.85	110.5	0.035
19	880	857	821	809	29.76	841.4	843.5	771	957	37.25	58.0	0.019
20	974	890	823	796	75.61	870.7	860.0	757	1091	76.80	110.0	0.038
21	764	739	707	707	28.66	729.0	727.5	673	801	32.26	50.0	0.019
22	890	865	835	843	27.13	858.2	858.5	798	925	30.03	49.5	0.015
23	863	805	729	689	67.00	771.4	762.5	655	916	72.17	129.0	0.041
24	983	932	889	882	46.85	921.5	914.5	842	1061	50.06	64.0	0.023
25	1072	978	883	820	94.68	938.1	919.0	771	1201	105.02	165.0	0.048
26	1018	956	901	883	58.67	939.6	925.5	849	1097	60.70	93.5	0.028
27	1119	1061	1010	1009	54.74	1049.7	1047.0	951	1195	58.65	85.5	0.024
28	902	877	855	870	23.38	876.1	879.5	819	949	28.00	43.0	0.014
29	741	716	675	656	33.16	697.2	696.0	635	775	37.93	62.5	0.024
30	743	697	667	676	38.49	695.6	691.0	639	802	36.60	53.5	0.023
31	1103	1104	1102	1154	0.85	1115.8	1107.0	1049	1270	46.71	66.0	0.018
32	918	890	846	829	36.26	870.7	870.5	794	965	41.35	67.5	0.021
33	1038	1013	993	1029	22.64	1018.4	1018.0	951	1121	38.40	55.0	0.016
34	1111	1102	1078	1089	16.95	1094.7	1092.5	1019	1184	38.94	63.0	0.015
35	889	896	880	882	8.06	886.5	884.5	841	941	22.19	32.0	0.011
36	1095	945	861	846	118	936.8	906.5	781	1322	116.43	151.5	0.051
37	994	955	926	940	33.76	953.9	947.5	890	1072	40.07	59.0	0.018
38	1223	1052	930	884	147	1022.4	981.5	816	1470	153.40	210.0	0.063
39	855	772	689	654	82.87	742.7	722.5	627	919	83.39	147.5	0.048
40	988	909	848	832	70.08	894.0	883.0	793	1092	69.27	100.5	0.033
41	1403	1310	1222	1212	90.58	1286.6	1273.5	1099	1642	120.95	183.0	0.040
42	1434	1309	1179	1111	128	1258.1	1245.0	1009	1714	156.65	232.0	0.053
43	1234	1052	929	884	153	1024.9	991.5	809	1503	160.26	231.0	0.065
44	1205	1028	899	837	153	992.2	944.0	771	1426	157.88	231.0	0.067
45	1040	1029	1016	1047	12.34	1033.0	1035.0	971	1143	34.21	54.0	0.014
46	864	828	789	783	37.37	815.9	817.0	749	903	39.43	60.0	0.021
47	943	894	847	827	47.87	877.8	872.5	792	1009	51.18	82.5	0.025
48	887	831	791	803	48.54	828.1	822.5	751	937	44.65	66.5	0.023
49	781	771	758	781	11.39	772.8	774.0	727	829	22.30	34.0	0.013
50	881	857	829	838	26.37	851.0	851.0	790	914	29.31	47.0	0.015
51	649	613	578	575	35.53	603.4	598.0	551	677	33.24	52.0	0.024
52	1222	1051	930	871	147	1018.2	977.0	817	1423	150.69	217.5	0.062
53	1222	1046	926	873	149	1016.7	969.0	802	1428	151.53	227.0	0.062

Figure 2

Variation of Thickness Distribution
With Experiment Number

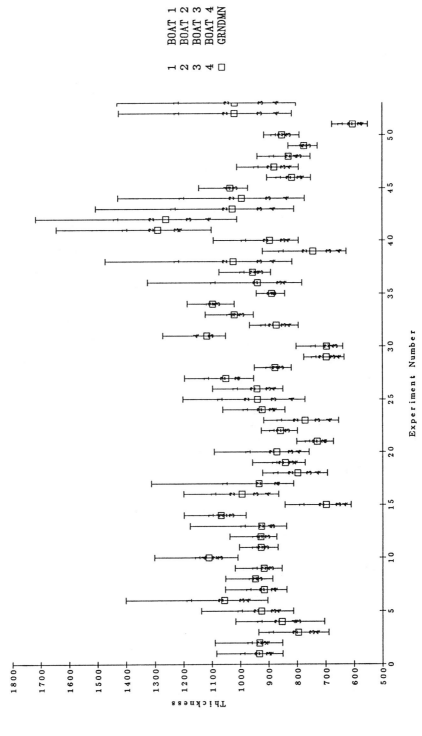

Table 2 summarizes the regression models derived in this study. Note the similarity in structure for location as well as variation effects. Note also that two interactions appear to control the variability independent of the location.

In performing this analysis we postulated a full 2nd order Taylor Series Model and refined it using stepwise regression which obeyed hierarchy. (Hierarchy rules require that one remove a main effect term only when all higher-order terms involving it have been removed.) This contrasts directly with the methods advocated by G. Taguchi. The quotes below are from a recent paper he published in The ASI Journal (6):

> "...In order to prove that there are no interactions, one assigns only main effects to the orthogonal array and finds the optimum conditions. One makes the assumption that there is not interaction and estimates the process average. If there is little interaction, the estimated value of the process average and its confirmatory experiment value will match well. If the effect of interactions is large, these two values will differ greatly and it becomes necessary to take additional measures, such as redoing the experiment on site or bracing oneself for the necessity of improving the design according to marketplace information, or redesigning by rethinking the quality characteristic so that there will no longer be interactions...
>
> ...The author suggests that, even if there are interactions, only main effects should be assigned to an orthogonal array. The process average is then estimated, and one compares this process average with the value of the confirmatory experiment...
>
> ...When an interaction is significant, this indicates that the optimum level of a control factor will differ, depending on the levels of other control factors. This also means the optimum conditions we have found will not be useful downstream where conditions differ. This is true because if there are interactions between control factors, there will also be interactions with the manufacturing scale and interactions between laboratory and marketplace conditions. Experience shows that optimum conditions determined after finding interactions do not stand up under large-scale production conditions...
>
> The greatest general difference between Taguchi Methods and traditional design of experiments is the belief that it is useless and counter-productive to assign interactions, whether or not there are interactions. Orthogonal arrays should be used to help us discover experimental failures when interactions exist..."

How in the world anyone would be able to predict a significant interaction is beyond this author. Furthermore, how do we define the "quality characteristic" variability so that interactions are no longer important? The authors experience is that interactions among control factors remain important regardless of the scale of the production. Perhaps new adjustments become necessary if one scales up a process to much larger equipment or output, but control factor interactions do **NOT** disappear.

This general ignoring of interactions in both optimization and some screening experiments, the general disregard for the importance of randomizing experimental tests, and the tendency to "pick the winner" among existing experimental conditions are the **THREE** main problems with the "Taguchi Method". Twenty years of doing experimental designs has shown this author that interactions are common, that processes drift with time while we probe them, and that experiments done to satisfy a matrix seldom contain the optimum point.

Modeling a collection of outputs as functions of inputs requires multiple regression analysis. Box and Draper (7) have recently reminded us that all regression models are wrong, but that some might be useful. How useful depends on their ability to predict responses within the limits of our experimental space. Table 3 shows the results of optimization runs done within the software using all the metrics for variability studied here. Regardless of their complexity, they seem to converge on the same general experimental region. Are the differences observed of any consequence?

Within the MULREG program a utility allows us to predict responses from our models, given a set of input parameters. Table 4 summarizes several of these. The ranges of values chosen for use in generating Table 4 represent the ranges seen from the various optimization runs. The confidence intervals show that the minor differences observed among the metrics are inconsequential.

Another utility uses the regression models to create contour plots showing us graphically the region of minimum variability predicted by each. Figures 3 and 4 are the contourplots for all the metrics discussed here. Note how all converge to one region.

Table 2

Regression Models -- Location and Variability

Term	MEAN	MEDIAN	TARGET IS BEST	-LOG10 (STDEV)	LOG10 STDEV	2*SD/MN *100	2*SDBT/GMN *100	THKIQR	IQR/MED *100	-LOG10 IQR
1	0.0329	0.0331	23.939	-1.757	-1.565	0.290	2.302	0.112	0.335	-1.932
~CLP	-0.0045	-0.0046	-0.241	-0.060	0.014	-0.018	0.142	-0.006	-0.017	-0.046
~TF	0.0027	0.0028	1.042	-0.054	-0.055	0.023	-0.498	-0.005	0.027	-0.052
~RAB	-0.0009	-0.0009	-1.781	0.002	0.091	-0.024	0.340	-0.004	-0.040	-0.032
~CTMP	0.0007	0.0009	0.148	-0.040	-0.005	-0.011	0.169	-0.011	-0.020	-0.060
~ZN1	-0.0010	-0.0010	-3.227	-0.163	0.155	-0.061	0.709	-0.022	-0.072	-0.155
~ZN3			1.271	0.025	-0.066	0.020	-0.373	0.011	0.041	0.075
~CLP*CTMP			-9.923	-0.327	0.482	-0.095	1.121	-0.037	-0.103	-0.312
~TF*CTMP	0.0018	0.0017	6.607	0.337	-0.329	0.090	-0.771	0.041	0.107	0.355
~RAB*CTMP	-0.0021	-0.0020	-4.884	-0.312	0.247	-0.083	0.818	-0.038	-0.096	-0.327
~ZN1*ZN3			-8.469	-0.324	0.409	-0.091	0.815	-0.045	-0.126	-0.368
TRANSFORM	INVSQRT	INVSQRT	NONE	NONE	NONE	INVSQRT	LOG	INVSQRT	INVSQRT	NONE
VAR. WT.	NONE	NONE	CTMP	CTMP	RAB	CTMP	CTMP	CTMP	CTMP	CTMP
ADJ R**2	0.8156	0.8573	0.462	0.296	0.462	0.450	0.502	0.403	0.568	0.403
REL. PRES	0.7760	0.8200	0.314	0.089	0.313	0.282	0.373	0.236	0.449	0.183

Table 3

Optimization Summary -- Variability Metrics

Factor, Resp., Formula	Range	Initial Setting	TARGET IS BEST	-LOGS	LOG10SD LOGTHK	2*SD/MN *100	2*SDBT/GMN *100	THKIQR	IQR/MED *100	-LOG10 IQR
Factors										
CLP	300 to 550	425	300.440	300.000	300.000	300.010	307.260	300.120	300.140	300.000
TOTFLW	200 to 500	350	500.000	499.620	500.000	491.000	490.320	498.780	499.260	499.990
RAB	3 to 7	5	3.006	3.001	3.240	3.003	3.014	3.001	3.002	3.019
CTMP	780 to 800	800	799.920	800.000	799.940	799.990	799.930	799.990	799.950	800.000
ZN10FF	-30 to -10	-30	-29.230	-28.601	-30.000	-29.184	-29.900	-29.993	-29.874	-29.999
ZN30FF	10 to 30	30	27.622	28.630	26.538	26.821	26.339	29.824	27.031	29.749
Resp.										
BTUNIF			0.067	0.064	0.078	0.078	0.083	0.046	0.065	0.046
IQRMED			1.270	1.250	1.330	1.326	1.347	1.135	1.265	1.137
LGIQR			-0.522	-0.502	-0.576	-0.568	-0.590	-0.386	-0.517	-0.389
LSDLTH			-3.258	-3.281	-3.207	-3.207	-3.176	-3.404	-3.263	-3.403
SDMNUN			1.803	1.784	1.871	1.869	1.899	1.642	1.791	1.644
THKTGT			58.248	58.701	57.231	57.213	56.596	61.256	58.367	61.222
THKIQR			12.803	12.582	13.396	13.335	13.579	11.371	12.743	11.396
THKLOGS			-0.429	-0.415	-0.468	-0.465	-0.486	-0.315	-0.420	-0.317

Table 4

Predictions from Regression Models (IQR/MED * 100)

Predictions and 95% simultaneous confidence intervals
for mean responses of IQRMED using model TKIQRMD
  RAB = 3, CTMP = 800, CLP = 300, TF = 490

ZN3		ZN1=-30	ZN1=-29	ZN1=-28
	Lower	0.657995	0.677811	0.698022
25	Predicted	1.382611	1.427586	1.474793
	Upper	4.563445	4.741089	4.938965
	Lower	0.535777	0.561639	0.588548
30	Predicted	1.145710	1.195903	1.249468
	Upper	3.963159	4.089269	4.238291

Predictions and 95% simultaneous confidence intervals
for mean responses of IQRMED using model TKIQRMD
  RAB = 3, CTMP = 800, CLP = 300, TF = 500

ZN3		ZN1=-30	ZN1=-29	ZN1=-28
	Lower	0.640432	0.659353	0.678640
25	Predicted	1.353949	1.397522	1.443234
	Upper	4.541693	4.719994	4.918446
	Lower	0.523517	0.548327	0.574109
30	Predicted	1.124059	1.172821	1.224826
	Upper	3.931724	4.059521	4.210180

7. Summary:

What then is the engineer to do?

This study has shown that where some correlation exists between the variability of a process and the location of a process, one may use that metric to describe that variability that makes the most sense in his environment and is the most convenient.

Without doubt the logS statistic advocated by Dr. George Box is adequate to the task described here. In this case, however, because engineers in the semiconductor industry are more familiar with metrics connoting a +/- variation around a central value, we elect to use any one of the several generally based on the coefficient of variation. If one employs a "fudge-factor" to generate values similar to those found using the standard deviation, that metric based on the IQR and the median is probably the one of choice.

If you wish to interpret decibel values from the SN_T metric, then do so. Most importantly, however, instead of the highly-fractionated or saturated designs advocated by Taguchi use more conservative 1/2-, or 1/4-fraction factorials so that you can estimate interactions. Randomize experimental trials to protect yourself and your results from nuisance variation due to a change in the process as you probe it. Finally, model your responses using multiple regression to help identify optimal regions without trying to "pick a winner" among the experimental trials actually done.

Figure 2

BTUNIF, IQRMED, THKTGT, THKLOGS
CLP = 300, CTMP = 800, ZN10FF = -30, ZN30FF = 30

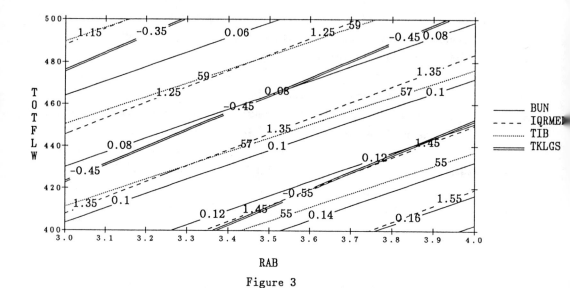

RAB

Figure 3

LGIQR, LGSDLGTHK, SDMNUNIF, THKIQR
CLP = 300, CTMP = 800, ZN10FF = -30, ZN30FF = 30

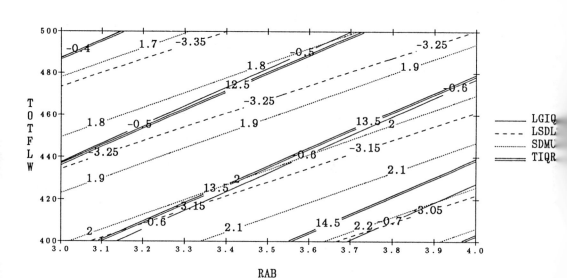

RAB

8. References:

1. Box, G.E.P., Hunter,W.G., and Hunter,J.S., "Statistics for Experimenters", John Wiley & Sons, New York, 1978.

2. Montgomery, D.C., "Design and Analysis of Experiments", John Wiley & Sons, New York, 1984.

3. See for example: Taguchi, G., and Phadke, M.S., "Quality Engineering Through Design Optimization", Conference Record Vol. 3., IEEE Globecom 1984 Conference, Atlanta, GA, 1106-1113; Taguchi, G., and Wu, Y., Introduction to Off-Line Quality Control, Central Japan Quality Control Association (1986); and Taguchi, G., Introduction to Quality Engineering: Designing Quality into Products and Processes. Asian Productivity Organization, UNIPUB, Kraus International Publications, White Plains, NY. See also Phadke, M.S., "Quality Engineering Using Design of Experiments", Proceedings of the American Statistical Association, section on Statistical Education, 11 - 20 (1982).

4. Box, G.E.P, "Studies in Quality Improvement: Signal to Noise Ratios, Performance Criteria, and Transformation", Report No. 26, University of Wisconsin Center for Quality and Productivity Improvement, Madison, WI., 1987.

5. Reece, J.E., and Good, M.A., "Process Variability and Response Surface Methodology", Proceedings of the 1988 Rocky Mountain Quality Conference, Denver, CO.

6. Taguchi, G., "The Development of Quality Engineering", ASI Journal 1, (1988).

7. Box, G.E.P, and Draper, N.R. "Empirical Model-building and Response Surfaces", John Wiley & Sons, New York, 1987.

9. About the Author:

Dr. Jack E. Reece received the Ph.D. in Organic Chemistry from the University of N. Carolina in 1966. He has practiced statistical experimental design techniques since 1970 in all areas of process development and manufacturing in both the manufacture of specialty coatings and semiconductor devices. Currently he is a Senior Principal Process Engineer for a major semiconductor manufacturer and defense contractor. In that position he assists in the integration of applied statistics methods, including experimental design and SPC, in all manufacturing stages and trains co-workers to use the techniques.

In addition he does free-lance consulting and training in the general areas of statistical design of experiments for clients in the computer, semiconductor, and defense contracting community.

Lt Col Mark J. Kiemele
Tenure Associate Professor
United States Air Force Academy

COMPUTER NETWORK PERFORMANCE ANALYSIS*

4. Introduction

The verification and validation of simulation models is a difficult task [8, 18]. In the absence of real data from the system being simulated, the task seems even close to insurmountable. Action has been taken, however, to gain a high confidence level in the simulator described in Chapter III.

Verification of simulator performance with regard to voice/data transaction arrival patterns was accomplished by Clabaugh [3]. He found that the simulator behavior was within a 95% confidence interval of the true means, where the true means were determined by the use of standard Poisson generators and user run time specifications. Clabaugh also provides additional verification of simulator output by configuring the simulator to correspond to an analytic model and comparing the output results [3]. These verification efforts apply to the modified model as well, since none of the modifications impact Clabaugh's verification work. This chapter documents another step taken toward the verification and validation of the integrated network simulator.

*reprinted from the author's dissertation, "Adaptive Topological Configuration of an Integrated Circuit/Packet-Switched Computer Network," Texas A&M University, 1984.

The motivation for this analysis stems not only from the desire to move in the direction of model verification but also from the need to obtain and analyze performance data from an integrated circuit/packet-switched computer network, for such data is scarce. Fortunately, these two encompassing goals do not diverge.

4.1 Goals and Scope of the Analysis

The specific goals of the analysis are as follows:

(1) Design an experiment whereby the performance data can be efficiently and economically obtained.

(2) Determine the effective ranges of network parameters for which realistic and acceptable network performance results.

(3) Investigate the sensitivity of performance measures to changes in the network input parameters. Specifically, determine how network performance is affected by changes in the network traffic load, trunk line or link capacity, and network size.

The scope of the analysis is restricted to those parameters that are closely related to the network topology and the workload imposed on that topology. With regard to performance sensitivity to workload and link capacity, the following four parameters are investigated:

CS: Circuit Switch Arrival Rate (voice calls/min)

PS: Packet Switch Arrival Rate (packets/sec)

SERV: Voice Call Service Rate (sec)

SLOTS: Number of Time Slots per Link (a capacity
indicator)

The sensitivity to network size is also investigated by
varying the number of nodes and links in a network. In
all cases, the performance measures observed are mean
packet delay (MPD), fraction of voice calls blocked (BLK),
and average link utilization (ALU).

4.2 Design of Experiment

The 10-node network topology shown in Figure 4 was
considered sufficiently complex to provide practical
performance data without exhausting the computing budget.

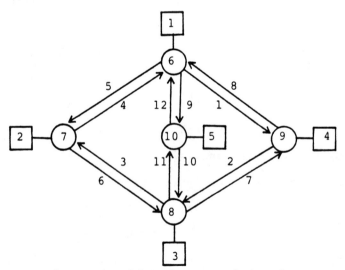

Figure 4. 10-Node network topology

The circuit switch (circular) nodes correspond to the
major computing centers of Tymshare's TYMNET, a
circuit-switched network [14]. The trunk lines are
full-duplex (FDX) carriers, each of which is modeled by
two independent, half-duplex (HDX) channels.

A preliminary sizing analysis [13] indicated that the
effective range of input parameters to be investigated is
as follows:

CS: 0-8 (calls/minute at each node)

PS: 0-600 (packets/sec at each node)

SERV: 60-300 (sec/call)

SLOTS: 28-52 (a link capacity indicator)

The fixed parameter settings were such that each slot
represents a capacity of about 33 Kbits/sec.

The experimental design selected for this analysis is
a second order (quadratic), rotatable, central composite
design [10, 11]. Such a design for k (number of
parameters) = 3 is illustrated in Figure 5. This design
was chosen because it reduces considerably the number of
experimentation points that would otherwise be required if
the classical 3^k factorial design were used. The
"central composite" feature of the design replaces a
3^k factorial design with a 2^k factorial system
augmented by a set of axial points together with one or
more center points. A "rotatable" design is one in which
the prediction variance is a function only of the distance

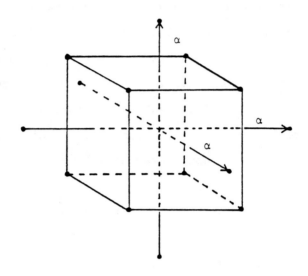

Figure 5. Central composite design (k=3)

from the center of the design and not on the direction.
Myers [10] shows that the number of experimental points
required for k = 4 is 31. That is, there are 16 factorial
points, 8 axial points, and 7 replications at the center
point. In order for the design to be rotatable, Myers
[10] shows that α must be chosen as $(2^k)^{1/4}$. In
this case, α, the distance from the center point to an
axial point, is 2. The seven replications at the center
point will allow an estimate of the experimental (pure)
error to be made; thus, a check for model adequacy is
possible [12]. Myers [10] recommends seven replications

at the center point simply because it results in a
near-uniform precision design. That is, it is a design
for which the precision on the predicted value $\hat{y}$, given by

$$\frac{N \text{ var } (\hat{y})}{\sigma^2} \quad , \text{ where}$$

    N = total number of experimental points, and
    σ^2 = the error variance,

is nearly the same at a distance of 1 from the center
point as it is at the center point [10].

In this design, each parameter is evaluated at five
different levels. The five levels for each of the four
parameters are as follows:

CS:	0	2	4	6	8
PS:	0	150	300	450	600
SERV:	60	120	180	240	300
SLOTS:	28	34	40	46	52

The center point is defined as CS/PS/SERV/SLOTS =
4/300/180/40.

Analysis of the results of these 31 runs indicated
that the original range of data was too large. Ten of
these experimental data points resulted in network
performance that would be totally unacceptable, e.g., a
MPD of more than 10 seconds. These "outliers" were
eliminated and 14 additional points in the moderate to
heavy loading of the network were added. Two observations

were taken at 8 of these 14 additional points for the purpose of estimating the error variance at points other than the center of the design. Table I shows the 43 observations that form the data set used in the subsequent regression analysis. The first 21 observations in Table I represent those data points retained from the original 31-run design. The remaining 22 observations represent the data points used to augment the original design.

Table I. Experimental data

OBS	CS	PS	SERV	SLOTS	MPD	ALU	BLK
1	2	150	120	34	0.117	0.262	0.000
2	2	150	120	46	0.116	0.194	0.000
3	2	150	240	34	0.113	0.372	0.000
4	2	150	240	46	0.116	0.275	0.000
5	2	450	120	34	0.430	0.553	0.016
6	2	450	120	46	0.260	0.396	0.000
7	2	450	240	34	0.433	0.676	0.031
8	2	450	240	46	0.449	0.481	0.003
9	6	150	120	34	0.339	0.520	0.012
10	6	150	120	46	0.115	0.386	0.000
11	6	450	120	46	0.462	0.594	0.016
12	0	300	180	40	0.123	0.231	0.000
13	4	0	180	40	0.000	0.325	0.000
14	4	300	180	52	0.153	0.427	0.000
15	4	300	180	40	0.350	0.561	0.016
16	4	300	180	40	0.527	0.571	0.005
17	4	300	180	40	0.502	0.561	0.011
18	4	300	180	40	0.316	0.581	0.020
19	4	300	180	40	0.381	0.562	0.017
20	4	300	180	40	0.798	0.608	0.036
21	4	300	180	40	0.616	0.560	0.007
22	5	400	210	43	1.613	0.739	0.070
23	5	400	210	43	2.754	0.746	0.082
24	5	400	150	37	1.656	0.725	0.067
25	5	400	150	37	1.345	0.725	0.084
26	5	400	150	43	0.192	0.615	0.027
27	5	400	210	40	4.814	0.785	0.106
28	5	400	210	40	5.624	0.782	0.125
29	4	400	210	37	1.604	0.747	0.084
30	4	400	210	37	2.730	0.745	0.101
31	4	400	210	40	2.440	0.695	0.063
32	4	400	210	40	1.199	0.687	0.057
33	4	400	180	40	0.910	0.648	0.035
34	5	400	180	40	1.970	0.722	0.074
35	5	400	180	40	1.130	0.731	0.077
36	5	400	180	37	4.084	0.775	0.110
37	5	400	180	37	4.019	0.778	0.139
38	4	400	180	37	4.817	0.712	0.057
39	4	400	180	37	1.135	0.701	0.065
40	3	400	210	37	0.599	0.652	0.017
41	3	400	210	43	0.293	0.541	0.003
42	3	400	150	37	0.454	0.570	0.003
43	3	400	150	43	0.280	0.476	0.005

4.3 Regression Analysis and Model Selection

The Statistical Analysis System (SAS) [15, 16, 17]
has been used to perform a multiple regression analysis
for each of the three performance measures. The
regression variables are the four input parameters. SAS
procedures REG [16] and RSREG [16] were used to analyze
the data. The assumption of a quadratic response surface
allows for the estimation of 15 model parameters,
including the intercept.

Although the model selection procedure can be based
on a number of possible criteria [4, 6, 9], the
approach taken is as follows. RSREG was used first to
check the full (quadratic) model for specification error
(lack of fit test) and to determine significance levels
for the linear, quadratic, and crossproduct terms. The
models for MPD, BLK, and ALU exhibited a lack of fit that
was significant at the .14, .05, and .82 levels,
respectively. It was found, however, that these
significance levels were heavily influenced by the
observations at the extremes of the heavy-loading region.
For example, by eliminating 13 of the observations in the
heavy-loading range of data, the lack of fit for BLK could
be raised to the .70 significance level. Hence, although
the .05 level for the BLK model borders on statistical
significance, the lack of fit was not deemed sufficient to
justify a more complex model.

7-65

The RSREG results also indicated that some terms were
insignificant and possibly could be deleted from the
model. Using the RSREG results as a guide, several
subsets of the full quadratic models were investigated
using SAS procedure REG. In these models, all linear
terms were retained because even though a lower order term
in a polynomial model may not be considered significant,
dropping such a term could produce a misleading model
[6]. Several crossproduct and quadratic terms were
deleted, however, and the final regression models selected
are shown in Table II. Despite their appearance of being
insignificant, several of the crossproduct and quadratic
terms were retained in the model simply because their
retention resulted in a smaller residual mean square than
if they had been deleted. The interpretation of these
regression models is presented graphically in the
following section in terms of a sensitivity analysis.

4.4 Sensitivity Analysis

This section presents data to describe how the
various performance measures change with respect to
changes in traffic load, link capacity, and network size.
All of the graphs presented are obtained from the
quadratic response surfaces (regression models) developed
in the previous section.

Table II. Regression models

DEP VARIABLE: MPD

SOURCE	DF	SUM OF SQUARES	MEAN SQUARE	F VALUE	PROB>F
MODEL	10	61.561986	6.156199	6.874	0.0001
ERROR	32	28.658932	0.895592		
C TOTAL	42	90.220918			
ROOT MSE		0.946357	R-SQUARE	0.6823	
DEP MEAN		1.218093	ADJ R-SQ	0.5831	
C.V.		77.69169			

VARIABLE		DF	PARAMETER ESTIMATE	STANDARD ERROR	T FOR HO: PARAMETER=0	PROB > |T|
INTERCEP		1	12.556905	13.142988	0.955	0.3465
X1	CS	1	-1.654903	1.326989	-1.247	0.2214
X2	PS	1	0.014620	0.012349	1.184	0.2452
X3	SERV	1	-0.026635	0.010541	-2.527	0.0167
X4	SLOTS	1	-0.531166	0.597025	-0.890	0.3803
X11		1	0.232675	0.078122	2.978	0.0055
X12		1	0.001153381	0.0009976058	1.156	0.2562
X13		1	0.013125	0.00331975	3.954	0.0004
X14		1	-0.048214	0.025537	-1.888	0.0681
X24		1	-0.000395179	0.0003018693	-1.309	0.1998
X44		1	0.009018565	0.007302992	1.235	0.2259

DEP VARIABLE: ALU

SOURCE	DF	SUM OF SQUARES	MEAN SQUARE	F VALUE	PROB>F
MODEL	11	1.123895	0.102172	970.909	0.0001
ERROR	31	0.003262242	0.0001052336		
C TOTAL	42	1.127158			
ROOT MSE		0.010258	R-SQUARE	0.9971	
DEP MEAN		0.581233	ADJ R-SQ	0.9961	
C.V.		1.764929			

VARIABLE		DF	PARAMETER ESTIMATE	STANDARD ERROR	T FOR HO: PARAMETER=0	PROB > |T|
INTERCEP		1	0.022183	0.150510	0.147	0.8838
X1	CS	1	0.068950	0.014713	4.686	0.0001
X2	PS	1	0.001859007	0.0001334741	13.928	0.0001
X3	SERV	1	0.001788392	0.0006903726	2.590	0.0145
X4	SLOTS	1	-0.0094644	0.006917257	-1.368	0.1811
X11		1	-0.000929364	0.0008468003	-1.098	0.2809
X13		1	0.0003897799	.00003664077	10.638	0.0001
X14		1	-0.00127589	0.0002929912	-4.355	0.0001
X24		1	-.0000259887	.00000329748	-7.881	0.0001
X33		1	-.0000024518	.00000146947	-1.669	0.1053
X34		1	-.0000214946	.00000932481	-2.305	0.0280
X44		1	0.0001583866	.00008483328	1.867	0.0714

Table II.   (Continued)

```
DEP VARIABLE: BLK
                      SUM OF          MEAN
    SOURCE      DF    SQUARES         SQUARE      F VALUE      PROB>F
    MODEL       12    0.062669    0.005222451     26.743       0.0001
    ERROR       30    0.005858444 0.0001952815
    C TOTAL     42    0.068528
         ROOT MSE      0.013974       R-SQUARE      0.9145
         DEP MEAN      0.038163       ADJ R-SQ      0.8803
         C.V.          36.61764

                      PARAMETER      STANDARD     T FOR H0:
    VARIABLE   DF     ESTIMATE        ERROR     PARAMETER=0   PROB > |T|

    INTERCEP    1     0.197892       0.205845       0.961        0.3441
    X1   CS     1    -0.028562       0.020039      -1.425        0.1644
    X2   PS     1    0.0003439953  0.0002038975     1.687        0.1020
    X3   SERV   1    0.0001307622  0.0005157113     0.254        0.8016
    X4   SLOTS  1    -0.012008      0.00887909      -1.352       0.1863
    X11         1    0.006553606    0.00115578       5.670       0.0001
    X12         1    .00005438269   .00001502268     3.620       0.0011
    X13         1    0.0003339836   0.0000496682     6.724       0.0001
    X14         1    -0.00175778    0.0004020761    -4.372       0.0001
    X22         1    3.79530E-07    1.75917E-07      2.157       0.0391
    X24         1    -.0000158292   .00000450668    -3.512       0.0014
    X34         1    -.0000194136   .00001274852    -1.523       0.1383
    X44         1    0.0002843414   0.0001092323     2.603       0.0142
```

4.4.1   Sensitivity to Traffic Load

The workload imposed upon an integrated network is
described by the voice arrival rates (CS) at each circuit
switch node and the data packet arrival rates (PS) at each
packet switch node, as well as the length of service for
each voice call (SERV).   It is assumed that, for a given
arrival at any particular node, all of the other nodes of
the same type are equally likely to be the destination
node for that arrival.   That is, the workload is said to
be uniformly distributed between node pairs.   Unless
otherwise stated, it is also assumed that SERV = 180

seconds. Besides these three parameters, the link capacity (SLOTS) parameter also affects the network traffic load. If the CS and PS arrival rates are also assumed to be fixed, then the smaller values of SLOTS represent a heavier network load while the larger SLOTS values correspond to a lighter network load.

Figures 6, 7, and 8 depict the MPD, BLK, and ALU performance measures, respectively, as a function of CS for four different SLOTS/PS combinations. Similarly, Figures 9, 10, and 11 show MPD, BLK, and ALU, respectively, as a function of PS for four load levels defined by combinations of SLOTS and CS. The performance measure sensitivity to the traffic load parameters CS, PS, and SLOTS (for SERV=180) is seen by observing the relative slopes and ordinate values of the four curves in each of the Figures 6 - 11. For example, examination of the curves in Figures 6 and 9 indicates that, within the range of data shown, MPD is more sensitive to CS than to either SLOTS or PS. Additionally, the sensitivity of MPD to both CS and SERV is shown in the three dimensional plot of Figure 12, where the higher levels of one input parameter are seen to magnify the effects of the other parameter.

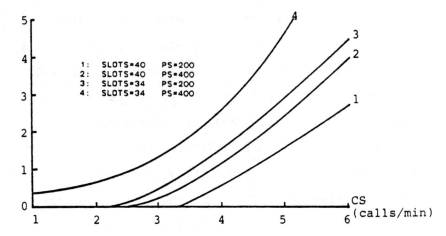

Figure 6.   Mean packet delay (MPD) vs.
            voice arrival rate (CS)

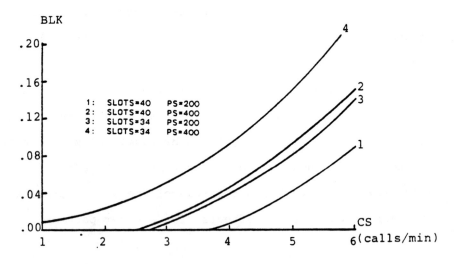

Figure 7.   Fraction of calls blocked (BLK)
            vs. voice arrival rate (CS)

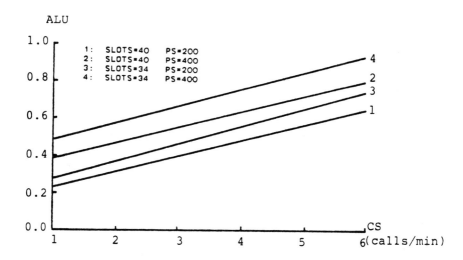

Figure 8.   Average link utilization (ALU)
vs. voice arrival rate (CS)

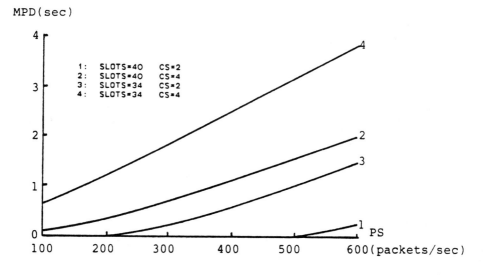

Figure 9.   Mean packet delay (MPD) vs.
data arrival rate (PS)

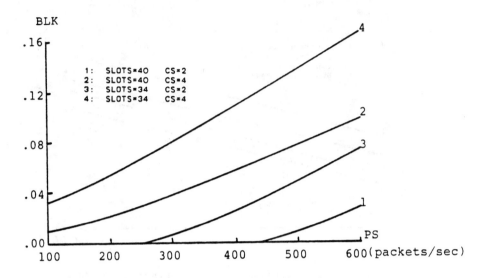

Figure 10.   Fraction of calls blocked (BLK)
vs. data arrival rate (PS)

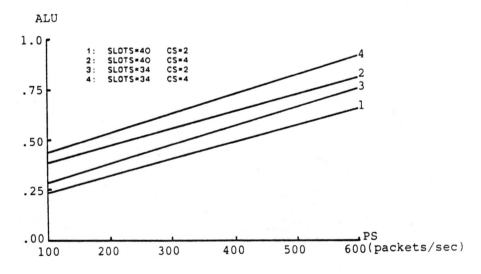

Figure 11.   Average link utilization (ALU)
vs. data arrival rate (PS)

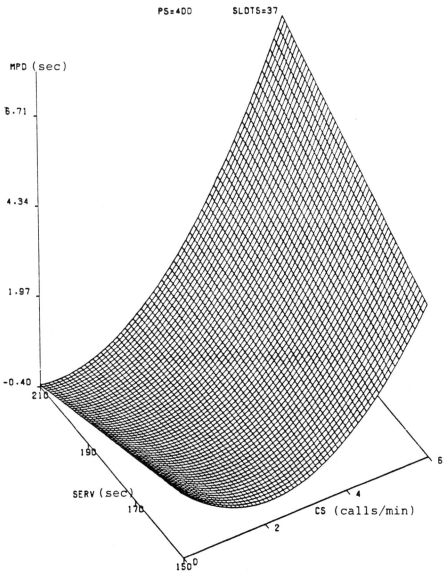

Figure 12. Mean packet delay (MPD) vs. voice arrival
rate (CS) vs. voice service time (SERV)

4.4.2 Sensitivity to Link Capacity

The trunk line carrying capacity is an important design parameter in integrated networks. Hence, in addition to the plots of the performance measures versus number of SLOTS, which are shown in Figures 13, 14, and 15, the confidence intervals for each of the performance measures are also presented. Figures 16, 17, and 18 give the 95% confidence limits for a mean predicted value of MPD, BLK, and ALU, respectively. The voice and data arrival rates for these graphs correspond to a fairly heavy traffic load (CS = 4, PS = 400, SERV = 180).

4.4.3 Sensitivity to Network Size

In addition to checking the performance of the simulator for varying traffic loads and link capacities on a fixed network topology, the analysis also examines a fixed traffic load on varying sized topologies. In particular, a throughput requirement of 2000 data packets per second with a voice call arrival rate of 20 calls per minute was imposed upon three different sized networks. A 10-node, 20-node, and 52-node network were each subjected to the fixed traffic load.

The 10-node network is the TYMNET topology shown previously in Figure 4. Six links interconnect the backbone nodes.

MPD(sec)

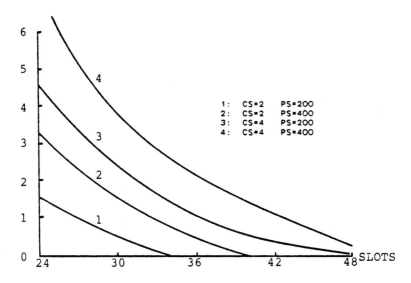

Figure 13. Mean packet delay (MPD) vs.
link capacity (SLOTS)

BLK

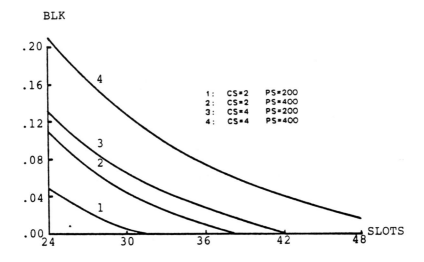

Figure 14. Fraction of calls blocked (BLK)
vs. link capacity (SLOTS)

7-75

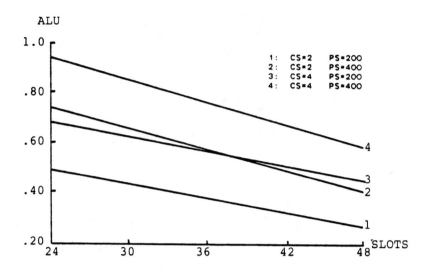

Figure 15.   Average link utilization (ALU)
vs. link capacity (SLOTS).

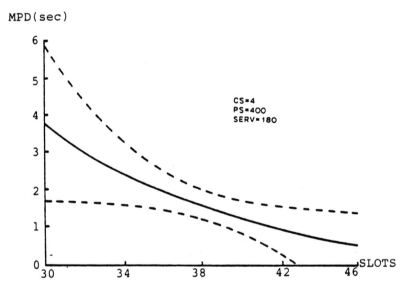

Figure 16.   95% Confidence limits for
mean packet delay (MPD)

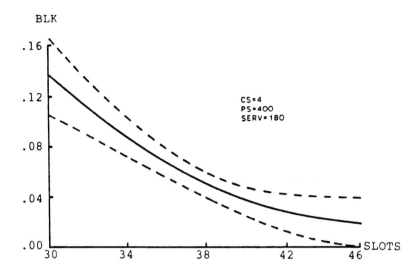

Figure 17.   95% Confidence limits for
             fraction of calls blocked (BLK)

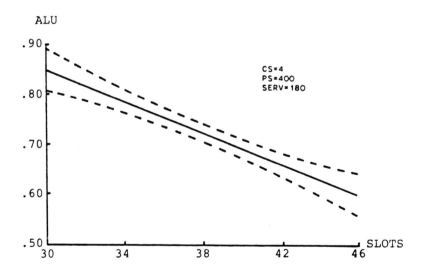

Figure 18.   95% Confidence limits for
             average link utilization (ALU)

7-77

The 20-node network consists of 10 packet switches and 10 circuit switches. The 10 circuit switches forming the backbone of the network are 10 of the major computing centers in the CYBERNET network [14]. The nodes on the subnet are interconnected by 12 trunk lines. The backbone of the 20-node network is depicted in Figure 19.

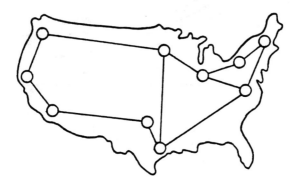

Figure 19.   20-Node network backbone

The 52-node network is comprised of 26 packet switches and 26 circuit switches, with the circuit switch nodes corresponding to a 26-node substructure of the ARPANET network. This 26-node subset of ARPANET is commonly used in the literature for comparative analyses [1, 2, 5, 7]. The subnet is interconnected with 33 links. Figure 20 shows the communications subnet of the 52-node network.

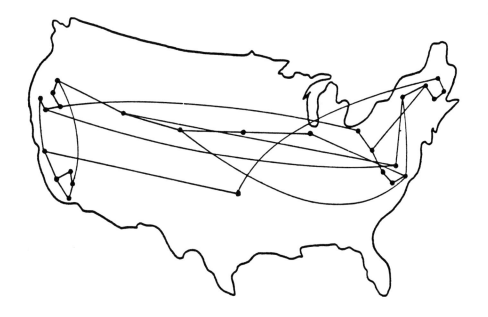

Figure 20.   52-Node network subnet

All links in the three topologies have a fixed
capacity of 40 slots.   Additionally, each of the three
communications subnets is node-biconnected (i.e., there
are at least two node-disjoint paths between any pair of
nodes).

Table III summarizes the results obtained when
subjecting the three different network topologies to the
given workload.   The performance measures given for the
10-node network are averages obtained from a total of 10
simulation runs.   Correspondingly, the data for the
20-node and 52-node networks are averages of 7 simulations

Table III. Sensitivity to network size

Topology	Subnet Link/Node Ratio	MPD	BLK	ALU
10-Node	1.2	.987	.034	.654
20-Node	1.2	.451	.033	.416
52-Node	1.27	.313	.006	.289

and 1 simulation, respectively. Despite the fact that there is a reduction in MPD as the number of nodes increases, the relatively smaller decrease in MPD when going from the 20-node network to the 52-node network (as compared to going from the 10-node network to the 20-node network) results from the fact that the backbone of the integrated network is circuit-switched. This implies that there is a fixed switching delay (assumed to be 50 ms in this study) at each node, and each packet incurs this delay at each intermediate node on its route. Hence, if the subnet link/node ratio remains fairly constant, the MPD is expected to decrease to a certain point and then begin to increase as the number of nodes in the network increases. This phenomenon is due to the fact that as more nodes are added to the network, a greater proportion of the delay can be attributed to switching delays, even though the queueing delay steadily decreases. However, as technology improves the switching delays, the effect of this phenomenon on total MPD is reduced.

4.5 Summary

Simulation model validation is a never-ending
-process, but a well designed sensitivity analysis of the
simulator can increase the user's confidence in the model
as well as his knowledge of the model. In this regard, a
sensitivity analysis is an important step in the direction
of model validation. This chapter has outlined an
approach to analyzing the performance characteristics of
an integrated computer-communication network simulation
model and has presented the results of the analysis.

The investigation has centered on the integrated
network traffic load parameters of packet arrival rate
(PS), voice call arrival rate (CS), and voice call service
times (SERV), as well as the design parameters of link
capacity (SLOTS) and network size. Network performance
has been measured in terms of mean packet delay (MPD), a
strict data measure; fraction of voice calls blocked
(BLK), a pure voice criterion; and average link
utilization (ALU), a gauge which tends to combine both the
data and voice attributes of an integrated network.

The analysis has determined a range of input
parameters for which second-order response surfaces can be
used to adequately describe realistic network performance.
This range is summarized as follows:

```
  CS:  1-6 calls/min

  PS:  100-600 packets/sec

SERV:  150-210 sec

SLOTS:  24-48 slots
```

In light of the fact that an integrated network with a
circuit-switched subnet has not yet been implemented, the
term "realistic" performance is admittedly dubious.
However, performance criteria for current packet-switched
and circuit-switched networks can and have been used as
guidelines (e.g., a packet delay of 10 seconds is clearly
unacceptable, for the message would automatically time
out), although flexibility in the guidelines has been
preserved so as to not stifle the range of model
applicability.

Furthermore, the graphs presented in this chapter
consistently support the theme that the performance
measures are more sensitive to the CS (voice) arrival rate
than to the other parameters investigated. The "slices"
(i.e., graphs) of the response surfaces that correspond to
an increased CS level generally have higher ordinate
values and steeper slopes than the "slices" that
correspond to an increased network loading due to
variations in the other parameters. In essence, voice
arrival rates tend to dominate the network in the sense
that the virtual circuits established by successful call
initiations provide the framework of paths by which data

packets can "piggyback" the digitized voice.

Additionally, heavily loaded networks tend to intensify the effect of any parameter. Increasing the number of nodes in a network (along with a corresponding increase in the number of links) for a fixed traffic load will decrease link utilization rates and call blocking, but may or may not decrease mean packet delay, depending on what proportion of the delay is switching delay.

The simulation model analyzed in this research is a tool that can be used by integrated network designers and managers alike. As a result of this analysis, the user of the simulation model now has a more precise understanding of the relationship between integrated network performance and the network parameters that influence such performance. In light of this, the model is a more viable tool now than it was prior to the analysis.

REFERENCES

1.   Boorstyn, R. R., and Frank, H.   Large-scale network topological optimization.   IEEE Trans. on Comm. Com-25, 1 (Jan 1977), 29-47.

2.   Chou, W., and Sapir, D.   A generalized cut-saturation algorithm for distributed computer communications network optimization.   IEEE 1982 Int. Conf. on Comm. (ICC-82), Philadelphia, PA (June 13-17, 1982), 4C.2.1-4C.2.6.

3.   Clabaugh, C. A.   Analysis of flow behavior within an integrated computer-communication network.   Ph.D. dissertation, Texas A&M University (May 1979).

4.   Draper, N. R. and Smith, H.   Applied Regression Analysis (2nd ed.).   John Wiley & Sons, Inc., New York, NY, 1981.

5.   Frank, H., and Chou, W.   Topological optimization of computer networks.   Proc. of IEEE 60, 11 (Nov 1972), 1385-1397.

6.   Freund, R. J., and Minton, P. D.   Regression Methods.   Marcel Dekker, Inc., New York, NY, 1979.

7.   Gerla, M., and Kleinrock, L.   On the topological design of distributed computer networks.   IEEE Trans. on Comm. Com-25, 1 (Jan 1977), 48-60.

8.   Graybeal, W. T., and Pooch, U. W.   Simulation: Principles and Methods.   Winthrop Publishers, Inc., Cambridge, MA, 1980.

9.   Montgomery, D. C., and Peck, E. A.   Introduction to Linear Regression Analysis.   John Wiley & Sons, Inc., New York, NY, 1982.

10.   Myers, R. H.   Response Surface Methodology.   Allyn and Bacon, Inc., Boston, MA, 1971.

11.   Naylor, T. H. (Ed.).   The Design of Computer Simulation Experiments.   Duke University Press, Durham, NC, 1969.

12.   Ostle, B., and Mensing, R. W.   Statistics in Research (3rd ed.).   The Iowa State University Press, Ames, IO, 1975.

13. Pooch, U. W., and Kiemele, M. J. A simulation model
for evaluating integrated circuit/packet-switched
networks. Presented at Winter Simulation Conf.,
Arlington, VA (Dec 12-14, 1983).

14. Pooch, U. W., Greene, W. H., and Moss, G. G.
Telecommunications and Networking. Little, Brown and
Company, Boston, MA, 1983.

15. SAS Institute Inc. SAS User's Guide: Basics, 1982
Edition. SAS Institute Inc., Cary, NC, 1982.

16. SAS Institute Inc. SAS User's Guide: Statistics,
1982 Edition. SAS Institute Inc., Cary, NC, 1982.

17. SAS Institute Inc. SAS/Graph User's Guide, 1981
Edition. SAS Institute Inc., Cary, NC, 1981.

18. Shannon, R. E. Systems Simulation: The Art and
Science. Prentice-Hall, Inc., Englewood Cliffs, NJ,
1975.

OPTIMIZING THE ANODIZE PROCESS AND PAINT ADHESION FOR SHEET METAL PARTS

Tom Bingham
Supplier Improvement Manager
Boeing Commercial Airplanes

INTRODUCTION

Sheet metal used for aircraft parts must be finished in such a way that it is resistant to the problems associated with corrosion. The anodizing process and painting of sheet metal are important procedures to insure acceptable corrosion resistance. The problems addressed in this experiment were optimization of the anodizing process and maximization of paint adhesion in order to assure acceptable corrosion resistance on metals exposed to salt spray tests. One of the problems to be overcome was the dilemma caused by conflicting affects of seal times and temperature on anodizing results and paint adhesion. The historical solution was to balance the process so both salt spray tests and paint adhesion failures are in spec. This balancing required slowing down the anodize time and increasing the temperature. Both of these settings produced an expensive solution to the problem. Therefore, it was decided to obtain information through experimental design which could ultimately lead to a more economical solution.

APPROACH TO THE PROBLEM

A team of qualified people from diverse backgrounds and levels of expertise was formed to brainstorm the problem. A fishbone diagram generated by the team yielded over 50 possible factors relevant to the anodize process and paint adhesion. The team then reduced the possible factors to 9 probable factors. Two experiments were then designed to model both paint adhesion and salt spray failures. The factors and the levels of interest are shown below.

Experimental Variables

Factor	Test Values	
	1	2
A: Deox Time	6 min	20 min
B: Anodize Time	35 min	55 min
C: Anodize Teamp	91°F	99°F
D: Free C_rO_3	32%	4.7%
E: Seal Time	23 min	28 min
F: Seal Temp	191°F	199°F
G. Ph	3.2	3.8
H: Primer Thickness	.3 mils	1.5 mils
I: Topcoat Thickness	.3 mils	1.5 mils

Controlled in Anodize Process (A–G)

Paint Adhesion Specific Factors (H–I)

The response of interest for the anodize process is the number of pits that appeared on the finished metal product after two weeks of exposure to the salt spray test.

RESULTS

The results indicated that anodize time and temperature had average response values as shown in the following graphic.

Salt Spray
Pits

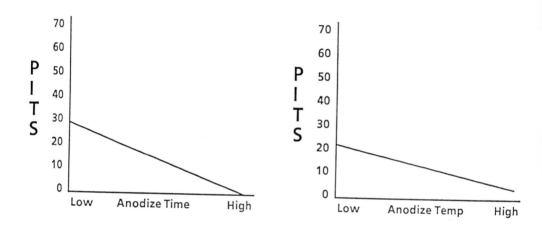

At first glance, it appeared that anodized time and temperature would play a moderate role in optimizing the anodize process. The optimal settings still seemed to favor high time and high temperature which were not the best for reducing costs. However, further investigation through the use of interaction graphs of time with CrO_3 and temperature with CrO_3 provided the following results.

Salt Spray
Pits

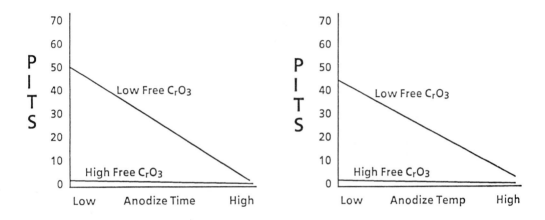

The interaction graphs revealed that when high free CrO_3 is used the time and temperature become unimportant for the anodize process. These two factors can now be set at levels which will speed up the process and reduce costs. The discovery of this very important interaction provided the knowledge required to produce a superior quality product at a lower cost.

Cadet First Class Jim Hecker
Cadet First Class Phil Herre
United States Air Force Academy

STATEMENT OF THE PROBLEM:

The purpose of this experiment is to find the best design and construction of a paper helicopter through the use of experimentation.

OBJECTIVE OF THE EXPERIMENT:

Find a combination of the input factors which will maximize the hangtime.

START DATE: 22 JAN 1989

END DATE: 27 JAN 1989

QUALITY CHARACTERISTIC:

Our response variable will be hangtime--time in the air. Hangtime is a quantitative variable (seconds), with an anticipated range of one to four seconds. The person who launches (drops) each helicopter from a height of ten feet will also time the fall in order to cut down the variability of reaction time.

FACTORS:

Three controllable variables are considered--1) ratio of blade length to width, 2) type of paper, and 3) the number of paper clips attached.

The ratio of blade length to width will have two levels: 2:1 and 1:2 (length to width). Paper is also a catagorical variable with two levels: filler paper and construction paper. The paper clip is a quantitative

variable with either one or two as the levels of interest.
We expect all the interactions to be worthy of
investigation, except for the three-way interaction of
ratio-paper-clips.

The noise variables (uncontrollable and unavoidable)
that may have an effect are:

    1) wind or drafts--To control for these, we
performed the experiments in a closed room.

    2) fall height--The helicopter does not start to
spin when released, creating a "fall height" where the
helicopter does not spin. From observation, we assume the
fall height to be random for each helicopter, therefore not
an important factor.

    3) reaction time--In timing the hangtime, there
may be some variance caused by the reaction time of hitting
the timer when the helicopter is released and when it hits
the ground. To control for this problem, the same person
launched the helicopters and timed the fall in order to
minimize and randomize reaction time.

EXPERIMENTAL DESIGN:

Due to the minimal cost, both in time and price, of a
single run, we chose a full-factorial design for this
experiment. A full-factorial has the advantage of testing
all possible combinations of the factors and accounting for
all possible interaction effects.

Since we had three factors at two levels, this meant we
needed 8 runs. In order to check our results and to have a
variance interval, we duplicated each run, meaning there
were 16 total runs.

Here is how the design looked:

RUN	RATIO	PAPER	R-P	CLIP	R-C	P-C	R-P-C	Y_1	Y_2	s^2
1	+	+	+	+	+	+	+	1.28	1.31	.00045
2	+	+	+	-	-	-	-	1.60	1.58	.00020
3	+	-	-	+	+	-	-	1.15	1.10	.00125
4	+	-	-	-	-	+	+	1.19	1.20	.00005
5	-	+	-	+	-	+	-	2.03	2.05	.00020
6	-	+	-	-	+	-	+	**2.50**	**2.42**	**.00320**
7	-	-	+	+	-	-	+	1.32	1.40	.00320
8	-	-	+	-	+	+	-	1.64	1.64	.00000

RATIO   + : 1:2   PAPER   + : filler   CLIPS   + : 2
        - : 2:1   - : construction   - : 1
(length to width)

ANALYSIS TECHNIQUES:

The primary method we used to analyze our results is a 2-way ANOVA table (ANalysis Of VAriance). We found the average value at each level of each factor, added the squared differences of each level from the grand mean (1.588) per factor, multiplied by the number that went into each average (4 runs each level x 2 runs) and divided by two. The error was calculated by adding the variances of each run. See Appendix 1 for the results.

Appendix 2 is a graph of the marginal means--the averages at each level per factor. The main factors--ratio, paper, and clips--clearly stand out, as do the three two-way interactions. The three-way interaction is insignificant, as expected.

Appendix 3 is a Pareto Diagram, which shows the half effects--half the difference of the averages. This plot is not as clear as the marginal means, however it does show the importance of the main factors and the two-way interactions in comparison to the three-way interaction.

Since the variance per run was so small it did not appear to be worthwhile to investigate dispersion effects.

CONCLUSIONS:

Based on a simple technique for full-factorials called "pick-the-winner", supported by the marginal means plot, we concluded that the best combination is **run six**--2:1 ratio, filler paper, one clip. Five confirmation runs were made of this design, giving an average hangtime of 2.378 seconds and a variance of 0.00868. These results are much better than those from any other combination.

RECOMMENDATIONS:

It is our professional opinion that in order to make a paper helicopter with the greatest hangtime, a light paper with a small weight attached to the bottom with the blades twice as long as wide will be the best design.

We recommend for anyone else who tries this experiment to use a greater release height in order to create more variance in the results. Also, a three factor, three level design would not be that hard to accomplish: simply create another ratio, like a 1:1 length to width; find a third kind of paper, like onion-skin typing paper; and a third level for the additional weight, i.e. 3 clips or even no clips.

APPENDIX 1: ANOVA Table

	Ratio	Paper	Clip	R-P	R-C	P-C	R-P-C
Average +	1.301	1.846	1.455	1.477	1.630	1.543	1.578
Average -	1.875	1.330	1.721	1.705	1.546	1.634	1.599

Source	SS	df	MS	F_O
Ratio	1.318	1	1.318	1318
Paper	1.065	1	1.065	1065
Clip	0.283	1	0.283	283
R-P	0.219	1	0.219	219
R-C	0.028	1	0.028	28
P-C	0.033	1	0.033	33
R-P-C	0.002	1	0.002	2 *
Error	0.009	8	0.001	--
Total	2.957	15	--	--

$F_{CRIT} = F_{(0.05,1,8)} = 5.32$

* Not significant because $F_O < F_{CRIT}$

APPENDIX 2: Marginal Means Plot

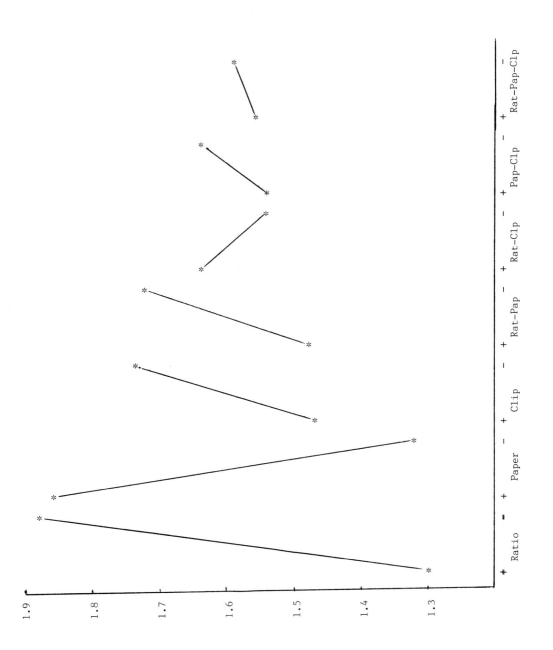

APPENDIX 3:   Pareto Diagram

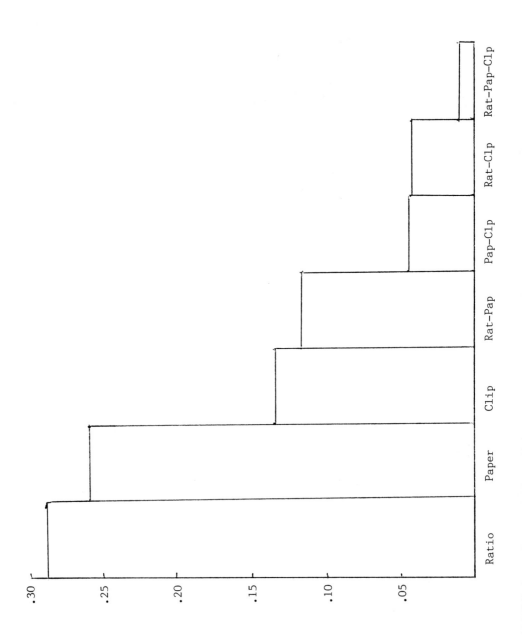

APPENDIX 4:   The Optimal Design

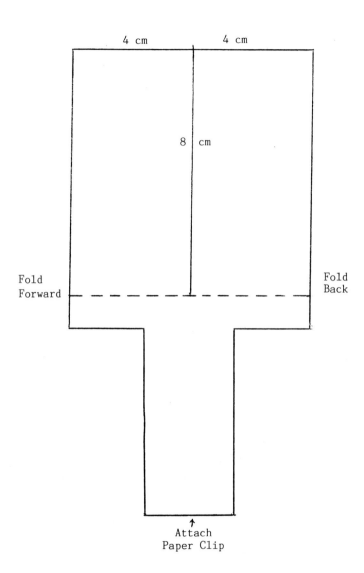

4 cm   4 cm

8 cm

Fold   Fold
Forward   Back

↑
Attach
Paper Clip

OPTIMIZING LUNAR CONCRETE

Presented to the American Concrete Institute,
Committee 125

Engineering Design Project
USAF Academy
Col. David O. Swint, Professor

Lt. Anderies   Lt. Kleinsmith Lt. Milam   Lt. Peddycord
Lt. Bennett   Lt. Krause   Lt. Mingus   Lt. Platt
Lt. Bramer   Lt. Leante   Lt. Neulander   Lt. Shackleford
Lt. Finn   Lt. Martin   Lt. Novak   Lt. Smith
Lt. Haggard   Lt. Meyer   Lt. Noyes

Problem Statement

The Spring 1989 Engineering 410 class at the US Air
Force Academy took on the project of making lunar concrete.
The task at hand was to design, fabricate, test and analyze
concrete mixtures, mixed under ambient and semi-evacuated
environments which optimized and maximized engineering
properties. The design, construction and testing of the
specimens were performed in accordance with American Society
for Testing Materials (ASTM) standards to ensure quality
control and to minimize errors (Swint, 1989).

Background

Of the various types of lunar construction materials
examined, concrete was found to be one of the most feasible
(Lin, 1987). Dr T.D. Lin has advocated the use of lunar
soils and rocks as aggregates because of their quality
physical characteristics (Lin, 1987). According to Dr Lin,
lunar materials contain sufficient amounts of silicate,
aluminum, and calcium to produce the "cementitious material"
required. Finally, when compared to three other major
construction materials (aluminum alloy, mild steel, and
glass), concrete requires less energy, and thus less cost,
for its production. It is also temperature, radiation, and
abrasion resistant (Lin, 1987).

A foreseeable problem with lunar construction of
concrete lays in the hydration process required in making
the concrete (Swint, 1989). Due to the evacuated conditions
on the lunar surface, curing of the concrete would have to
be accomplished in a pressure chamber to ensure that at
least water vapor pressure is maintained. This would
prevent the necessary amount of water for optimal concrete
strength from being drawn out into a lower pressure
environment.

Problem Solving Strategy

With the feasibility of lunar concrete as a
construction material established, one is left with solving
the problem of determining the best concrete to be used. To
accomplish this task, the Engineering 410 group had to

determine the best of 80 possible concrete combinations. The group used a D-optimal design to test 18 different combinations of cement, environment, plasticizer, wetting agent, and additive variables. From the results of the testing and the prediction equation, the characteristics of the remaining 62 combinations were estimated. The generation of these 18 test runs will be discussed in the section entitled "Generation of Test Runs."

Concrete Variables

Cement -- Two types of cement were used for the experiment, Calcium Aluminate and Portland Type III. The Portland Type III contains approximately 65% calcium oxide, 23% silica, and 4% alumina (Lin, 1986). Calcium aluminate is a high alumina cement (38% calcium oxide, 10% silica, 52% alumina oxide) and approximates well a cement that could be made from lunar soils. According to Dr Lin, these proportions are satisfactory for the Portland cement to be considered a cementitious material (Lin, 1986).

Environment -- The two mixing environments used during this experiment were ambient and semi-evacuated (mixed at 200 Torr) conditions. This was done to confirm the effect of mixing conditions on the resulting porosity of the concrete.

Plasticizer/Wetting Agent (admixture) -- The concrete combinations were made with and without each of these variables. According to Dr Lin, these variables would be two of the determining factors in the ultimate strength of the concrete produced.

Reinforcements -- The reinforcements used were steel wire, aluminum wire, polymer, simulated lunar soil, and no additive. Though aluminum is known to react with the calcium oxide present in the cements, the group was asked to investigate its feasibility as an additive. If it was a feasible additive, using aluminum already in space would eliminate the cost of shipping it to the moon.

Generation of Test Runs

With the five factors and their possible variations mentioned above, the group had 80 possible combinations of ingredients to test (see Appendix C). However, due to time and monetary constraints, it was not feasible to test all eighty combinations. After consulting with Lt. Col. Stephen R. Schmidt, the class used his work on Experimental Design theory for the project (Schmidt, 1988). Lt. Col. Schmidt consulted Dr Lin to insure that the most probable interactions, those between environment, plasticizer, and wetting agent, were included in the design generation. These interactions combined with the other variables were used to generate the 18 combinations of ingredients that would provide the information necessary to predict the properties of the other 62 combinations (see "Prediction Process" section). The actual combinations generated can be seen in Appendix A.

Mixing Conditions

For each of the 18 combinations discussed above, one beam and three column specimens were produced. For every concrete batch made, certain mixing conditions were standardized.

Sand -- The sand used in the concrete mixtures is Ottawa sand because its particle size distribution is similar to that of a lunar soil sample (Lin, 1987). The grain size distribution was the same proportion as that used by Dr Lin, 4% of #30 sieved sand, 31% of #40 sieved sand, 45% of #50 sieved sand and 20% of #100 sieved sand (to the nearest percent). It was sieved according to ASTM procedures, C136 (Lin, 1987).

Water -- As with Dr Lin's experiment, the distilled water was used.

Mix -- Though the recommended ASTM mixture for Portland cement is 2.75:1:0.485 of sand-cement-water, the sand-cement ratio was adjusted to achieve the same level of workability and attain the same water-cement ratio ad the Calcium Aluminate cement (Lin, 1987). The resulting ratio used by Dr Lin was 1.75:1:0.485. Due to work from previous classes that produced different results, it was decided that a ratio likely to provide better results would be 1.75:1:0.435. This ratio was used in all concrete combinations poured.

Additives -- The additives were added by volume after removing 2% of the sand from the mixtures. This was done because of the different densities of the various additives.

Mixing Time -- Each specimen was mixed for 3 minutes before being placed in the molds. According to the Fall 1988 class, a 3-minute mix resulted in the strongest mix and removed a high percentage of air pockets from the specimens.

Curing Time -- Dr Lin's test results confirmed a 3 to 4-day cure of the Calcium Aluminate cement for optimal strength (Lin, 1987). For this reason, the Engineering 410 class allowed the Calcium Aluminate cement mixes to cure for 4 days. The Portland cement cured for 7 days according to standard testing procedures. The first day of curing took place in the molds in a 100% humidity environment, the molds were removed after 24 hours and the remaining days of curing took place in a 100% humidity environment also.

Dimensions for the cylinder and beam specimens are:

Cylindrical Column: Diameter = 1", Length = 2"
Square Cross Section Beam: Side dimension = 1.57", Length = 6.69"

Testing Conditions

From each test run, the specimens were tested for compressive and flexural strength. A three-point load was placed on the beams to be tested (according to ASTM C 78-64) using the Satec Systems 120 WHVL Universal Testing Machine. The load rate used was 0.0025 in/min, consistent with the

Spring and Fall 1988's load rate, and was applied until beam failure.

The columns were capped with sulfur (for uniform application of load), and strain gauges were applied with epoxy onto the columns after surface preparation. The wires were connected to a Vishay/Ellis 20 Strain Indicator and a load rate was set at 0.0012 in/min. The lower load rate (as compared with the beam loading) was done so the group could obtain the necessary information from the testing device. As with the beam, the columns were loaded to ultimate failure.

From these tests it was possible to determine maximum flexural strength (modulus of rupture), maximum compressive strength, modulus of elasticity and Poisson's ratio.

Experimental Data

Two sets of specimens were prepared for each of the 18 experimental conditions shown in Appendix A. Each specimen was tested for compressive strength resulting in two response values (A and B) for each row of the design matrix. The average compressive strength per experimental condition is displayed in Appendix A column labeled "Compressive Actual".

Analysis

Once all the specimens were tested, the information from the testing was used to determine the performance of the specimens in terms of four different properties: maximum compressive strength, maximum flexural strength, Poisson's Ratio, and Young's Modulus of Elasticity. Maximum compressive strength is a measure of the compressive load per unit area a specimen can withstand before catastrophic failure occurs. Maximum flexural strength is the product of the bending moment times the distance from the neutral axis to the outer fibers divided by the area moment of inertia. Poisson's Ratio is the ratio of lateral strain to axial strain in the specimen. Young's Modulus of Elasticity is the ratio of stress versus strain which can be determined using the compressive load versus longitudinal deformation curve. It measures the stiffness of the specimen.

To determine the values for these four characteristics, a computer program, CNCRT.PAS was written. The input values for this program were: maximum compressive load, two points defining the load versus deformation curve, maximum flexural load, and the code defining the concrete mixture. The equations used in this program are as follows.

MAXIMUM COMPRESSIVE STRENGTH (SEE APPENDIX A)

$$f'_c = P_{max} / Area$$

POISSON'S RATIO

$$\mu = \frac{\text{LATERAL STRAIN}}{\text{AXIAL STRAIN}}$$

FLEXURAL STRENGTH

$$f_r = \frac{My}{I} = \frac{Pl}{bh^3}$$

YOUNG'S MODULUS OF ELASTICITY

$$E = \frac{\text{STRESS}}{\text{STRAIN}} = \frac{\sigma}{\varepsilon}$$

Using the program, these four properties for both runs of the 18 combinations were obtained.

Prediction Process
At this point in the experiment, the group decided to carry on only with the compressive strength for the prediction equation. The reason this was done is that although standard mixing and testing procedures were followed, the group was not confident about the measurements for the three properties of Poisson's Ratio, maximum flexural strength and Young's Modulus of Elasticity. The group was however confident about the maximum compressive strength because of the simple procedure used to obtain the results. Having the values necessary to determine the maximum compressive strength, the prediction equation below was used to estimate the value of the compressive strength for the other 62 untested combinations.

$$\hat{Y} = \overline{T} + (\overline{C}_i - \overline{T}) + (\overline{E}_i - \overline{T}) + (\overline{P}_i - \overline{T}) + (\overline{A}_i - \overline{T})$$

$$+ [(\overline{P_iW_i} - \overline{T}) - (\overline{P}_i - \overline{T}) - (\overline{W}_i - \overline{T})]$$

$$+ [(\overline{W_iE_i} - \overline{T}) - (\overline{W}_i - \overline{T}) - (\overline{E}_i - \overline{T})]$$

$$+ [(\overline{E_iP_i} - \overline{T}) - (\overline{E}_i - \overline{T}) - (\overline{P}_i - \overline{T})]$$

(Schmidt, 1988)

In this equation, "$\hat{Y}$" represents the predicted compressive strength for any concrete sample with a given combination of ingredients. "$\overline{C}_i$" represents the average response when C is at level i. The "i" subscript allows for variation of the type of cement used. For example, $\overline{C}_1$ represents the average response (f'_c) for specimens made with Portland Type III cement, while $\overline{C}_2$ represents the average response (f'_c) for samples made with Calcium Aluminate cement. The other factors are interpreted in a similar fashion. The marginal averages can be located in Appendix B.

Just as the single factors provide marginal averages (average response), the double factor combinations provide information on interactions. An interaction is very similar to a marginal average except that instead of averaging the value of the property for experimental specimens containing one specific ingredient, one must average the value of the property for the experimental specimens containing the two ingredients in the interaction. For example, for "$\hat{Y}$" representing compressive strength, $\overline{E_1P_1}$ is calculated by finding the average of the compressive strength for all of the 18 combinations containing both E_1 and P_1 (see Appendix B). "$\overline{T}$" is the average value of compressive strength for all 18 of the experimental runs. Applying the prediction equation to any of the 80 possible combinations of ingredients, one would be able to determine the estimated compressive strength for a concrete specimen containing the given combination. For instance, if one wishes to determine the compressive strength for the specimen with the ingredients Calcium Aluminate (C_2), wetting agent (W_1), plasticizer (P_1) and no additives (A_1) mixed under ambient conditions (E_1), the equation would be as follows.

$$Y = \overline{T} + (\overline{C}_2 - \overline{T}) + (\overline{E}_1 - \overline{T}) + (\overline{W}_1 - \overline{T}) + (\overline{P}_1 - \overline{T}) + (\overline{A}_1 - \overline{T})$$

$$+ [(\overline{P_1W_1} - \overline{T}) - (\overline{P}_1 - \overline{T}) - (\overline{W}_1 - \overline{T})]$$

$$+ [(\overline{W_1E_1} - \overline{T}) - (\overline{W}_1 - \overline{T}) - (\overline{E}_1 - \overline{T})]$$

$$+ [(\overline{E_1P_1} - \overline{T}) - (\overline{E}_1 - \overline{T}) - (\overline{P}_1 - \overline{T})]$$

In order to speed up the prediction process, two programs were written to implement the equation described above. The first program, 410INIT.FOR, accepts the property values for the 18 runs and generates all the marginal averages, interactions, and $\overline{T}$ value for the given property, i.e., compressive strength (see Appendices A & B).
This data is then inserted into the text of the second program, 410.FOR. This program iterates through all 80 combinations using the prediction algorithm above and gives the predicted $\hat{Y}$ value for each combination (see Appendix C). In this way it is possible to approximate the remaining, untested 62 conditions with the results of only 18 tests.

Optimal Combinations
Using the mathematical model, the group was able to predict the compressive strength for all of the 80 combinations. In determining the optimal combinations, the graphical results of the prediction algorithm were analyzed (see Appendix D). The high points marked the greatest value of compressive strength. From this graph, runs 61, 66 and

71 were of interest and chosen for confirmation testing (see Appendix D).

Confirmation Tests
        Once the optimal runs were chosen, they were remixed and tested to confirm the results with those of the prediction algorithm. The combinations of interest were as follows:

1) #61 -- Calcium Aluminate cement, semi-evacuated mixing conditions, wetting agent, plasticizer, no reinforcement.

2) #66 -- Calcium Aluminate cement, semi-evacuated mixing conditions, wetting agent, no reinforcement.

3) #71 -- Calcium Aluminate cement, semi-evacuated mixing conditions, plasticizer, no reinforcement.

        One mix, consisting of three cylinders and a beam, was done for each of the combinations of interest. The mixing and testing procedures were carried out exactly as they had been for the original 18 specimens.
        The confirmation results (see Appendix D) exceeded our predictions (from 647 psi to 833 psi). Although the difference in predicted and confirmation results is in a direction advantageous to our objective, the group is still concerned as to what has caused this difference. One strong possibility is that an important interaction was not included in this design. To overcome this problem it is recommended that a follow on experiment be conducted allowing for all two-way interactions to be estimated.

Conclusions
        The purpose of this experiment was twofold. It was designed to find the optimum concrete mixture for lunar construction. It also used Lt. Col. Schmidt's prediction equation as a statistical tool in this process. The confirmation data supports the algorithm's predictions and thus shows that the design of experiments approach is a vital tool to reduce the experimenter's work. The group realizes that for actual lunar construction, many more inputs must be analyzed to obtain the optimal combination. However, the group also realizes that the use of the prediction equation is a must to decrease time and money while still obtaining the optimum concrete mixture.

List of References

"Design and Control of Concrete Mixtures." Portland Cement
    Association. 12th ed., 1979.

ENGR 410 class, Fall 1988. Final Report.

ENGR 410 class, Spring 1988. Display.

Flexural Strength of Concrete (Using Simple Beam with Third-
    Point Loading), ASTM C 78-64, pp 40-42.

Lin, T.D. "Concrete for Lunar Base Construction." The
    Portland Cement Association, 1987.

Lin, T.D., H. Love and D. Stark. "Physical Properties of
    Concrete Made With Apollo 16 Lunar Soil Sample."
    Construction Technology Laboratories, 1987.

Rader, Stanley P., Maj (USAF) Consultant. Fall, 1988, Spring
    1989.

Schmidt, Stephen R., Lt.Col. (USAF) and Robert G. Launsby.
    Understanding Industrial Design Experiments, 1988.

Shah, S.P. "Alternative Reinforcing Materials for Concrete
    Construction." International Journal for Development
    Technology, vol 1, 3-15 (1983).

Student Manual for Strain Gauge Technology. Raleigh, NC:
    Measurements Group, Inc., 1983.

Simmerer, Stephen J. (Capt, USA). "Lunar Construction: The
    Countdown Begins." The Military Engineer, Jul 1987,
    pp 354-357.

Swint, David O., Col. (USAF). Instruction and discussions.
    Spring, 1989.

APPENDIX A

Mixing Assignments

C$_1$ - Portland Type III cement   P$_1$ - Plasticizer - 1ml
C$_2$ - Calcium Aluminate cement   P$_2$ - No Plasticizer

E$_1$ - Ambient mixing   A$_1$ - No Reinforcement
E$_2$ - Semi-Evacuated   A$_2$ - Steel
    A$_3$ - Aluminum
W$_1$ - Wetting Agent - 0.07 ml   A$_4$ - Polymer
W$_2$ - No Wetting Agent   A$_5$ - Simulated Lunar Soil

COMPRESSIVE STRENGTH (psi)

Run	C	E	W	P	A	Response A	Response B	Actual	Compressive Strength Predicted	Deviation (%)
1	1	1	2	2	1	4132.94	3428.83	3780.89	3893.40	− 2.98
2	2	1	1	1	1	5067.50	6042.79	5555.15	4962.02	+ 10.68
3	1	2	2	1	1	5207.55	4004.34	4605.95	5064.10	− 9.95
4	2	2	1	2	1	6944.25	6009.69	6476.97	6119.87	+ 5.51
5	1	2	1	1	3	3101.61	2792.21	2946.91	2214.17	+ 24.86
6	1	1	1	2	3	2272.73	2639.43	2456.08	2188.99	+ 10.87
7	1	1	2	1	4	2317.30	4278.08	3297.69	2702.07	+ 18.06
8	1	2	2	2	5	3997.97	4507.27	4252.62	4030.06	+ 5.23
9	2	1	2	2	2	4010.00	3670.00	3840.00	3665.31	+ 4.55
10	2	2	1	1	4	4260.26	5458.38	4859.32	4989.60	− 2.68
11	2	1	2	1	3	1914.95	1960.79	1937.87	2125.84	− 9.70
12	2	1	1	2	5	4318.83	5672.28	4995.56	5100.78	− 2.11
13	2	2	2	2	4	4870.00	4210.00	4540.00	4993.30	− 9.98
14	2	2	2	1	2	5692.65	5692.65	5692.65	4836.01	+ 15.05
15	1	1	1	1	2	1706.14	2294.38	2000.26	2534.73	− 26.72
16	2	2	1	1	5	3496.32	4219.52	3857.92	5125.97	− 33.13
17	1	2	1	2	2	3808.30	3342.25	3575.28	3692.58	− 3.28
18	2	2	2	2	3	2591.04	3179.28	2885.16	3317.47	− 14.98

$\overline{T}$ = 3975.35

APPENDIX B

Variable	Compressive Strength
$\overline{C}_1$	3364.46
$\overline{C}_2$	4464.06
$\overline{E}_1$	3482.94
$\overline{E}_2$	4369.28
$\overline{W}_1$	4080.38
$\overline{W}_2$	3870.31
$\overline{P}_1$	3861.52
$\overline{P}_2$	4089.17
$\overline{A}_1$	5104.74
$\overline{A}_2$	3777.05
$\overline{A}_3$	2556.51
$\overline{A}_4$	4232.34
$\overline{A}_5$	4368.70

INTERACTIONS

Interaction	Compressive Strength
$\overline{W}_1, \overline{E}_1$	3751.76
$\overline{W}_1, \overline{E}_2$	4343.28
$\overline{W}_2, \overline{E}_1$	3214.11
$\overline{W}_2, \overline{E}_2$	4395.28
$\overline{E}_1, \overline{P}_1$	3197.74
$\overline{E}_1, \overline{P}_2$	3768.13
$\overline{E}_2, \overline{P}_1$	4392.55
$\overline{E}_2, \overline{P}_2$	4346.01
$\overline{W}_1, \overline{P}_1$	3843.91
$\overline{W}_1, \overline{P}_2$	4375.97
$\overline{W}_2, \overline{P}_1$	3883.54
$\overline{W}_2, \overline{P}_2$	3859.73

APPENDIX C

RUN #	C	E	W	P	A	VALUE
1	1	1	1	1	1	3862.417
2	1	1	1	1	2	2534.727
3	1	1	1	1	3	1314.186
4	1	1	1	1	4	2990.017
5	1	1	1	1	5	3126.379
6	1	1	1	2	1	4737.217
7	1	1	1	2	2	3409.527
8	1	1	1	2	3	2188.986
9	1	1	1	2	4	3864.817
10	1	1	1	2	5	4001.179
11	1	1	2	1	1	3574.465
12	1	1	2	1	2	2246.775
13	1	1	2	1	3	1026.234
14	1	1	2	1	4	2702.065
15	1	1	2	1	5	2838.427
16	1	1	2	2	1	3893.400
17	1	1	2	2	2	2565.710
18	1	1	2	2	3	1345.169
19	1	1	2	2	4	3021.000
20	1	1	2	2	5	3157.362
21	1	2	1	1	1	4762.402
22	1	2	1	1	2	3434.713
23	1	2	1	1	3	2214.172
24	1	2	1	1	4	3890.003
25	1	2	1	1	5	4026.365
26	1	2	1	2	1	5020.270
27	1	2	1	2	2	3692.580
28	1	2	1	2	3	2472.039
29	1	2	1	2	4	4147.870
30	1	2	1	2	5	4284.232
31	1	2	2	1	1	5064.096
32	1	2	2	1	2	3736.406
33	1	2	2	1	3	2515.865
34	1	2	2	1	4	4191.696
35	1	2	2	1	5	4328.059
36	1	2	2	2	1	4766.098
37	1	2	2	2	2	3438.408
38	1	2	2	2	3	2217.867
39	1	2	2	2	4	3893.698
40	1	2	2	2	5	4030.060

APPENDIX C

RUN #	C	E	W	P	A	VALUE
41	2	1	1	1	1	4962.018
42	2	1	1	1	2	3634.328
43	2	1	1	1	3	2413.787
44	2	1	1	1	4	4089.618
45	2	1	1	1	5	4225.980
46	2	1	1	2	1	5836.818
47	2	1	1	2	2	4509.128
48	2	1	1	2	3	3288.587
49	2	1	1	2	4	4964.418
50	2	1	1	2	5	5100.780
51	2	1	2	1	1	4674.066
52	2	1	2	1	2	3346.376
53	2	1	2	1	3	2125.835
54	2	1	2	1	4	3801.666
55	2	1	2	1	5	3938.029
56	2	1	2	2	1	4993.001
57	2	1	2	2	2	3665.311
58	2	1	2	2	3	2444.770
59	2	1	2	2	4	4120.601
60	2	1	2	2	5	4256.963
61	2	2	1	1	1	5862.004
62	2	2	1	1	2	4534.313
63	2	2	1	1	3	3313.773
64	2	2	1	1	4	4989.604
65	2	2	1	1	5	5125.966
66	2	2	1	2	1	6119.871
67	2	2	1	2	2	4792.181
68	2	2	1	2	3	3571.640
69	2	2	1	2	4	5247.471
70	2	2	1	2	5	5383.833
71	2	2	2	1	1	6163.697
72	2	2	2	1	2	4836.007
73	2	2	2	1	3	3615.466
74	2	2	2	1	4	5291.297
75	2	2	2	1	5	5427.659
76	2	2	2	2	1.	5865.699
77	2	2	2	2	2	4538.009
78	2	2	2	2	3	3317.468
79	2	2	2	2	4	4993.299
80	2	2	2	2	5	5129.661

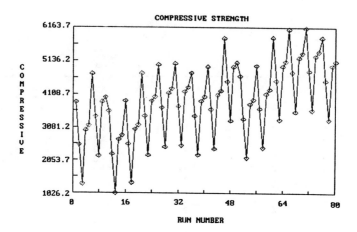

CONFIRMATION RESULTS

Test#	Design Matrix C E W P A					Response 1	2	3	Average (psi)	Compressive Strength Predicted	Deviation
61	2	2	1	1	1	4588.6	8225.1	7270.2	6694.6	5862.0	+ 832.6
66	2	2	1	2	1	6369.8	7109.8	7345.3	6941.6	6119.9	+ 821.7
71	2	2	2	1	1	5765.3	8641.5	6026.24	6811.0	6163.7	+ 647.3

Design of Experiments
Silane Doping in GaAs

Statistics Seminar Project
University of Northern Colorado

Alan Arnholt
Steve Smith
Robert Kaliski

I. Objective

In the growth of gallium arsenide (GaAs), a material used to fabricate electronic devices, the level of impurities in the crystal has a great influence on the electrical properties. The level of these trace impurities is known as the doping level. Since the doping level determines some of the electrical properties, we desire to have control of this level to tailor the electrical properties to the device or circuit requirements.

II. Response

For the operation of heterostructure bipolar transistors, there are two different doping levels of silicon in GaAs. These levels are $3x10^{16}$ cm^{-3} and $5x10^{18}$ cm^{-3}. The goal of this experimental design is to model doping level response, select the proper conditions to obtain the two doping levels and then confirm the results. The response will be the actual doping level measured on a profiling instrument. This instrument can measure the doping level both accurately and precisely in a range of 10^{14} to 10^{19} cm^{-3}.

III. Factors

In the case of silicon doping in GaAs, three factors may influence the doping level: the silane flow, the arsine flow and the trimethylgallium (TMGa) flow. There are other factors that could also modify the response, but the machinery is calibrated at these fixed levels. In addition, engineering is only interested in the perturbation of three factors and cannot afford to recalibrate the reactor. The three factors are quantitative and are highly controllable (+/- 2% of set level), and literature and engineering experience suggest that no interactions exist.

IV. Experimental Constraints

The growth parameters selected are to be used in the growth of heterostructure in March. Hence engineering prefers to have the data in three weeks (end of February) with 15 runs and a maximum of 30 runs including confirmation runs. Although cost is not a concern, each run requires four hours from start to completion.

V. Experimental Plan

Since engineering believes no interaction exists, we will use a fractional factorial design, the Taguchi $L_4(2^3)$, with two replications. The factors with both high and low levels are listed below.

Factors	−	level	+	Total range
silane flow	15		50	0-100
TMGa flow	20		35	5-50
arsine flow	100		140	90-150

The analysis will include a Pareto chart of the three main effects and also a determination of variance. A regression analysis will follow using linear terms. We plan to check the center point for validity of the linear model. If the linear model is inappropriate, we will then examine the response of the axial points and determine the appropriate regression equation. From the empirical regression, we finally will compute the correct levels and make at least two confirmation runs for each target level.

VI. Experiment and Analysis

Results:

	A=Arsine	B=TMGa	C=Silane	Y1	Y2	S	Ln(S)
run 1.	−1	−1	−1	0.70	0.66	0.0282	35.578
" 2.	−1	+1	+1	2.0	1.6	0.282	37.881
" 3.	+1	−1	+1	4.0	3.7	0.212	37.593
" 4.	+1	+1	−1	1.6	1.2	0.282	37.881

Note: Response is in terms of 1×10^{17} cm^{-3}.   T=1.93

Avg.[Ln(S)]=37.233

Marginal Means:

	A	B	C	
Avg +	2.63	1.60	2.83	
Avg −	1.24	2.27	1.04	
d	1.39	−0.67	1.79	

Factor C is the largest location effect.

	A	B	C	
Avg Ln(S.D.) +	37.731	37.881	37.737	
Avg Ln(S.D.) −	36.730	36.586	36.730	
d	1.007	1.295	1.007	

From our limited data, no factor appears to stand out as a disperion reduction factor.

PARETO DIAGRAM

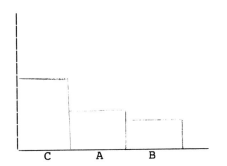

A - Arsine
B - TMGa
C - Silane

MARGINAL MEANS PLOT

(10^{17})

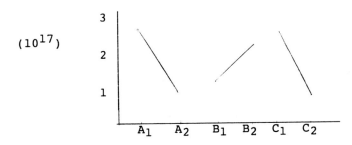

$A_1=100$ $A_2=140$
$B_1=20$ $B_2=35$
$C_1=15$ $C_2=50$

TEST FOR SIGNIFICANCE OF REGRESSION

1) H_0: $A = B = C \ne 0$
   H_1: $A \ne 0$ or $B \ne 0$ or $C \ne 0$

2)   a. F-test
     b. $\alpha = 0.05$
     c. DF = (3,4)
     d. CV for $F_{\alpha=0.05,3,4}$ = 6.59

$$\mathbf{X}=\begin{bmatrix} 1 & -1 & -1 & -1 \\ 1 & -1 & 1 & 1 \\ 1 & 1 & -1 & 1 \\ 1 & 1 & 1 & -1 \\ 1 & -1 & -1 & -1 \\ 1 & -1 & 1 & 1 \\ 1 & 1 & -1 & 1 \\ 1 & 1 & 1 & -1 \end{bmatrix} \quad \mathbf{X'}=\begin{bmatrix} 1 & 1 & 1 & 1 & 1 & 1 & 1 & 1 \\ -1 & -1 & 1 & 1 & -1 & -1 & 1 & 1 \\ -1 & 1 & -1 & 1 & -1 & 1 & -1 & 1 \\ -1 & 1 & 1 & -1 & -1 & 1 & 1 & -1 \end{bmatrix} \quad \mathbf{Y}=\begin{bmatrix} 0.7 \\ 2.0 \\ 4.0 \\ 1.6 \\ 0.66 \\ 1.6 \\ 3.7 \\ 1.2 \end{bmatrix}$$

$$\mathbf{X'X}=\begin{bmatrix} 8 & 0 & 0 & 0 \\ 0 & 8 & 0 & 0 \\ 0 & 0 & 8 & 0 \\ 0 & 0 & 0 & 8 \end{bmatrix} \quad (\mathbf{X'X})^{-1}=\begin{bmatrix} 0.125 & 0 & 0 & 0 \\ 0 & 0.125 & 0 & 0 \\ 0 & 0 & 0.125 & 0 \\ 0 & 0 & 0 & 0.125 \end{bmatrix}$$

$$\underline{\mathbf{B}} = (\mathbf{X'X})^{-1}\mathbf{X'Y}=\begin{bmatrix} 1.93 \\ 0.63 \\ -0.33 \\ 0.89 \end{bmatrix} \qquad \underline{Y} = 1.93 + 0.69A - 0.33B + 0.89C$$

$$\underline{\mathbf{B}}\mathbf{X'Y} =\begin{bmatrix} 1.93 & 0.69 & -0.33 & 0.89 \end{bmatrix} \begin{bmatrix} 15.46 \\ 5.54 \\ -2.66 \\ 7.14 \end{bmatrix} = 40.892$$

$\mathbf{Y'Y} = 41.1756$ $(\sum y_i)^2/n = 29.876$

Source	SS	DOF	MS	F_0
Regression	$S_R=\underline{\mathbf{B}}\mathbf{X'Y}-(\sum y_i)^2/n=11.016$	K=3	3.672	51.93
Error	$SS_E=\mathbf{Y'Y}-\underline{\mathbf{B}}\mathbf{X'Y}=0.2823$	n-k-1=4	0.0707	
Total	$SS_T=\mathbf{Y'Y}-(\sum y_i)^2/n=11.299$	n-1=7		

4) Reject H_0, $F_0 = 51.93 > F_{\alpha=0.05,3,4} = 6.59$

5) We found evidence that at least $A \ne 0$ or $B \ne 0$ or $C \ne 0$

     $R^2 = SS_R/SS_T = 0.975$

TEST ON INDIVIDUAL REGRESSION COEFFICIENTS

1) a. H_0: A = 0   b. H_0: B = 0   c. H_0: C = 0
      H_1: A $\neq$ 0 H_1: B $\neq$ 0 H_1: C $\neq$ 0

2) a. t-test
   b. α = 0.05
   c. DF = n-k-1
   d. CV for $t_{0.05/2,4}$ = 2.776

3) Calculations

   $t_0 = \underline{B_j}/\sqrt{\sigma^2 \, c_{jj}}$

   a. $t_0 = 0.69/\sqrt{(0.0707)(0.125)}$ = 7.339

   b. $t_0 = -0.33/\sqrt{(0.0707)(0.125)}$ = -3.51

   c. $t_0 = 0.89/\sqrt{(0.0707)(0.125)}$ = 9.467

4) Reject H_0 for A, B, and C since $|t_0| > t_{\alpha/2,n-k-1}$

   a. $|t_0|$ = 7.339 > t = 2.776

   b. $|t_0|$ = 3.51 > t = 2.776

   c. $|t_0|$ = 9.467 > t = 2.776

5) We found evidence that A, B, and C are all non-zero.

From the experimental data, all three factors are significant while no factors have any greater influence on the variability. Since the variability of each factors is approximately equal, we will not concern ourselves with the standard deviation any further.

Using the half-effects of the marginal means, the regression equation is

$$\underline{Y} = (\ 1.93 + 0.69*A - 0.33*B + 0.89*C)\ 10^{17}\ cm^{-3}.$$

With two runs at the center point A=B=C=0, the responses are 2.0 and 2.2 (10^{17} cm^{-3}). The average of the center points is 2.1. This value does not appear to be deviate enough from 1.93 to justify the addition of curvature to the regression model.

To reach the lower target of 0.3, we need to set A=-1 and B=1 by inspection. Solving the regression equation for C, we obtain a coded value for C of -0.67. Uncoding this value by using the inverse of the standard coding transformation yields a silane setting of 21. Two confirmation runs were performed with the three factors set at the specified levels. The responses are 0.27 and 0.30 (10^{17} cm^{-3}) which are acceptable.

VII. Conclusions

To achieve the lower target value we recommend the arsine be set at 90, the TMGa set at 50 and the silane set at 21. Since the upper target cannot be met with the present machine configuration, we recommend the the engineering staff make changes to the machine to achieve this doping level or change their target level.

Submitted by: Randy Boudreaux
Cray Research, Inc.

Bonding Gold Wire to Gallium Arsenide Wafers:
Overcoming Pre-Conceived Notions

Statement of the Problem:

The opening of a new Cray plant in Colorado Springs posed an interesting problem. Gold wire bonding to Gallium Arsenide wafers had been optimized at the Wisconsin plant and using the same parameters should have optimized the process in Colorado. The solution was not to be that simple. Several months passed while "best guess" and "one factor at a time experimentation" were used to find an improved process. After hearing and reading the materials in this text it was decided to try a designed experiment matrix as shown in Chapter 3. Several lessons learned from this first experiment are:

1. Proper experimental designs can be used with little training provided adequate guidance is available.

2. Do not consume all of your resources during the initial experiment - things may go wrong, or the information gained may lead to the need for further experimentation.

3. Preconceived notions are a barrier to process improvements but they can be overcome through properly designed experiments.

4. Designed experimentation is a much more powerful tool than previously used "best guess" or one-factor-at-a-time approaches.

THE DESIGN:

Prior to the actual experimentation, the brainstorming session generated five input factors to be tested along with two responses of interest as shown in Table 1.

Input Factors	Levels
Force	60, 150
Time	40, 100
Ultrasonics	40, 150
Temperature	125, 175
Z-Speed	50, 150

Response	Type
Bond/No Bond	Attribute
Bond Strength	Variable

The L_8 design matrix and response values for Bond/No Bond are show in Table 2

RESULTS:

As you can see, some errors were made in Z-speed settings for runs 2, 3, and 8. For runs 2 and 8, the problem was associated with machine limitations. Run 3 should have been set at 150, but was inadvertently set incorrectly. These problems will limit our ability to obtain unambiguous results for each factors's effectiveness; however, there still exists a very important finding from comparing the results for each of the 8 runs. For example, prior to collecting the data, it was generally accepted that runs 1, 2 and 6 would be poor performers and were only run because they showed up in the design matrix. Run 6 obviously was a surprise - among other things, preconceived notions had previously prevented this successful combination from being tested.

CONCLUSION:

Not only did the experimental matrix provide the best combination of input factors, resulting in 100% good bonds, these bonds all exceeded the minimum bond strength requirements. In fact, the bond strength testing device could not knock them off. The knowledge gained in this one experiment was worth many times that of four months of one–at–a–time experiments.

TABLE 2
DESIGN MATRIX AND RESPONSE VALUES FOR BONDING EXPERIMENT

RUN	FORCE	TIME	U/S	TEMP	Z-SPEED	1	2	3	4	5	6	7	8	9	10	11	12	13	14	15	16	17	18	19	20	21	22	23	24
1	60	40	40	125	50	0	0	0	0	0	0	0	0	0	0	0	0	0	0	0	0	0	0	0	0	0	0	0	0
2	60	40	150	175	150 / 131	0	0	0	0	0	0	0	0	0	0	0	0	0	0	0	0	0	0	0	0	0	0	0	0
3	60	100	40	125	150 / 50	1	0	0	0	0	0	0	1	0	0	0	1	0	1	0	0	0	0	1	1	0	1	1	0
4	60	100	150	175	50	0	0	0	0	0	0	0	0	0	0	0	0	0	0	0	0	0	0	0	0	0	0	0	0
5	150	40	40	175	150	0	0	0	0	0	0	1	1	0	0	0	0	0	0	0	0	0	1	0	1	1	0	0	0
6	150	40	150	125	50	1	1	1	1	1	1	1	1	1	1	1	1	1	1	1	1	1	1	1	1	1	1	1	1
7	150	100	40	175	50	1	1	1	0	1	1	1	1	1	1	1	1	1	1	1	1	1	1	1	1	1	1	0	1
8	150	100	150	125	150 / 131	1	1	1	1	0	1	1	1	0	1	1	1	1	1	1	0	1	1	1	1	0	1	1	1

CONSTANT VELOCITY JOINTS ON A FRONT WHEEL DRIVE
ANALYSIS OF MULTIPLE RESPONSES FROM DESIGNED EXPERIMENTS

DON F. WILSON
QINAS, INC.
LITTLETON, COLORADO

ABSTRACT: This case study presents methods of visualizing and analyzing multiple responses from a designed experiment. The case study is a product development example - elastomeric boots for constant velocity joints on front wheel drive automobiles. Methods presented include use of linear graphs to represent interactive factors, blocking to eliminate noise factors, use of percent contribution to identify strong relationships, analysis of non-parametric (qualitative) data, and compromises to balance performance of several key objectives.

BACKGROUND: The use of constant velocity (CV) joints on front wheel drive automobiles presents a unique challenge for development of elastomeric products. The CV joints are permanently sealed with a boot to contain the grease and to prevent contaminants from entering. If undetected by the customer, the tearing or loosening of the elastomeric boot may require replacement of an entire CV Joint, costing the customer in excess of $350.00 per CV joint.

The boot must be able to flex at angles up to 45 degrees; operate at speeds up to 1900 RPM without more than 7mm of radial growth; maintain resistance to attack by ozone, oxidation, and grease; perform at temperatures below -40 degrees and over 200 degrees; and provide cut resistance to road hazards.

At the time of this case study, the product dimensions had been determined by the customer and most of the key functional parameters of the CV boots had been met. The primary focus of this study was to improve the resistance of the boot to ozone attack, low temperature properties, and radial growth characteristics at high RPM, while maintaining the performance of previous rubber formulations on other functional tests.

DETERMINING OBJECTIVES: The primary objective of the experiment was to substantially improve the dynamic ozone characteristics of the formulation without adversely affecting the other requirements:

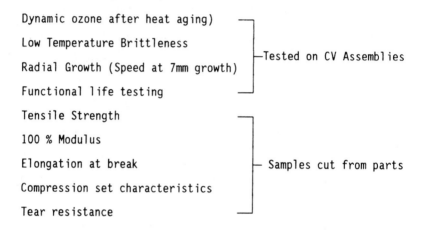

Dynamic ozone after heat aging)
Low Temperature Brittleness
Radial Growth (Speed at 7mm growth) ⎤—Tested on CV Assemblies
Functional life testing

Tensile Strength
100 % Modulus
Elongation at break ⎤— Samples cut from parts
Compression set characteristics
Tear resistance

PRIMARY OBJECTIVE: The customer tests the ozone resistance of the formulation on the CV boot, on an accelerated life test preceded by heat aging the samples to simulate internal deterioration (inner noise) of the product. The specification was to show no cracking under 7X magnification after 168 hours of running the CV joint at a 34 degree angle at 23 mph (350 RPM) in an atmosphere of 50 ppm of ozone. At the time of this study, minute cracks were visible at 168 hours (no cracks at 144 hours). Several attempts to "patch" the formulation with one-at-a-time changes had failed. Parts had been marginally passing the radial growth test and low temperature flex requirements. Any change to the rubber formulation might cause problems with these tests.

MEASURING THE OBJECTIVE: The dynamic ozone test at the customer's facility could not be directly duplicated in the development lab because 1) the expense of the test equipment was high, 2) the time necessary to acquire the equipment would delay development, and 3) correlation problems between labs existed. To circumvent these problems, a "predictor" test had been developed in the supplier's development lab to show relative improvements in ozone performance of formulations.

BRAINSTORMING POTENTIAL CAUSES / "KNOBS": The primary cause of deterioration of the rubber boot is the ozone level in the atmosphere. However, the intent of the experiment is not to control the root cause (ozone), but to make the boots "robust" or insensitive to the "outer" noise of ozone. The major factors to improve ozone resistance relate to either reducing the stress level of the rubber in operation or using protective chemicals in the formulation.

PRIORITIZING THE INVESTIGATIONS: Since the product design had been finalized, and modulus requirements were necessary, the focus of the experiment was on modifying the rubber formulation to improve chemical resistance to ozone .

The rubber formulation consists of thirteen ingredients, but the chemist believed only four ingredients were potential major contributors to improving the ozone performance.

SELECTING FACTORS FOR THE EXPERIMENT: The four factors were to be characterized by both type and amount, creating seven Control Factors for the experiment:

```
Neoprene Type       (N)
Antiozonant Type    (Z)
Antiozonant Level   (ZL)
Antioxidant Type    (X)
Antioxidant Level   (XL)
Wax Type            (W)
Wax Level           (WL)
```

One Noise Factor was considered to be dominant in the ozone testing - the variability of the ozone concentration from test to test.

Interactions of the three protective ingredients were considered to be strong possibilities, creating seven suspected interactions:

```
ZxZL   (Antiozonant Type x Antiozonant Level)
XxXL   (Antioxidant Type x Antioxidant Level)
WxWL   (Wax Type x Wax Level)
ZxX    (Antiozonant Type x Antioxidant Type)
ZxW    (Antiozonant Type x Wax Type)
XxW    (Antioxidant Type x Wax Type)
ZLxXL  (Antiozonant Level x Antioxidant Level)
```

SELECTING LEVELS FOR EACH FACTOR: Two types of neoprene (GNA, GW) were selected to investigate differences in performance on the three major responses: Dynamic Ozone, Low Temperature, and Radial Growth. Technical literature indicated that stress levels might be significantly different in the neoprene types to improve performance characteristics in all three areas.

The selection of the protective ingredients were based on technical literature and previous lab screenings. The levels of the antioxidant and antiozonant were selected to produce "adequate" to "improved" protection, but below the maximum solubility of the materials in the formulation. The levels of the wax types were varied from 0 to 4 parts per hundred (pph) of base polymer to examine a published theory that wax may improve static ozone resistance, but inhibit dynamic ozone resistance.

CHOOSING THE APPROPRIATE DESIGN: With the seven factors listed above and selecting two levels per factor, all combinations of seven factors at two levels each would require 128 formulations. Running the 128 trials would produce information on all interactions of the factors. However, resources for the experiment were limited. The dynamic ozone chamber in the development lab would test only 8 samples at a time, and each test required approximately 2 weeks (32 weeks for 128 combinations). The cost of testing each formulation was $250.00, or $32,000 for 128 combinations.

The minimum use of resources to examine the effects of each factor would be 8 trials, using a one-at-a-time strategy, or an L8 design (a fully saturated fractional factorial). A one-at-a-time strategy would not permit analysis of interactions and would be very sensitive to individual test results. The L8 Design would confound all two way interactions with the factor effects (3 2-way interactions with each factor), but factor effects would be based on averages of 4 data points each. Larger fractional factorial designs (L16, L32, L64) would reduce confounding of interactions at the expense of spending more resources. Selection of non-geometric orthogonal designs (L12, L20, L24, L28) would prevent total confounding of interactions with factors, but would not permit analysis of the interactions.

The chemist decided to select a design that would prevent confounding of the seven factors and seven interactions that were suspected, but allow-ing the other 14 potential interactions to be confounded with factors and/or interactions. Using the linear graph method developed by Genichi Taguchi, the experiment can be accomplished using the L16 design:

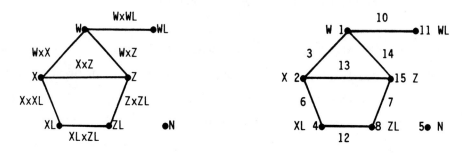

REVIEWING THE DESIGN: The linear graph does not show the confounding of 14 other potential two way interactions, or higher order interactions. Using an interaction table, the following confounding situations were identified:

FACTOR			POTENTIAL 2 WAY INTERACTIONS	
1	W	WAX TYPE	NxXL	
2	X	ANTIOX. TYPE		
3	WxX	WAX x ANTIOX	ZLxWL	
4	XL	ANTIOX LVL	WxN	WLxZ
5	N	NEOPRENE	WxXL	
6	XxXL			
7	ZxZL		XxN	
8	ZL	ANTIOZ LVL		
9	(not assigned)		WxXL	XxWL
10	WxWL		XxZL	NxZ
11	WL	WAX LEVEL	XLxZ	
12	XLxZL			
13	XxZ		NxZL	
14	WxZ		NxWL	
15	Z	ANTIOZ.	XLxWL	

The two-way interactions that were confounded were not considered to be any of the "vital few" effects that would significantly alter ozone performance.

A review of the formulation combinations showed all combinations to be feasible for the experiment:

Design No.	WAX TYPE	ANTIOX TYPE	ANTIOX LEVEL	NEOPRENE TYPE	ANTIOZ LEVEL	WAX LEVEL	ANTIOZ TYPE
1	PARAFIN	X1	1 PHR	GNA	1 PHR	0 PHR	Z1
2	PARAFIN	X1	1 PHR	GNA	3 PHR	4 PHR	Z2
3	PARAFIN	X1	3 PHR	GW	1 PHR	0 PHR	Z2
4	PARAFIN	X1	3 PHR	GW	3 PHR	4 PHR	Z1
5	PARAFIN	X2	1 PHR	GNA	1 PHR	4 PHR	Z2
6	PARAFIN	X2	1 PHR	GNA	3 PHR	0 PHR	Z1
7	PARAFIN	X2	3 PHR	GW	1 PHR	4 PHR	Z1
8	PARAFIN	X2	3 PHR	GW	3 PHR	0 PHR	Z2
9	POLY.	X1	1 PHR	GW	1 PHR	4 PHR	Z2
10	POLY.	X1	1 PHR	GW	3 PHR	0 PHR	Z1
11	POLY.	X1	3 PHR	GNA	1 PHR	4 PHR	Z1
12	POLY.	X1	3 PHR	GNA	3 PHR	0 PHR	Z2
13	POLY.	X2	1 PHR	GW	1 PHR	0 PHR	Z1
14	POLY.	X2	1 PHR	GW	3 PHR	4 PHR	Z2
15	POLY.	X2	3 PHR	GNA	1 PHR	0 PHR	Z2
16	POLY.	X2	3 PHR	GNA	3 PHR	4 PHR	Z1

DETERMINING THE SEQUENCE OF EXPERIMENTING: The test samples could be prepared in random sequence (to reduce the chances of a nuisance variable affecting the results). However, the variability of the dynamic ozone test (8 tests/setup) was likely to affect the test results substantially. Factor 9 was used to block the order of testing to isolate the effect of the ozone test variation.

Trial Order	Design No.	WAX TYPE	ANTIOX TYPE	ANTIOX LEVEL	NEOPRENE TYPE	ANTIOZ LEVEL	TEST BLOCK	WAX LEVEL	ANTIOZ TYPE
1	3	PARAFIN	X1	3 PHR	GW	1 PHR	1st	0 PHR	Z2
2	10	POLY.	X1	1 PHR	GW	3 PHR	1st	0 PHR	Z1
3	7	PARAFIN	X2	3 PHR	GW	1 PHR	1st	4 PHR	Z1
4	14	POLY.	X2	1 PHR	GW	3 PHR	1st	4 PHR	Z2
5	12	POLY.	X1	3 PHR	GNA	3 PHR	1st	0 PHR	Z2
6	5	PARAFIN	X2	1 PHR	GNA	1 PHR	1st	4 PHR	Z2
7	1	PARAFIN	X1	1 PHR	GNA	1 PHR	1st	0 PHR	Z1
8	16	POLY.	X2	3 PHR	GNA	3 PHR	1st	4 PHR	Z1
9	15	POLY.	X2	3 PHR	GNA	1 PHR	2nd	0 PHR	Z2
10	13	POLY.	X2	1 PHR	GW	1 PHR	2nd	0 PHR	Z1
11	6	PARAFIN	X2	1 PHR	GNA	3 PHR	2nd	0 PHR	Z1
12	11	POLY.	X1	3 PHR	GNA	1 PHR	2nd	4 PHR	Z1
13	9	POLY.	X1	1 PHR	GW	1 PHR	2nd	4 PHR	Z2
14	8	PARAFIN	X2	3 PHR	GW	3 PHR	2nd	0 PHR	Z2
15	4	PARAFIN	X1	3 PHR	GW	3 PHR	2nd	4 PHR	Z1
16	2	PARAFIN	X1	1 PHR	GNA	3 PHR	2nd	4 PHR	Z2

PERFORMING EXPERIMENTS/OBTAINING RESULTS: As anticipated, the dynamic ozone results (hours to first cracks observed) changed dramatically from the first test group to the second, averaging 342 hours and 174 hours, respectively. The chemist, on noting the difference, requested a supplier (DuPont) to perform a "second opinion" test at their laboratories, using a different apparatus. Results were compiled in a data table:

Trial Order	Design No.	DYNAMIC OZONE	DUPONT OZONE	OZONE RATING	MODULUS 100%	TENSILE PSI	ELONG %	HI SPEED SPIN	LOW TEMP
7	1	384	4	2B	380	1870	325	2250	-49
16	2	168	5	2B+	390	1930	350	2300	-47
1	3	360	5	2B+	300	2150	375	2100	-45
15	4	192	2	1B+	315	2070	350	2100	-46
6	5	288	7	3B	360	1970	300	2250	-48
11	6	168	5	2B+	420	1930	325	2350	-48
3	7	240	7	3B	300	2020	375	2050	-46
14	8	168	5	2B+	340	2060	400	2200	-47
13	9	168	5	2B+	295	2140	325	2150	-44
2	10	408	3	2B-	340	2180	375	2150	-46
12	11	216	2	1B+	370	2030	325	2150	-48
5	12	408	4	2B	415	2060	300	2300	-49
10	13	168	7	3B	310	2200	450	2100	-47
4	14	336	6	3B-	325	2170	425	2100	-45
9	15	144	7	3B	390	2080	300	2300	-48
8	16	312	6	3B-	395	2050	350	2250	-47

ANALYZING DATA: The data for each response was analyzed using DESIGN CUBE software. Analysis of Variance (ANOVA) tables and Level Effects reports were generated for each response. The analysis for the Dynamic Ozone responses are shown on the following page. To determine significant effects, factors and interactions with less importance were "pooled" into error terms.

The results on Dynamic Ozone show the TEST BLOCK (the difference between the first and second test groups) to be the largest effect on ozone performance (79.9% contribution), followed by ANTIOXIDANT TYPE (10.1%), and WAX LEVEL (3.6%). The pooling process may have created "significant" results out of insuficcient information by pooling the smaller sources of variation preferentially. Thus, the Probability column (1-alpha) may not be accurate. Reviewing the tables of Level Means and Level Effects shows the preferred Antioxidant is X1. The preferred wax level is 0 PHR. The interaction plots show a preference for 3 PHR of Antioxidant if X1 is selected.

The Dynamic Ozone results, however, were questionable, considering the large variation between tests. Therefore, a "second opinion" test was conducted by DuPont on a different method of dynamic ozone testing. All samples were run for 336 hours in the ozone chamber and removed. Unfortunately, the Dupont measurement criteria was not a continuous quantitative scale as with the "time to cracking" criteria. The data was qualitative, using a visual rating scale, from 1B- (very slight cracks under 7X magnification) to 3B+ (almost ready to break). The data from the DuPont Test was reviewed in two ways: 1) By sorting the trials from low to high ratings and 2) by assigning a numerical rating scale to the visual ratings (1B = 1.... 3B+ = 8).

Grp.	Factor	Name	Level 1	Level 2	2 way interactions
I	1 W	WAX TYPE	PARAFIN	POLY.	2x3 4x5 6x7 8x9 10x11 12x13 14x15
II	2 X	ANTIOX.	X1	X2	1x3 4x6 5x7 8x10 9x11 12x14 13x15
II	3 WxX				1x2 4x7 5x6 8x11 9x10 12x15 13x14
III	4 XL	ANTIOX LVL	1 PHR	3 PHR	1x5 2x6 3x7 8x12 9x13 10x14 11x15
III	5 N	NEOPRENE	GNA	GW	1x4 2x7 3x6 8x13 9x12 10x15 11x14
III	6 XxXL				1x7 2x4 3x5 8x14 9x15 10x12 11x13
III	7 ZxZL				1x6 2x5 3x4 8x15 9x14 10x13 11x12
IV	8 ZL	ANTIOZ LVL	1 PHR	3 PHR	1x9 2x10 3x11 4x12 5x13 6x14 7x15
IV	9 B	TEST BLOCK	1st	2nd	1x8 2x11 3x10 4x13 5x12 6x15 7x14
IV	10 WxWL				1x11 2x8 3x9 4x14 5x15 6x12 7x13
IV	11 WL	WAX LEVEL	0 PHR	4 PHR	1x10 2x9 3x8 4x15 5x14 6x13 7x12
IV	12 XLxZL				1x13 2x14 3x15 4x8 5x9 6x10 7x11
IV	13 XxZ				1x12 2x15 3x14 4x9 5x8 6x11 7x10
IV	14 WxZ				1x15 2x12 3x13 4x10 5x11 6x8 7x9
IV	15 Z	ANTIOZ.	Z1	Z2	1x14 2x13 3x12 4x11 5x10 6x9 7x8

Trial Order	Design Order	1	2 3	4 5	6 7	8 9	10 11 12	13 14 15	DYNAMIC OZONE
7	1	1	1 1	1 1	1 1	1 1	1 1 1	1 1 1	384
16	2	1	1 1	1 1	1 1	2 2	2 2 2	2 2 2	168
1	3	1	1 1	2 2	2 2	1 1	1 1 2	2 2 2	360
15	4	1	1 1	2 2	2 2	2 2	2 1 1	1 1 1	192
6	5	1	2 2	1 1	2 2	1 1	2 2 1	1 2 2	288
11	6	1	2 2	1 1	2 2	2 2	1 1 2	2 1 1	168
3	7	1	2 2	2 2	1 1	1 1	2 2 2	2 1 1	240
14	8	1	2 2	2 2	1 1	2 2	1 1 1	1 2 2	168
13	9	2	1 2	1 2	1 2	1 2	1 2 1	2 1 2	168
2	10	2	1 2	1 2	1 2	2 1	2 1 2	1 2 1	408
12	11	2	1 2	2 1	2 1	1 2	1 2 2	1 2 1	216
5	12	2	1 2	2 1	2 1	2 1	2 1 1	2 1 2	408
10	13	2	2 1	1 2	2 1	1 2	2 1 2	2 2 1	168
4	14	2	2 1	1 2	2 1	2 1	1 2 2	1 1 2	336
9	15	2	2 1	2 1	1 2	1 2	2 1 2	1 1 2	144
8	16	2	2 1	2 1	1 2	2 1	1 2 1	2 2 1	312

Grand Average 258.00   Averages   258.00   0.00

ANOVA TABLE

ERROR (e)	SOURCE	NAME	df	SUM OF SQUARES	MEAN SQ.	% CONTR.	F	Prob.
	1 W	WAX TYPE	1	2304.00	2304.00	1.54	18.67	.9960**
	2 X	ANTIOX.	1	14400.00	14400.00	10.12	116.67	.9998**
e	3 WxX		1	0.00	0.00	0.00		
e	4 XL	ANTIOX LVL	1	144.00	144.00	0.00		
e	5 N	NEOPRENE	1	144.00	144.00	0.00		
	6 XxXL		1	1296.00	1296.00	.83	10.50	.9858 *
e	7 ZxZL		1	144.00	144.00	0.00		
	8 ZL	ANTIOZ LVL	1	2304.00	2304.00	1.54	18.67	.9960**
	9 B	TEST BLOCK	1	112896.00	112896.00	79.91	914.67	.9999**
	10 WxWL		1	576.00	576.00	.32	4.67	.9341
	11 WL	WAX LEVEL	1	5184.00	5184.00	3.59	42.00	.9993**
e	12 XLxZL		1	144.00	144.00	0.00		
	13 XxZ		1	1296.00	1296.00	.83	10.50	.9858 *
e	14 WxZ		1	144.00	144.00	0.00		
e	15 Z	ANTIOZ.	1	144.00	144.00	0.00		
	Replications		0	0.00	0.00	0.00		
	Error		7	864.00	123.43	1.31		
	Total		15	141120.00		100.00		
	Standard Error				11.11			

CV BOOT COMPOUNDING STUDY
L16 DESIGN

FACTOR	NAME	LEVEL	LEVEL MEANS	LEVEL EFFECTS	TOTAL EFFECTS
1 W	WAX TYPE	1 PARAFIN	246.00	-12.00	24.00**
		2 POLY.	270.00	12.00	
2 X	ANTIOX.	1 X1	288.00	30.00	-60.00**
		2 X2	228.00	-30.00	
3 WxX		1	258.00	0.00	0.00
		2	258.00	0.00	
4 XL	ANTIOX LVL	1 1 PHR	261.00	3.00	-6.00
		2 3 PHR	255.00	-3.00	
5 N	NEOPRENE	1 GNA	261.00	3.00	-6.00
		2 GW	255.00	-3.00	
6 XxXL		1	249.00	-9.00	18.00 *
		2	267.00	9.00	
7 ZxZL		1	261.00	3.00	-6.00
		2	255.00	-3.00	
8 ZL	ANTIOZ LVL	1 1 PHR	246.00	-12.00	24.00**
		2 3 PHR	270.00	12.00	
9 B	TEST BLOCK	1 1st	342.00	84.00	-168.00**
		2 2nd	174.00	-84.00	
10 WxWL		1	264.00	6.00	-12.00
		2	252.00	-6.00	
11 WL	WAX LEVEL	1 0 PHR	276.00	18.00	-36.00**
		2 4 PHR	240.00	-18.00	
12 XLxZL		1	261.00	3.00	-6.00
		2	255.00	-3.00	
13 XxZ		1	267.00	9.00	-18.00 *
		2	249.00	-9.00	
14 WxZ		1	255.00	-3.00	6.00
		2	261.00	3.00	
15 Z	ANTIOZ.	1 Z1	261.00	3.00	-6.00
		2 Z2	255.00	-3.00	

Grand Average   258.00

DYNAMIC OZONE
INTERACTION PLOTS

ANTIOXIDANT TYPE x ANTIOXIDANT LEVEL

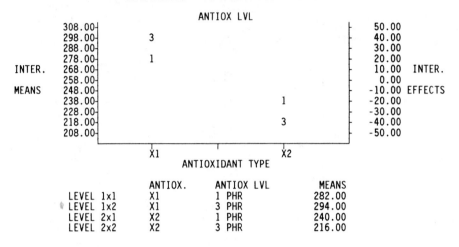

	ANTIOX.	ANTIOX LVL	MEANS
LEVEL 1x1	X1	1 PHR	282.00
LEVEL 1x2	X1	3 PHR	294.00
LEVEL 2x1	X2	1 PHR	240.00
LEVEL 2x2	X2	3 PHR	216.00

WAX TYPE x WAX LEVEL INTERACTION

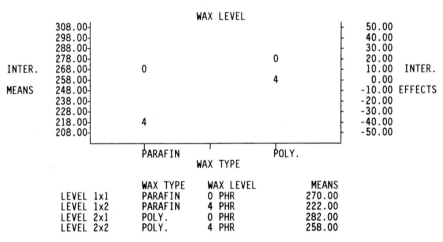

	WAX TYPE	WAX LEVEL	MEANS
LEVEL 1x1	PARAFIN	0 PHR	270.00
LEVEL 1x2	PARAFIN	4 PHR	222.00
LEVEL 2x1	POLY.	0 PHR	282.00
LEVEL 2x2	POLY.	4 PHR	258.00

The results from the DuPont Test tended to confirm Antioxidant X1 as the largest contributor in the formulation for ozone protection, but did not confirm the effects of the Wax Level. ANOVA Analysis using the Numerical rating scale appeared to show an effect from the ANTIOXIDANT TYPE (59.5% contribution), the ANTIOZONANT TYPE (9.5%), the ANTIOZONANT LEVEL (9.5%), and an ANTIOXIDANT TYPE / ANTIOZONANT TYPE interaction (9.5%).

DUPONT OZONE RESULTS
SORTED BY VISUAL RATING

Des No	WAX TYPE	ANTIOX TYPE	ANTIOX LEVEL	NEOPRENE TYPE	ANTIOZ LEVEL	TEST BLOCK	WAX LEVEL	ANTIOZ TYPE	DUPONT VISUAL RATING	DUPONT NUMER. RATING
11	POLY.	X1	3 PHR	GNA	1 PHR	2nd	4 PHR	Z1	1B+	2
4	PARAFIN	X1	3 PHR	GW	3 PHR	2nd	4 PHR	Z1	1B+	2
10	POLY.	X1	1 PHR	GW	3 PHR	1st	0 PHR	Z1	2B-	3
12	POLY.	X1	3 PHR	GNA	3 PHR	1st	0 PHR	Z2	2B	4
1	PARAFIN	X1	1 PHR	GNA	1 PHR	1st	0 PHR	Z1	2B	4
6	PARAFIN	X2	1 PHR	GNA	3 PHR	2nd	0 PHR	Z1	2B+	5
3	PARAFIN	X1	3 PHR	GW	1 PHR	1st	0 PHR	Z2	2B+	5
9	POLY.	X1	1 PHR	GW	1 PHR	2nd	4 PHR	Z2	2B+	5
8	PARAFIN	X2	3 PHR	GW	3 PHR	2nd	0 PHR	Z2	2B+	5
2	PARAFIN	X1	1 PHR	GNA	3 PHR	2nd	4 PHR	Z2	2B+	5
14	POLY.	X2	1 PHR	GW	3 PHR	1st	4 PHR	Z2	3B-	6
16	POLY.	X2	3 PHR	GNA	3 PHR	1st	4 PHR	Z1	3B-	6
7	PARAFIN	X2	3 PHR	GW	1 PHR	1st	4 PHR	Z1	3B	7
13	POLY.	X2	1 PHR	GW	1 PHR	2nd	0 PHR	Z1	3B	7
5	PARAFIN	X2	1 PHR	GNA	1 PHR	1st	4 PHR	Z2	3B	7
15	POLY.	X2	3 PHR	GNA	1 PHR	2nd	0 PHR	Z2	3B	7

ANALYSIS OF MULTIPLE RESPONSES: Each response can be analyzed separately by ANOVA to determine the effect of each factor on each response. To gain a better understanding of the relationships of each factor to each response, Multiple Response Summary Tables can be constructed to show a matrix of the factors and product characteristics on a single page. The Summary Tables may display the Level Means, Level Effects, or Percent Contribution. The Percent Contribution Summary Table shows the relative effect of each factor on each product characteristic and can be used to quickly identify the "vital few" factors, versus the "trivial many". The Percent Contribution Summary Table for this experiment shows the importance of antioxidant selection for Ozone Protection, and the importance of Neoprene selection for Modulus, Tensile, Elongation, High Speed Spin, and Low Temperature properties. The Wax selection appears important to maintain tensile strength. Factors of lesser significance are also identified.

The Percent Contribution Table does not indicate the preferred level for each factor - only the relative importance. Summarizing information by Level Effects shows which level is preferred for any particular response. The Level Effect is obtained by subtracting the Grand Average from the Mean Response of each Factor Level. Thus, a prediction can be made by adding the Level Effects for any combination of Factor Levels to the Grand Average. For example, the Tensile prediction for the combination of Poly Wax with GNA Neoprene would be: 2056.88 (Grand Average) + 56.88 (Poly Wax Level Effect) - 66.88 (GNA Neoprene Level Effect) = 2046.88, assuming other Factor effects to be insignificant.

The Level Effects Summary Table shows the preferred Levels of Neoprene Type to be GW Neoprene for Tensile and Elongation Properties, but GNA Neoprene for High Speed Spin and Low Temperature. The selection of Neoprene Type is dependent on the relative importance of the response criteria. For factors that have no substantial relationship to any responses, the factor level with the lower cost would be the choice.

FACTOR	NAME	LEVEL	DYNAMIC OZONE	DUPONT OZONE	MODULUS 100%	TENSILE PSI	ELONG. %	HI SPEED SPIN	LOW TEMP BRITTLE.
W	WAX TYPE	1 PARAFIN 2 POLY.	1.63	0.00	.28	36.56	.52	.48	.79
X	ANTIOX.	1 X1 2 X2	10.20	59.52	.28	.11	8.38	.48	.79
WxX		1 2	0.00	2.38	.28	.75	8.38	.48	.79
XL	ANTIOX LVL	1 1 PHR 2 3 PHR	.10	2.38	.00	.75	2.09	1.93	.79
N	NEOPRENE	1 GNA 2 GW	.10	0.00	81.05	50.55	52.36	69.56	63.78
XxXL		1 2	.92	2.38	.05	2.76	.52	1.93	.79
ZxZL		1 2	.10	0.00	.05	2.34	13.09	1.93	7.09
ZL	ANTIOZ LVL	1 1 PHR 2 3 PHR	1.63	9.52	12.64	.00	2.09	7.73	0.00
B	TEST BLOCK	1 1st 2 2nd	80.00	2.38	.05	.04	0.00	1.93	0.00
WxWL		1 2	.41	2.38	.00	.53	.52	0.00	3.15
WL	WAX LEVEL	1 0 PHR 2 4 PHR	3.67	0.00	4.81	.99	.52	7.73	7.73
XLxZL		1 2	.10	0.00	.14	.22	.52	.48	3.15
XxZ		1 2	.92	9.52	.00	.11	.52	.48	3.15
WxZ		1 2	.10	0.00	.28	2.34	8.38	.48	0.00
Z	ANTIOZ.	1 Z1 2 Z2	.10	9.52	.05	1.95	2.09	4.35	3.15

☐ = STRONG CONTRIBUTOR TO RESPONSE

☐ = LESSER CONTRIBUTOR TO RESPONSE

CV BOOT COMPOUNDING STUDY
SUMMARY TABLE - LEVEL EFFECTS

FACTOR	NAME	LEVEL	DYNAMIC OZONE	DUPONT OZONE	MODULUS 100%	TENSILE PSI	ELONG. %	HI SPEED SPIN	LOW TEMP BRITTLE.
1 W	WAX TYPE	1 PARAFIN	-12.00	0.00	-2.19	-56.88	-3.12	6.25	-.12
		2 POLY.	12.00☺	0.00	2.19	56.88●	3.12	-6.25	.12
2 X	ANTIOX.	1 X1	30.00●	-1.25●	-2.19	-3.12	-12.50	-6.25	.12
		2 X2	-30.00	1.25	2.19	3.12	12.50☺	6.25	-.12
3 WxX		1	0.00	.25	-2.19	8.12	12.50☺	-6.25	.12
		2	0.00	-.25	2.19	-8.12	-12.50	6.25	-.12
4 XL	ANTIOX LVL	1 1 PHR	3.00	.25	-.31	-8.12	6.25	12.50	.12
		2 3 PHR	-3.00	-.25	.31	8.12	-6.25	-12.50	-.12
5 N	NEOPRENE	1 GNA	3.00	0.00	37.19	-66.88	-31.25	75.00●	-1.12
		2 GW	-3.00	0.00	-37.19	66.88●	31.25●	-75.00	1.12
6 XxXL		1	-9.00	.25	.94	-15.62	-3.12	12.50	.12
		2	9.00	-.25	-.94	15.62	3.12	-12.50	-.12
7 ZxZL		1	3.00	0.00	.94	-14.38	15.62☺	-12.50	-.38
		2	-3.00	0.00	-.94	14.38	-15.62	12.50	.38
8 ZL	ANTIOZ LVL	1 1 PHR	-12.00	.50	-14.69	.62	-6.25	-25.00	0.00
		2 3 PHR	12.00☺	-.50☺	14.69	-.62	6.25	25.00☺	0.00
9 B	TEST BLOCK	1 1st	84.00?	.25	-.94	1.88	0.00	-12.50	0.00
		2 2nd	-84.00?	-.25	.94	-1.88	0.00	12.50	0.00
10 WxWL		1	6.00	-.25	.31	-6.88	3.12	0.00	.25
		2	-6.00	.25	-.31	6.88	-3.12	0.00	-.25
11 WL	WAX LEVEL	1 0 PHR	18.00☺	0.00	9.06	9.38	3.12	25.00☺	-.5
		2 4 PHR	-18.00	0.00	-9.06	-9.38	-3.12	-25.00	.5
12 XLxZL		1	3.00	0.00	-1.56	-4.38	-3.12	6.25	-.2
		2	-3.00	0.00	1.56	4.38	3.12	-6.25	.2
13 XxZ		1	9.00	-.50☺	-.31	-3.12	-3.12	-6.25	-.2
		2	-9.00	.50	.31	3.12	3.12	6.25	.2
14 WxZ		1	-3.00	0.00	2.19	-14.38	-12.50	6.25	0.0
		2	3.00	0.00	-2.19	14.38	12.50☺	-6.25	0.0
15 Z	ANTIOZ.	1 Z1	3.00	-.50☺	.94	-13.12	6.25	-18.75	-.2
		2 Z2	-3.00	.50	-.94	13.12	-6.25	18.75	.2
	Grand Average		258.00	5.00	352.81	2056.88	353.12	2193.75	-46.8

● = STRONG PREFERENCE FOR LEVEL SETTING

☺ = MILD PREFERENCE FOR LEVEL SETTING

CTOR	FACTOR	LEVEL	DYNAMIC OZONE	DUPONT OZONE	MODULUS 100%	TENSILE PSI	ELONG. %	HI SPEED SPIN	LOW TEMP BRITTLE.
W	WAX TYPE	1 PARAFIN	246.00	5.00	350.62	2000.00	350.00	2200	-47.00
		2 POLY.	270.00	5.00	355.00	2113.75	356.25	2188	-46.75
X	ANTIOX.	1 X1	288.00	3.75	350.62	2053.75	340.62	2188	-46.75
		2 X2	228.00	6.25	355.00	2060.00	365.62	2200	-47.00
WxX		1	258.00	5.25	350.62	2065.00	365.62	2188	-46.75
		2	258.00	4.75	355.00	2048.75	340.62	2200	-47.00
XL	ANTIOX LVL	1 1 PHR	261.00	5.25	352.50	2048.75	359.38	2206	-46.75
		2 3 PHR	255.00	4.75	353.12	2065.00	346.88	2181	-47.00
N	NEOPRENE	1 GNA	261.00	5.00	390.00	1990.00	321.88	2269	-48.00
		2 GW	255.00	5.00	315.62	2123.75	384.38	2119	-45.75
XxXL		1	249.00	5.25	353.75	2041.25	350.00	2206	-46.75
		2	267.00	4.75	351.88	2072.50	356.25	2181	-47.00
ZxZL		1	261.00	5.00	353.75	2042.50	368.75	2181	-47.25
		2	255.00	5.00	351.88	2071.25	337.50	2206	-46.50
ZL	ANTIOZ LVL	1 1 PHR	246.00	5.50	338.12	2057.50	346.88	2169	-46.88
		2 3 PHR	270.00	4.50	367.50	2056.25	359.38	2219	-46.88
B	TEST BLOCK	1 1st	342.00	5.25	351.88	2058.75	353.12	2181	-46.88
		2 2nd	174.00	4.75	353.75	2055.00	353.12	2206	-46.88
WxWL		1	264.00	4.75	353.12	2050.00	356.25	2194	-46.62
		2	252.00	5.25	352.50	2063.75	350.00	2194	-47.12
WL	WAX LEVEL	1 0 PHR	276.00	5.00	361.88	2066.25	356.25	2219	-47.38
		2 4 PHR	240.00	5.00	343.75	2047.50	350.00	2169	-46.38
XLxZL		1	261.00	5.00	351.25	2052.50	350.00	2200	-47.12
		2	255.00	5.00	354.38	2061.25	356.25	2188	-46.62
XxZ		1	267.00	4.50	352.50	2053.75	350.00	2188	-47.12
		2	249.00	5.50	353.12	2060.00	356.25	2200	-46.62
WxZ		1	255.00	5.00	355.00	2042.50	340.62	2200	-46.88
		2	261.00	5.00	350.62	2071.25	365.62	2188	-46.88
Z	ANTIOZ.	1 Z1	261.00	4.50	353.75	2043.75	359.38	2175	-47.12
		2 Z2	255.00	5.50	351.88	2070.00	346.88	2212	-46.62

DETERMINING SETTINGS: The original set of priorities indicated that the Dynamic Ozone, High Speed Spin, and Low Temperature Brittleness characteristics were the most important responses. Through use of the Summary Tables, the following formulation was derived:

MATERIAL	AMOUNT	TO IMPROVE:
Neoprene GNA		High Speed Spin, Low Temperature
Antioxidant X1	3 phr	Dynamic Ozone
Antiozonant Z1	3 phr	(Dynamic Ozone ?)
Wax - Poly.	1 phr	(Ozone / Tensile ?)

This combination had not been run as one of the 16 formulations in the experiment. Instead of "Picking the Winner" from among the 16 formulations, the analysis is predicting a different combination. Predictions of average performance with the above combination (conservative estimates using Grand Average + largest Level Effects):

Dynamic Ozone:	Improved, but can't predict hours
DuPont Ozone:	5.0 - 1.25 - .5 -.5 -.5 = 2.25 = 1B+ rating
High Speed Spin:	2194 + 75 + 25 = Approx 2300 RPM
Low Temp. Brittleness:	-46.88 - 1.12 = -48 deg. F.
Tensile:	2057 + 57 - 67 = 2047 PSI
Elongation:	353 - 31 - 15.5 - 12.5 - 12.5 + 12.5 = 294%
Modulus:	353 + 37 + 15 = 405 PSI

VERIFYING RESULTS: Results from the experimental trials should always be verified before action is taken. Mixing the above formulation gave results close to the prediction for Tensile, Elongation, Modulus, and High Speed Spin. The dynamic ozone improvements were predicted to be better than previous formulations but could not be verified in running hours because of the large variability in testing. Samples of CV boots were sent to the customer for functional testing and all parts passed the performance tests as predicted.

Through Statistical Process Control, the Tensile, Elongation, and Modulus tests were monitored to maintain performance. Subsequent process improvements were introduced to improve consistency of the manufactured product.

The methods incorporated in this experiment have been used frequently in the development and production of CV boots and other rubber products. Through the use of Design of Experiments and Statistical Process Control, a better understanding of the chemistry of the formulation and manufacture of the boots was achieved.

A STRATEGIC DEFENSE CASE STUDY

Peter L. Knepell
PhD, Operations Research
Geodynamics Corporation

Minutes of the First Meeting:

"O.K. folks, here's the architecture we need to study," explained Capt. Kirk, an enthusiastic strategic defense expert. "Don't look at the picture too long because it might change any minute. Some things do remain constant:

- a two-tiered interceptor system (one is space-based and the other is ground-based),
- command centers,
- space-based and ground-based sensors to support the command centers and interceptors, and
- communications links.

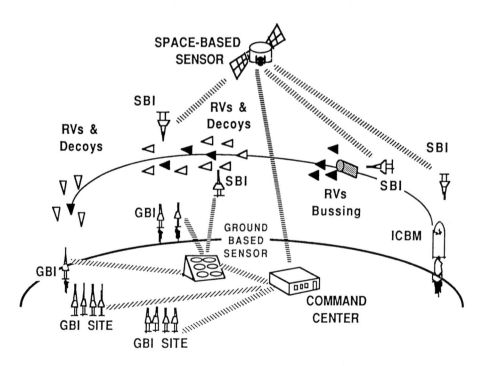

"Look, this study is quite simple" "It's a simple matter of input/output, like knobs on a radio." His knobs were labeled:

Number of SBI	Space-Based Interceptors (SBI)
Number of GBI	Ground-Based Interceptors (GBI)
RVs	Reentry Vehicles (alias: blivits or bombs)
Decoys	The objects used to look like RVs
Pk	The one-shot probability of kill of an interceptor against a target
Trelease	The time it takes someone to give the command to shoot at missiles
Discrimination factor	A measure of how well real missiles are picked over decoys
Tdiscrim	The time it takes to track and discriminate an RV so that an interceptor can shoot at it

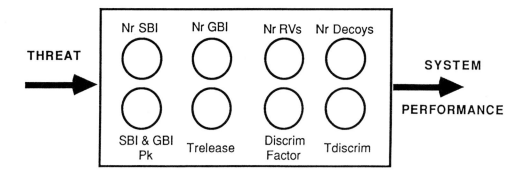

This stimulated a discussion on what measures will be used to gauge 'system performance.' The fundamental measure agreed upon was RVs destroyed. Then things got interesting and everyone got a chance to add knobs to the system.

"Not enough knobs" said Dee Coy, a renowned US expert on Russian threats through the year 2010. "You've got to consider the number of fast burn boosters and fast bussers." (Bussing is a term describing the act of casting several RVs off an ICBM in the proper direction -- it's not a rapid transit system.)

"Don't forget the ASAT (anti-satellite) threat," piped in Dr. Larry Leber, a national labs scientist. "A good ASAT attack could 'smear' SBIs all over the galaxy."

"I have a new SBI design that will make all ASATs ineffective," countered Dr. S B Iacocca.

"Yeah, but what about SEOs?" remarked the consultant from ERUDITE Corporation.

"Yep, don't forget SEOs."   "SEOs are tricky."   "Oh no, SEOs!"

It got very quiet, especially after I asked, "What are SEOs?"

"Survivability Enhancement Options," someone said disdainfully.

"Whose?" I asked undaunted -- after all I was conducting the "simple input/output" study.

"The SBI's of course. You've got to consider these in your study'" said Kirk.

"What exactly do the SBIs do?"

"THEY haven't decided yet. Anyway, if I knew, I couldn't tell you -- it's classified above your level."

"When will THEY decide?"

"Soon, probably after you finish the study."

Minutes of the Second Meeting:

I was much better prepared this time. "Gentlemen, this table shows all factors that you want me to consider.

PARAMETER	LEVELS	PARAMETER	LEVELS
ASAT Attack	4 types	Nr of Decoys	2 levels
SBI Design	2 types	Nr Fast-Burn Boosters	2 levels
Pk	2 levels		
Quantity	4 levels	Trelease	3 levels
GBI Design	2 types	Discrimination Factor	2 levels
Pk	2 levels		
Quantity	4 levels	Tdiscrim	2 levels

FULL FACTORIAL DESIGN = 4 x 2 x 2 x 4 x 2 x 2 x 4 x 2 x 2 x 3 x 2 x 2
= **49152 cases**

I am prepared to conduct a complete factorial analysis (49,152 cases). Since I can simulate an entire war in only 15 minutes, I'll be prepared to brief the results in 512 days (737,280 minutes) -- as long as the computer doesn't crash."

"I want an answer in three months," boomed Capt. Kirk.

"O.K., how about a fractional factorial design where I'll tell you which factors were dominant in affecting system performance," I offered.

"No good," he said. "I need to make clear statements about the architecture THEY are designing. All factors are important!"

"All right, please tell me which factors are more likely to remain fixed while we vary everything else." The following list was developed:

SBI & GBI design & Pk   Trelease
Number of decoys   Tdiscrim
Number of fast-burn boosters   Discrimination factor

"I see your problem with the number of factors and I have a suggestion," offered Dee Coy. "If we composed two different threat scenarios, which differ by the number of fast-burn boosters and decoy composition, we could combine two of your factors into one called 'strategic threat.' Then we can measure the effects separately since they occur at different times in the ICBM flyout."

This was an excellent suggestion except for one consideration: "Could there be an interaction effect between the number of fast-burn boosters and decoy composition?" I asked.

Dee Coy was very accommodating: "Given this architecture, there is no reason to believe that there is an interaction. I might also suggest that the effects of Pk, Trelease and Tdiscrim could be checked in just a few cases to give us a feel for their impact."

"So gentlemen, let's summarize this discussion. It appears that the following factors are foremost in your minds:

ASAT threat   SBI design
Number of SBI   GBI design
Number of GBI   Strategic threat

The Dr. Leber had a interesting suggestion: "Why not see how the SBI design performs against the ASAT threats."

"But I want a complete tradeoff analysis for varying the numbers of SBIs and GBI," said Kirk.

"Gentlemen, if we all agree that our main objective is to conduct an interceptor tradeoff analysis and our secondary objective is to see how changes in other parameters affect these trades, then I think I can have a study design very soon!"

Minutes of the Third Meeting:

I opened the meeting by restating our objective: "If we all agree that our primary mission is to conduct a study on the tradeoffs (i.e., effects) involved with varying the number of SBIs and GBIs in the architecture, I think I have a proposition we all can live with. In particular, I will discuss a study that will support tradeoff analyses for 10 different scenarios and some other limited excursions with less than 200 simulations. With that number of runs, I can meet our schedule."

Here's what I presented: For each of the 10 scenarios, I have a "design space" representing different numbers of interceptors. I would conduct 18 simulations at the lattice points shown.

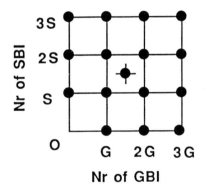

(I don't need to simulate the origin for obvious reasons.)

"There are only 16 lattice points shown. Why 18 runs?"

The reason: "The center point is simulated (replicated) three times to get a measure of variation. This will support the fit of a quadratic response surface. The contours of this surface (level curves) can then be plotted in the design space as shown."

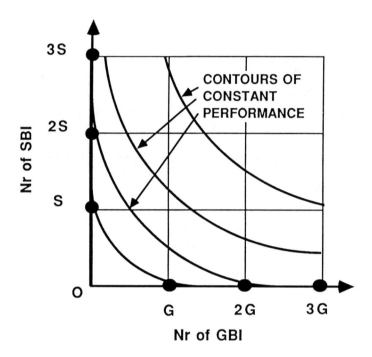

"What do these curves represent?" asked Capt.Kirk.

"One curve shows which combinations of SBIs and GBIs give a constant number of RVs killed. In fact, if you have some baseline number of RV kills that you want to compare the results to, I can generate that curve and highlight it in this display."

"So we could call these curves 'isomorts' for constant mortality," responded Kirk.

At this point, a vocal critic said: "This is statistical voodoo. You must replicate at all points at least 6 times and use the average to counteract variation."

This would increase the number of simulation runs to almost 1000! Discussions of statistical design, analysis of variance, etc. did not calm this persistent nay sayer. Nor did an appeal to the fact that these methods have been accepted by the most respected people in the field. After many unsuccessful attempts we entered into an interesting dialog:

"You agree that there is a chance for variation at each point in the design space?" I asked.

"Of course." he replied.

"And you agree that with sufficient data, a representative response surface can be created."

"Yes."

"But your fear is that with only one replication for 15 of the points, the response could be too high or too low."

"Correct."

"Could the results at these points all come in too high?"

"Yes, but that would be rare."

"Well then, what would you expect to happen across 16 points in the design space?"

"I guess some would come in too high and some too low."

"Just like we would see with replications at one point?"

"Yes," he agreed.

"Well then it seems like the 16 points will act like replications where some are too high, some are too low, but together, they should balance each other out to provide a close approximation of the true response surface."

"O.K., but how is this all put together to make a statistical statement?"

"That's voodoo - F statistics!" I answered cryptically.

Capt. Kirk looked at the data and, while he liked the contour plots, he suggested another display which is more appealing to his tastes.

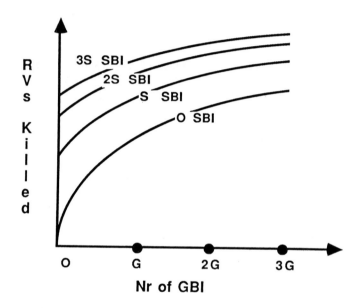

"These curves are essentially generated by cutting the response surface with four planes parallel to the GBI axis, where these planes cut through fixed levels of SBIs." I noted.

"Enough on display of data; what are your scenarios?" asked Kirk.

I suggested the following five cases of SBI type and ASAT threat be considered:

SBI TYPE

ASAT MIX	I	II
NO ASATS	CASE A	
FORCE 1	CASE B	
FORCE 2	CASE C	CASE D
FORCE 3		CASE E

The audience was in wide agreement. I then asked, "Which of these cases would represent a baseline, that we should examine while changing other factors?" For the first time a consensus choice was easy - Case D was chosen as a baseline for all other sensitivity studies. Then using Case D, I suggested that we run complete tradeoff studies while changing settings in:

Discrimination Factor	2 settings
Strategic threat	2 settings
GBI Design	one change

Thus with the five cases for ASAT-SBI factors and the five excursions noted above, I got agreement on the study design. Each scenario required 18 simulation runs for a total of 180 to provide tradeoff analyses for all scenarios. We then could conduct a handful of simulations to provide a feel for the effects of Pk, Trelease and Tdiscrim.

Epilog:

This study took six months, primarily due to a massive software development effort to simulate the ASAT threat and the SEOs. Results were widely appreciated for two reasons: people liked our display of results (the tradeoff contours and the tradeoff curves) and we answered a multitude of questions without confounding the results. Specifically, we:

- Characterized the effects and interactions when the number of interceptors was varied (tradeoff analysis)

- Provided a statement on the effects of:
  ASAT attack
  Interceptor design
  Strategic threat
  Discrimination factor

- Provided a feel for the effects of variations in:
  Pk
  Trelease
  Tdiscrim

What was not appreciated by the customer, but was fundamental to our success, was the importance of a study design to limit the number of simulations. Given time and resource constraints, we had to limit the number of simulation runs early and design our study around that limit. When the limit was discussed with the customer, he had a much greater appreciation of our need to use an efficient design as well as our need to limit the scope of the study. Future studies will now focus on some of the more salient results of our initial study.

IDENTIFYING A PLASMA ETCH PROCESS WINDOW
Using Response Surface Methodology

Dr. Jack E. Reece, Sr. Principal Process Engineer
David Daniel, Process Engineering Supervisor
Honeywell Solid State Electronics Division

Ms. Robin Bloom, Sr. Process Engineer
United Technologies Microelectronics Center
Colorado Springs, CO

1. The Problem:

Removing a top layer in the presence of an underlayer is a common requirement in the manufacture of integrated circuits. The process must etch the layer uniformly across the wafer and must maintain considerable selectivity to avoid damage any underlayer. Furthermore, to minimize the risk of damaging other parts of the circuit, operators must have a convenient means for detecting the end point when the process has removed the target layer.

For approximately one year preceding the start of this study, four similar processes run in this equipment had given satisfactory etch rate uniformity and selectivity. In the process a relatively low-powered pulsed plasma activated an etchant gas at low pressure to remove the target layer. Unexpectedly, operators began having difficulty determining end points for the etch process, presumably due to the breakdown of some shielding or circuit that had protected the detector from the pulsing RF in the reactor. In addition while the uniformity of the etch rate across a wafer had been marginally acceptable, design demands for new circuits required better control.

The machine vendor, working cooperatively with factory maintenance personnel, failed to find an assignable mechanical or electrical cause for the malfunctions. After three weeks with the $200,000 machine offline, the best he could suggest was that the engineers devise new processes using higher-intensity pulsed cathode power to enhance etch rate uniformity. All parties had somewhat agreed that an unstable plasma due to the pulsing at low power contributed to the uniformity problems and that pulsing the plasma in general caused the end point detection interference. But the vendor could not show that he had eliminated the electrical interference with the photocell end point detector nor that the problem would not return.

When the engineers on site suggested using a steady-state, rather than pulsed cathode power, the vendor was adamant that the machine absolutely would not etch the layers uniformly under such conditions -- it was designed to use the pulsed power. In fact, preliminary best guess probing experiments suggested that the vendor was probably correct.

Given that they were faced with devising a new process and that they had no real confidence that end point detection problems would cease to plague the machine's operation, the engineers elected to invest the time necessary to characterize its performance using fixed power first.

2. Experimental Approach:

2.1 Factor Settings and Design Matrix:

When restricted to constant power, the machine offered only three process control variables: power, pressure, and flow of etchant gas. Experience gained from the probing experiments and previous use of the equipment identified the logical limits for factor settings, so the engineers elected to use a Box-Wilson central composite design requiring 20 trials. Table 1 shows the factors and their limits for this experiment.

Table 1

Factor Settings

Factor, Units	Low	High
Pressure, mtorr	200	300
Power, watts	75	150
Etchant Flow, sccm	20	40

Simple two-level factorial designs support linear or planar models; to characterize any curvature present in a response as a function of the inputs requires at least 3 levels for each factor. The Box-Wilson Central Composite Design (1,4) augments the usual 2-level factorial array with *star points* and *center points* such that one explores 5 levels of each factor in relatively few trials.

> Among alternatives to this design are a 3-level full factorial which requires 27 trials and the 3-level Box-Behnken, which requires 15 (including replication). We chose the central composite design as it provides 5 points for estimating curvilinearity and requires only 20 trials (including replication).

Modern experimental design software simplifies the tasks of creating and interpreting designs such as these. The limits for low and high indicated in Table 1 were truly limits, but the software used here allowed us to assign these points to the "star" positions while it constructed the two-level matrix within them, saving the tedium of the arithmetic calculations for factor settings. Table 2 summarizes the design matrix generated.

2.2 Responses:

Responses included the etch rates of layers 1 and 2, the uniformity of each etch, and the selectivity between layers 1 and 2. Measurements at several points on the wafer before and after each trial provided the etch rates. The ratio of etch rates during each trial provided the selectivity.

Etch rate measurements at several points on each wafer gave ranges of values for the rates suggesting that one or more of the factors was contributing a dispersion effect in the process. In addition the range of etch rate values observed was proportional to the mean etch rate. To describe the dispersion of etch rates we chose a relative of the *coefficient of variation (1)* because it was familiar to those involved in the investigation.

Percent Uniformity = [(Max - Min)/2*Mean]*100   *(1)*

7-152

Table 2

Design Matrix of Factor Settings

# Reps	PRESSURE (MTORR)	POWER (WATTS)	GASFLOW (SCCM)
	220	90	24
	280	90	24
	220	135	24
	280	135	24
	220	90	36
	280	90	36
	220	135	36
	280	135	36
	200	113	30
	300	113	30
	250	75	30
	250	150	30
	250	113	20
	250	113	40
6	250	113	30

Calculations are simple and the result provides a concept of a +/- variation about a central value. However, this measurement has a natural limit of 0, and regression models readily predict meaningless <u>negative</u> values. Also, range is sensitive to outliers, so a more robust metric that provides approximately the same answers substitutes $2*\sigma$ for the range.

Taguchi (2) has published a variety of complex equations to model variability and calls them <u>signal to noise</u> ratios. His SN_T for "Target is Best" is also a variant of the coefficient of variation and works well when significant correlation exists between the standard deviation of a series of observations and the mean of those observations. Box, (3) advocates a simpler approach by determining the logarithm of the standard deviations of the observations and basing the analysis on that metric.

The software used in this study supports Box-Cox transformations of responses and performs the arithmetic totally transparent to the user -- if the user takes full advantage of the scaling options available, s/he sees only normal response values and appropriate regression coefficients, regardless of the transformation used. Supported transformations include $y**2$, y, $sqrt(y)$, $log_e y$, $1/sqrt(y)$, and $1/y$.

Randomization of experimental trials distributes environmental or equipment bias uniformly among all factor levels; and while the software supports orthogonal blocking, we elected not to use it in this study. The first 3 columns of Table 3 illustrate the random run order software-generated for this investigation. The last 5 columns contain the observations.

Table 3

Randomized Worksheet and Results
Etch Process Optimization

RUN	PRS	PWR	FLOW	ETCH RATE1	LAYER1 UNIF	ETCH RATE2	LAYER2 UNIF	SELECT
1	280	90	24	827	13.213	193	23.714	4.3
2	250	113	30	1417	8.395	446	9.099	3.2
3	250	113	40	1351	6.295	375	12.912	3.6
4	250	75	30	719	14.289	200	20.701	3.6
5	250	113	30	1365	10.304	372	7.907	3.7
6	220	90	36	978	9.099	279	11.508	3.5
7	280	135	36	1597	9.795	397	6.998	4.0
8	250	113	30	1475	6.607	405	10.399	3.6
9	250	113	20	1376	8.395	316	7.295	4.4
10	250	113	30	1370	8.091	348	9.795	3.9
11	220	90	24	905	12.106	256	11.092	3.5
12	220	135	24	2100	2.000	734	5.794	2.9
13	250	113	30	1400	6.295	367	12.106	3.8
14	220	135	36	1961	2.698	660	5.105	3.0
15	280	90	36	841	11.508	195	21.979	4.3
16	200	113	30	1824	0.800	551	7.096	3.3
17	250	150	30	2065	4.198	615	10.789	3.4
18	300	113	30	1085	11.298	245	25.704	4.4
19	280	135	24	1647	9.099	398	12.388	4.1
20	250	113	30	1449	5.297	462	10.990	3.1

3. Results and Analysis:

Each etch rate is the average of 5 measurements on each wafer, and the author used the range of those in equation (1) to model uniformity after transforming the values to the natural logarithm using software utilities. Selectivity is the ratio of the etch rates.

Backwards stepwise regression provided the Taylor expansion series coefficients displayed in Table 4. The author prefers this technique because it efficiently provides the simplest (most parsimonious) model which adequately describes the response as a function of the factors. In this case we set software parameters such that the α risk associated with retaining a coefficient was 0.10. Analysis of variance (ANOVA) and residual diagnostic routines within the software help the user decide on the adequacy of a model.

The software coded the predictor variables according to the relationship

$$(X_i - ((MAX\ X_i + MIN\ X_i)/2)/((MAX\ X_i - MIN\ X_i)/2)$$

which results in values of -1 and + 1 for the low and high predictor values, respectively. Therefore we may judge the relative importance of each factor on a response by the absolute value of its coefficient. For example, the models for LAYER1 ETCH and LAYER1 UNIF indicate that Pressure and Power are highly coupled or interactive. That is, the effect on the response of changing Pressure from its low to its high setting depends on the setting of Power.

7-154

Table 4

Regression Coefficients

MODEL TERM	ETCH RATE 1	LOG UNIF1	LOG UNIF2	SELECTIVITY LAYER1/LAYER2
CONSTANT	1382.68	8.095	10.251	3.590
PRS	-278.96	7.795	6.232	0.696
PWR	737.89	-5.800	-5.804	-0.237
FLW	-------	------	------	-0.164
PRS**2	-------	-6.320	------	------
PRS*PWR	-213.22	7.385	------	------
PWR**2	-------	------	3.106	------
FLW**2	-------	------	------	0.377
Adj-R**2	0.97	0.826	0.748	0.664
Rel. Press	0.95	0.605	0.581	0.491
RMS Error	71.16	2.503	2.496	0.269

Note that selectivity is a quadratic function of etchant gas flow while the other responses are independent of this factor. In the table above the coefficients for the uniformity models reflect our choice of a scaling option that displays approximate values for the raw metric regardless of the transformation. This makes interpretation of the model somewhat simpler than using logarithms. In UNIF2, the α risk associated with the PWR**2 coefficient was 0.11.

3.1 Interpretation and Conclusions:

An extremely convenient method for interpreting the Taylor series regression models is contour plotting. Unlike a conventional graph this technique assigns a factor to each axis while holding any others at fixed values. The contour lines are lines of equal response, quite analogous to the contour intervals on a map.

Figures 1 to 6 are computer-generated plots based on the Taylor series models. The author adjusted the maximum value of the Pressure axis to expand the scale of the graphs and focus attention on the area of interest. The '*' on each represent actual data points. The first 4 add contours at etchant flow = 30 sccm -- the center of the investigation. The box in graphs 1 to 4 indicates the best conditions found during probing experiments which seemed to support the vendor's contention that this machine would not operate properly using fixed cathode power. Graphs 1 to 4 show that reducing the pressure and increasing the power improve uniformity and increase potential throughput at the cost of selectivity. Figure 5 shows that increasing the etchant flow to the limit studied improves selectivity, but not enough; Figure 6 shows that decreasing the etchant flow to the minimum investigated improves it further and provides an acceptable result. The box in Figure 6 represents a proposed process window.

An investigator must not forget that the models derived contain experimental error -- the values indicated on a contourplot are not absolute, but have confidence intervals. An additional software utility permits estimation of these intervals. The final stage in any investigation requires predicting results and testing them in the process. Table 5 summarizes the predicted values and observed values for each response when the authors tested the modified process.

Figure 1
ETCH1
FLOW = 30

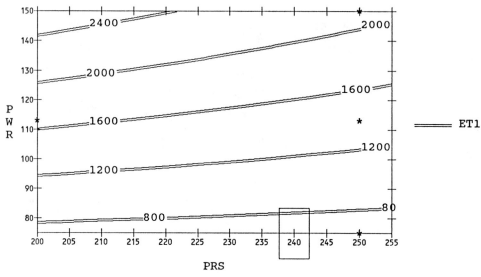

Figure 2
ETCH1, UNIF1
FLOW = 30

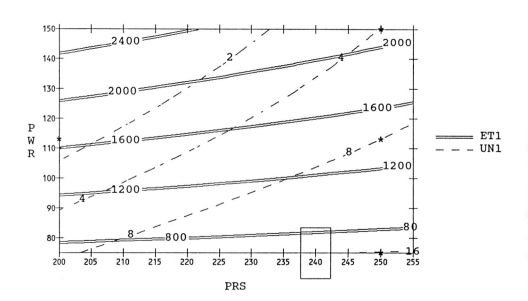

Figure 3
ETCH1, UNIF1, UNIF2
FLOW = 30

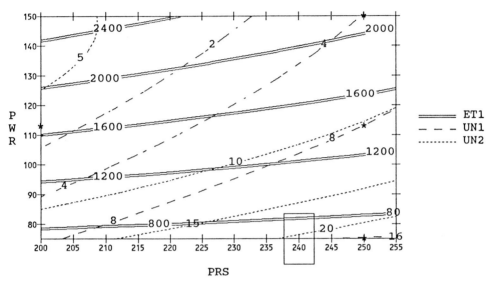

PRS

Figure 4
ETCH1, SELECT, UNIF1, UNIF2
FLOW = 30

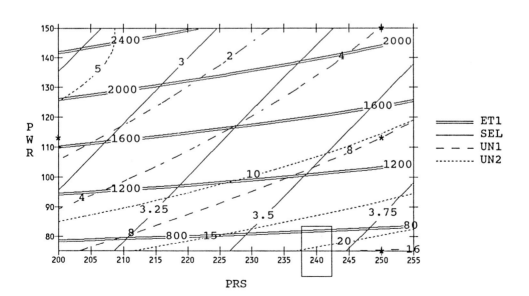

PRS

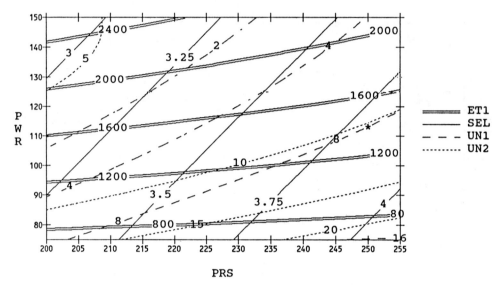

Figure 5
ETCH1, SELECT, UNIF1, UNIF2
FLOW = 40

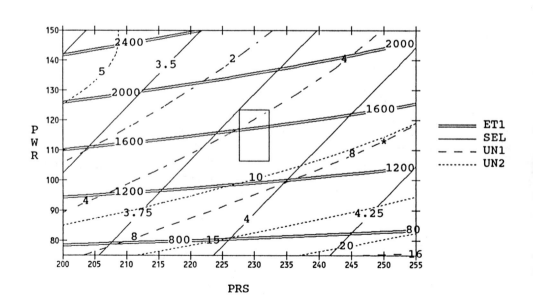

Figure 6
ETCH1, SELECT, UNIF1, UNIF2
FLOW = 20

Table 5

Predictions and Observed Values
for Responses in Etch Process
(95% Confidence Intervals)

FACTORS	ETCH RATE 1	UNIF LAYER1	UNIF LAYER2	SELECTIVITY LAYER1/LAYER2
PRESSURE	230	230.00	230.00	230.0
POWER	115	115.00	115.00	115.0
GASFLOW	20	20.00	20.00	20.0
Lower Bound	1477	3.30	6.00	3.0
RESPONSE->	1549	4.60	7.90	3.8
Upper Bound	1620	6.50	10.30	4.6
Observed	1517	4.60	7.70	3.5

This investigation required four days and included 20 experiments, their analysis and interpretation, and the testing of predictions. It identified a new process window which satisfied all process requirements which the vendor swore could not exist. The new process has performed as expected for 30 months, requiring nothing more than regular preventative maintenance and reducing scrap and rework while increasing the throughput at this step. The new process unexpectedly replaced the *four* previous ones, giving acceptable results regardless of device type or layer.

The application of experimental design and the contourplots derived from the models generated provide a far superior understanding of process windows available to the process engineer than any other experimental method.

Attempting to build these models from other than the orthogonal arrays provided by statistical experimental design usually leads to mathematical correlations among supposedly independent factors and potentially misleading models.

2. Acknowledgement:

The authors gratefully acknowledge the contributions of Mr. Tom Tjaden, Process Engineer, Honeywell Solid State Electronics Division, in helping to define the study and in interpreting its results and implications.

3. References:

(1) Box, G.E.P., Hunter, W.G., and Hunter, J.S., "Statistics for Experimenters", John Wiley & Sons, New York, 1978.

(2) See for example: Taguchi, G., and Phadke, M.S., "Quality Engineering Through Design Optimization", Conference Record Vol. 3., IEEE Globecom 1984 Conference, Atlanta, GA, 1106-1113; Taguchi, G., and Wu, Y., "Introduction to Off-Line Quality Control", Central Japan Quality Control Association (1986); and Taguchi, G., "Introduction to Quality Engineering: Designing Quality into Products and Processes". Asian Productivity Organization, UNIPUB, Kraus International Publications, White Plains, NY. See also Phadke, M.S., "Quality Engineering Using Design of Experiments", Proceedings of the American Statistical Association, section on Statistical Education, 11 - 20 (1982).

(3) Box, G.E.P, "Studies in Quality Improvement: Signal to Noise Ratios, Performance Criteria, and Transformation", Report No. 26, University of Wisconsin Center for Quality and Productivity Improvement, Madison, WI., 1987.

(4) Box, G.E.P, and Draper, N.R. "Empirical Model-building and Response Surfaces", John Wiley & Sons, New York, 1987.

Applying Experimental Design Techniques to Operating Systems and Software

Even though there are large numbers of documented designed experiments in the chemical and hardware segments of industry, few documented examples exist in the operating systems and software world. In view of the above, it is now fitting that we share the following case study with you.

A software engineer concerned with overall system performance when running a particularly large computer program decided to run a designed experiment. The experimental objective was to determine the best combination of operating system factors in order to minimize CPU time (regardless of the disk file scheme). Factors of interest to the experimenter were:

FACTOR	LOW	MIDDLE	HIGH
A. Buffered I/O Limit	- 1	0	+1
B. "Q" Limit	- 1	0	+1
C. Block Size	- 1	0	+1
D. File Limit	- 1	0	+1
E. Direct I/O Limit	- 1	0	+1
F. Working Set Quota	- 1	0	+1
G. Byte Limit	- 1	0	+1
H. "Bf"	A		B

In addition, a three different disk file schemes were appropriate to evaluate as a "noise factor" (X_1, X_2, X_3).

For the actual experiment, six different responses were considered worthy of consideration. In our example, however, we will only address one response (CPU time). Utilizing an 18 run computer-generated D–Optimal design, coupled with a "Taguchi style" outer array, the following design matrix with applicable data was:

TABLE I
Design Matrix

Experiment #	A	B	C	D	E	F	G	H	X₁	X₂	X₃
1	-1	-1	-1	-1	-1	-1	-1	A	127	155	269
2	-1	0	0	0	0	0	0	A	118	150	685
3	-1	1	1	1	1	1	1	A	115	169	2321
4	0	-1	-1	0	0	1	1	A	141	163	2237
5	0	0	0	1	1	-1	-1	A	110	134	207
6	0	1	1	-1	-1	0	0	A	131	160	791
7	1	-1	0	-1	1	0	1	A	139	271	1358
8	1	0	1	0	-1	1	-1	A	127	145	216
9	1	1	-1	1	0	-1	0	A	112	139	579
10	-1	-1	1	-1	0	0	-1	B	40	69	144
11	-1	0	-1	0	1	1	0	B	56	125	715
12	-1	1	0	1	-1	-1	1	B	63	92	652
13	0	-1	0	1	-1	1	0	B	41	73	574
14	0	0	1	-1	0	-1	1	B	68	125	889
15	0	1	-1	0	1	0	-1	B	49	82	129
16	1	-1	1	0	1	-1	0	B	51	100	580
17	1	0	-1	1	-1	0	1	B	50	80	786
18	1	1	0	-1	0	1	-1	B	57	87	153

<table>
<tr><td colspan="4">

TABLE II
Average Marginals for Statistic "$\bar{x}$"

	Low	Mid	High
A	337	339	279
B	363	265	327
C	333	276	347
D	315	321	319
E	252	331	373
F	247	290	418
G	128	288	540
H	417	- - -	220

</td><td colspan="4">

TABLE III
Average Marginals for the Statistic LnS *

	Low	Mid	High
A	5.48	5.41	5.26
B	5.55	5.26	5.34
C	5.43	5.27	5.45
D	5.44	5.31	5.39
E	5.27	5.41	5.47
F	5.24	5.31	5.59
G	3.94	5.75	6.45
H	5.56	- - -	5.20

* natural log of the sample std dev

</td></tr>
</table>

FIGURE I

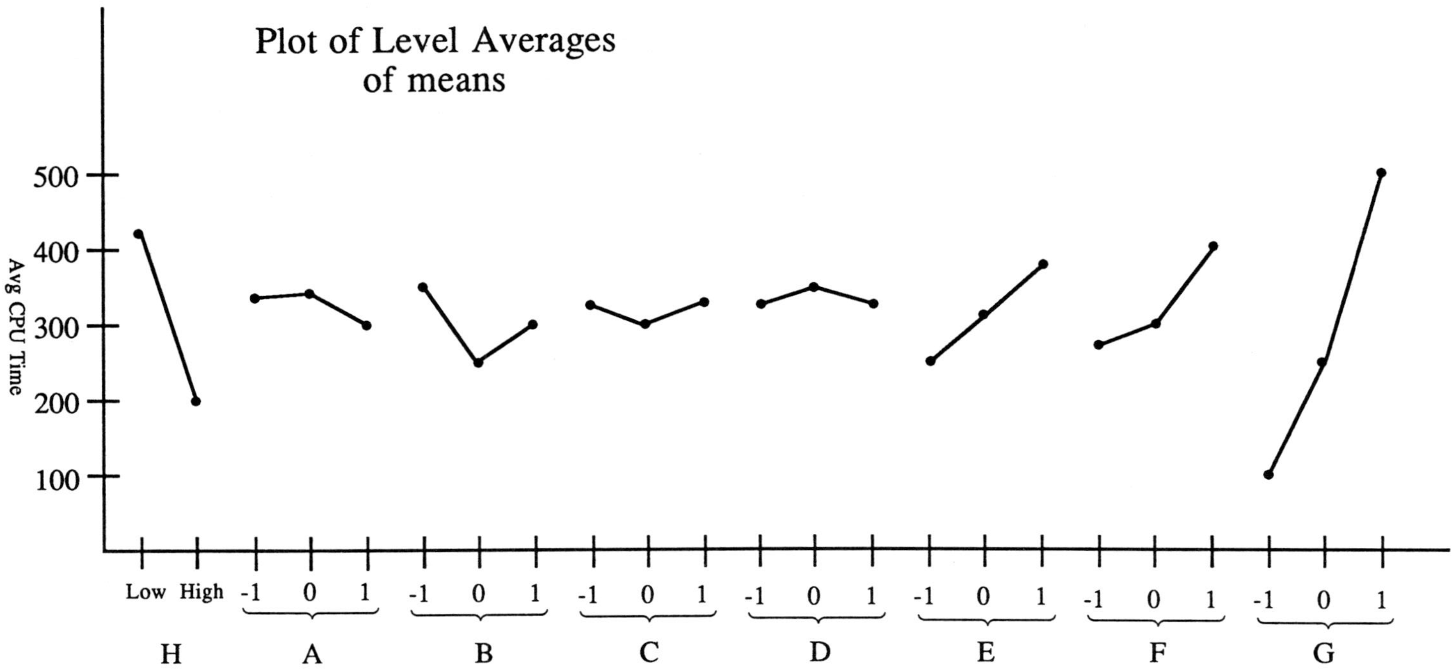

Plot of Level Averages
of means

Avg CPU Time

Low High

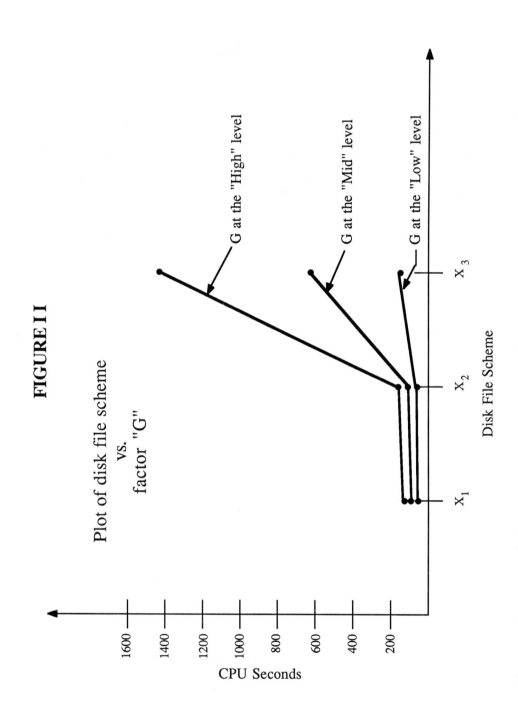

FIGURE I I

Plot of disk file scheme
vs.
factor "G"

G at the "High" level

G at the "Mid" level

G at the "Low" level

CPU Seconds

1600 1400 1200 1000 800 600 400 200

X_1 X_2 X_3

Disk File Scheme

An analysis of the data in Table III indicates that factor G may be a variance reduction factor with the low level being the best setting. Analysis of table II and figure I indicate that G, H, and possibly F have an important impact on the mean response. Figure II provides us with a graphical representation of the relationship between disk file scheme and factor G. Of particular interest is the relatively constant low CPU time when G is at the low level, regardless of the disk file scheme being used.

In summary, setting factor G at the low level will allow us to minimize CPU time.

Submitted by: **MAXCOR**

Gary Fenton

John Hopkinson

William Woodard

U.S. AIR FORCE ACADEMY

C1C James B. Hecker

C1C Phillip A. Herre

C1C Ray L. Plumley

C2C Raymond X. Sagui

"Reducing Variability in a Machining Process: An Industrial Application of Experimental Design"

Background:

An industrial bushing has a close tolerance requirement on the outer diameter of 14 mm $\pm .01$ mm (.5512" $\pm .00039$"). The machining process was producing approximately .6% of the product out of the tolerance limits. Some of the variability of the outer diameter was due to an "out of round" condition. A joint effort between MAXCOR and the Air Force Academy was initiated to reduce the amount of runout (as measured by Total Indicator Reading-TIR) by utilizing a designed experiment approach to optimizing the process settings.

Experiment Objective

The objective of the experiment was to see an improvement in the roundness of the bushings. Roundness is defined as the condition on a surface of revolution, such as a cylinder, where all points on the surface intersected by any plane perpendicular to a common axis are equidistant from the center.

Quality Characteristics

The two ends of the bushing may be distinguished by the presence of a dimple near one end (see Figure 1). The manufacturing specifications required that the entire surface (composed of an infinite number of diameters) meet the $14 \pm .01$ mm condition. To meet this condition, both the taper and the out of roundness must be minimized. Thus, both of these were used as criteria measures (response variables).

Figure 1: Outer Bushing Top View and Cross Section

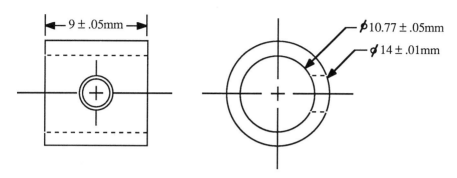

The inspection equipment used a two point contact so that as the part was rotated, both maximum and minimum diameters could be recorded. Both ends of the bushing were inspected, generating four pieces of data for each part. These four data pieces were then used to mathematically calculate the roundness (maximum minus minimum for each end) and the taper (maximum of one end minus maximum of the other end).

Table 1: Quality Characteristics

Response	Type	Anticipated Range
TIR, dimple	quantitative	0 - 0.0110 mm
TIR, non-dimple	quantitative	0 - 0.0110 mm
Taper	quantitative	-0.0075 - 0 mm

Factors

Perhaps the most difficult step of experimental design is deciding which factors should be included in the design. Many times there are so many potential factors that affect the results that the design becomes infeasible because of the size, or the time to run the entire design, or most common, the cost per experiment.

In this case study a brainstorming session with the engineers brought out the possible factors that go into the manufacturing of a single bushing. This initial list consisted of almost twenty different factors shown in Figure 2.

Figure 2: Fishbone Diagram of Possible Factors

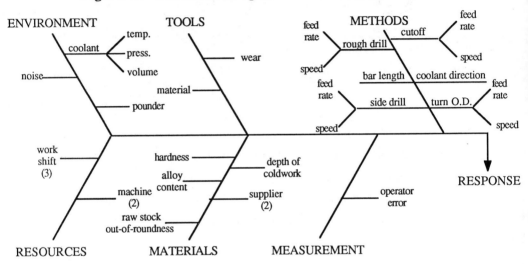

Some factors, such as alloy content, stock roundness, and bar feeding length, were beyond control so they were not included in the experiment. Others, like machine, worker shift, and environment were controlled by randomization or holding them constant. After two weeks of weighing the priority and appropriateness of each factor, ten two-level factors were chosen for the design.

Table 2: Experimental Factors

Factor	Levels	Type
Hardness (HD)	Lo - Hi	Qualitative
Rgh Drill Feed Rate (RDFR)	.0015 - .0025	Quantitative
Rgh Drill Speed (RDS)	2000 - 3000 rpm	Quantitative
Sd Hole Drill Feed Rate (SHFR)	0.7 - 1.7 ipm	Quantitative
Sd Hole Drill Speed (SHDS)	1800 - 2800 rpm	Quantitative
Turn O.D. Feed Rate (TFR)	.002 - .003	Quantitative
Turn O.D. Speed (TODS)	3000 - 5000 rpm	Quantitative
Cutoff Feed Rate (CFR)	.0002 - .0005	Quantitative
Cutoff Speed (CS)	2500 - 4000 rpm	Quantitative
Coolant Flow (CF)	Top - Bottom	Qualitative

Choosing the Factors for the Experiment

The hardness of the raw stock was considered the most important factor to the MAXCOR engineers. Because of all the drilling and cutting required to make a bushing, the drill bit can severely deform the metal if the metal hardness is low. The feed rate and speed factors were considered important because they cause movement of the raw stock during drilling and cutting. The place of coolant flow was also considered important because of affect on lubrication and cooling.

Run Limitations

The parts manufactured during the experiment were eventually to be sent to a customer, so it was important to select factor level settings that would not intentionally produce non-conforming products. The appropriate sample size for each run was selected by calculations based upon the historical standard deviation and the amount of shift that was to be detected, along with the alpha and beta risks that were reasonable. (See Chapter 2).

Appropriate Experimental Designs

Due to MAXCOR's familiarity with the Taguchi design, an L_{16} Taguchi sequence was used. Except for the factor of material hardness, which could not be easily changed due to equipment set-ups, the runs were randomized by making use of a random number table.

Experimental Implementation and Data Collection

MAXCOR's technicians were instucted to set-up the machinery in accordance with the randomized Taguchi L_{16} design. After each run, the quality characteristics were measured and recorded for each bushing. Throughout this entire process, management was present to make certain there was no deviation from the experimental design.

Analysis and Results

The analytical method used was a regression analysis on the three-response mean for each replicated run to determine main effects. For each of the three regressions, there was no significance at the 95% confidence level. Due to the small differences in factor settings, the resulting regression weights were not statistically significant. A set of "vital few" factors were suspected to exist despite the regression results. Therefore, it was necessary to rank order the regression weights to determine which factors appeared to be more significant than others. The coefficients of the orthogonally coded factors were used to set up the prediction equations; these coefficients are listed in Table 3.

A regression on the variance of the three responses for each replicated run was used to determine dispersion effects. Again, the regressions were not statistically significant. Likewise, these regression weights were used to determine the most likely significant factors. The coefficients of the orthogonally coded factors were used to set up the prediction equation. (See Table 3).

Using the coefficients listed in table 3, each factor was set to minimize the mean and variance of the response variables. This involved setting factors with a positive coefficient at the low level, and factors with a negative coefficient at the high level. This procedure was accomplished for each of the response variables. In some instances, the factor coefficients for the mean and variance responses were opposite in sign. Since MAXCOR desired to emphasize decreasing the variance in the responses, it was decided to give precedence to the coefficients of the variance response. The best factor settings for optimizing the mean and variance can be seen in Table 4.

Table 3:   Coefficients

Factor	Y1	Y2	Y3	S_1^2	S_2^2	S_3^2
HD					0.00505	
RDFR			0.024		-0.00462	
RDS						
SHFR	0.042	0.046	0.030	0.00641		
SHDS		0.031	0.037		-0.00385	
TFR				0.00459	0.00466	
TODS			-0.036		0.00444	
CFR	-0.038			-0.00453	-0.00571	
CS				0.00498		
CF				-0.00526	0.00539	-0.00478

Table 4:   Optimal Settings

Factor	Setting
HD	low
RDFR	high
RDS	high or low
SHFR	low
SHDS	high
TFR	low
TODS	low
CFR	high
CS	low
CF	high

Confirmation Runs

MAXCOR ran 200 confirmation runs at the recommended levels. To obtain an overall picture of the bushing, the TIR responses for both the dimpled and non-dimpled ends were combined. Although there was no significant change in the mean TIR response, the variation was decreased by 50% (see Appendices 1 and 2). Although the confirmation runs produced parts with reasonable taper, there is no way to discern any improvement in this response due to the lack of historical data.

MAXCOR MFG. - HISTORICAL DATA
TIR ON OUTER BUSHINGS

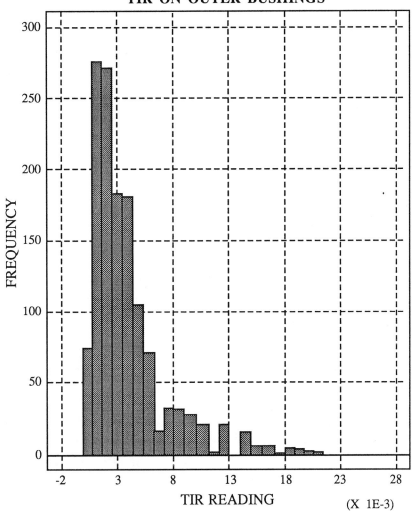

TIR Reading	
Sample Size	1345
Average	3.97435 E-3
Standard Deviation	3.50571 E-3

MAXCOR MFG. - CONFIRMATION DATA
TIR BOTH ENDS COMBINED

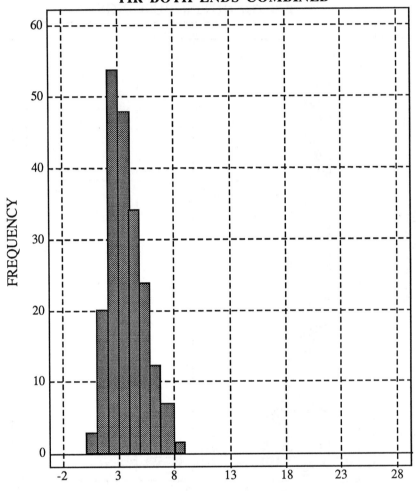

TIR READING

	TIR Reading
Sample Size	200
Average	37.995 E-4
Standard Deviation	16.081 E-4

THE EFFECT OF THE CE RIVET H PARAMETER ON HEAD PROTRUSION
by
Tom Gardner
James D. Riggs

INTRODUCTION

PROBLEM STATEMENT: Determine the effect of the CE rivet h parameter (see Figure 1) on head protrusion in skin panel rivet installations.

OBJECTIVE: Use a designed experiment and knowledge gained from previous experimentation to quantify the effect of the h parameter on rivet protrusion. This information will be applied to determine the required countersink dimension for each value of h.

BACKGROUND: The variation in head diameter of CE rivets has been found to be a major contributor to the variation in head protrusion in CE rivet skin installations. A request to rivet vendors to better control head diameter was rejected in favor of controlling the h parameter. Apparently, the head diameter is controlled by the h parameter. Examination of Figure 1 shows the geometric reasoning for this statement. A correlation analysis between h and head diameter showed marginal correlation, thus challenging the geometric foundation. This experiment is designed to establish the role of h on head protrusion.

RESULTS

The h parameter does effect rivet protrusion but the magnitude of the effect (0.0011") is much smaller than that of the combination of countersink and head diameter (0.0093"). See Figures 2 and 3. Figure 2 shows the expected increase in protrusion with an increase in h. Minimal correlation exists between h and the head diameter. The countersink required to control protrusion to 0.0020" can be determined from h only marginally. Head diameter is the factor from which the countersink should be determined.

This experiment shows that the countersink ranges over which the protrusion can be controlled to 0.0020" are different from the existing ranges shown below. This may be due to a change in rivet vendor - now Precision Form and formerly Allfast.

	CE 5	CE 6
Existing range	0.240" - 0.245"	0.290" - 0.295"
Experiment range	0.245" - 0.252"	0.292" - 0.294"

Figure 3 shows two effects. First, as the countersink increases, protrusion decreases as expected. Secondly, as the head diameter increases, the protrusion decreases when the countersink is small. When the countersink is larger, the protrusion increases as the head diameter increases. This effect is due to the mechanics compensating for small countersink diameter with increased pressure on the rivet gun.

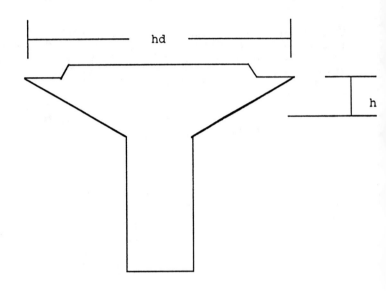

Figure 1.   The CE Rivet H Parameter

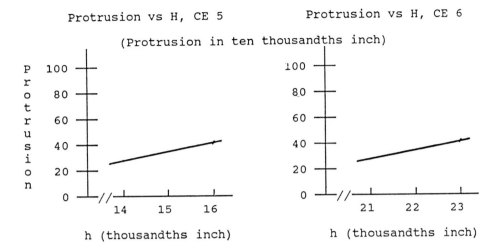

Figure 2. Effect of H on Protrusion

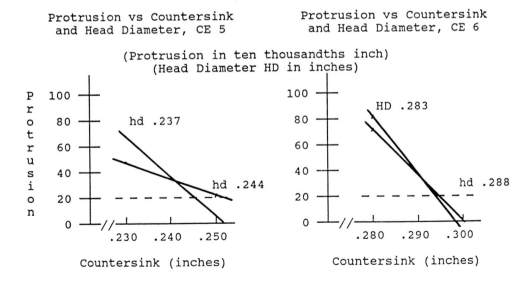

Figure 3. Effect of Countersink and Head Diameter on Protrusion

RECOMMENDATIONS

This experiment used rivets from a variety of batches which gives no clue as to the within batch variation. Both head diameter and h need to be measured from within several batches. The within variation can be determined as well as the correlation between head diameter and h. Also, the number of rivets that must be measured to find accurate averages for these factors can be found. If the correlation between head diameter and h is good enough, h can be used to characterize the batch and to determine the target countersink diameter. Otherwise, the head diameter must be used. In addition, if h can be correlated, recommendations to vendors for target h values and variances can be made.

The graphs in Figure 3 gives countersink ranges which will produce protrusions for the different head diameters for both the CE 5 and CE 6 rivets. Batches (boxes) on the floor need to have average head diameters and variation measured to allow the graph to give the target countersink diameter for each box. The goal is to find two or three countersink targets for each rivet type (CE 5 or 6). This will permit utilization of two or three countersink cutters per rivet type.

EXPERIMENT DESIGN

Previous experimentation in similar skin riveting installations shows the following factors, with levels, to be important.

Factor	Levels	Symbol
panel	0	
complexity	0	
shimming	0	
rivet head diameter	2	hd
h	3	csk
countersink diameter	3	h
rivet head protrusion	random	PROT

All samples are taken from a single panel in Acc 315 thus eliminating the panel-to-panel effect.

The complexity factor was eliminated as all the combinations of h and csk were available in the open (easy) areas of installation on the selected panel.

The shimming effect was eliminated by working only in areas where no shims are present.

The rivet head diameter, i.e., CE 5 and CE 6 is a fixed factor over which the experiment is blocked. The actual measurements are expected to be 0.228" to 0.248" and 0.279" to 0.299", respectively. The deviation from nominal for both the CE 5 (0.238") and CE 6 (0.288") rivets is used in the analysis.

The h parameter has the following levels within hd. The deviation from nominal (level 2 for each type) is used.

	Level	Actual
CE 5	1	0.0140"
	2	0.0148"
	3	0.0162"
CE 6	1	0.0209"
	2	0.0216"
	3	0.0230"

The countersink diameter is considered fixed within hd as follows. The deviation from nominal is used in the analysis and the nominal are the same as those of the head diameter.

	Level	Actual
CE 5	1	0.228"
	2	0.238"
	3	0.248"
CE 6	1	0.279"
	2	0.288"
	3	0.298"

The rivet head protrusion is the response variable. It has been shown that controlling this factor clearly reduces defects associated with skin panel installation. These include both rivet installation defects (such as exposed countersinks, gaps, shanked rivets, cracked butts) and skin defects (including scratches, tool marks, pillowing, canning).

MODEL

The informal model to be tested is

PROTRUSION = mean + hd + h + csk + hd*h + hd*csk + h*csk + h*h
            + csk*csk + hd*h*csk + hd*h*h + hd*csk*csk
            + RESIDUALS

The anticipated anova table is:

AA TABLE

Source	df	FC	RC
hd	1	$18n\phi(hd)$	σ_R^2
h	2	$18n\phi(h)$	σ_R^2
csk	2	$18n\phi(hd)$	σ_R^2
hd*h	2	$9n\phi(hd*h)$	σ_R^2
hd*csk	2	$9n\phi(hd*csk)$	σ_R^2
h*csk	4	$3n\phi(h*csk)$	σ_R^2
hd*h*csk	4	$n\phi(hd*h*csk)$	σ_R^2
REP	$18(n-1)$	0	σ_R^2
Total	$18n - 1$		

SAMPLE SIZE

To compute the number of samples needed per cell (h and csk in full factorial combinations) we first assume the df to be infinity, then recompute the sample size based on the previous estimate. The difference (in ten thousandths) we need to detect is 5, and σ_E is equal to 3 (determined from past experience). This gives us:

$$t_{0.025}(\infty) * \sigma_E * (2/n)^{1/2} = 5$$

$$\Rightarrow \quad 1.96 * 3 * (2/n)^{1/2} = 5$$

$$\Rightarrow \quad n = 2 * ((1.96 * 3)/5)^2$$

$$n = 2.77 \Rightarrow 3$$

If n = 3, we will take a total of 42 samples. Let df for t = 41.

$$\Rightarrow \quad n = 2 * ((2.02 * 3)/5)^2$$

$$n = 2.94 \Rightarrow 3$$

The sample size needed is 3 per cell, yielding a total of 42 samples.

RANDOMIZATION

Randomization is by runs within a block. Random numbers are generated and assigned to the factorial runs within a block. See Figure 4 for a diagram of the normal order of installation-first CE 6's, CE 5's, CE 6's, and then CE 5's.

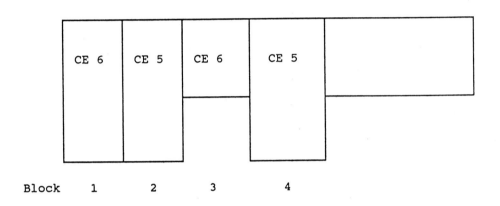

Block   1   2   3   4

Four blocks containing either CE 5 or CE 6 rivets are laid out on
the 65-45800-139 skin panel in ACC 315. The order of
installation is from left to right in the diagram.

Figure 4. Test Panel Layout

Within each block, the randomized run combinations are as follows:

	Block 1			Block 2	
Run	h	c	Run	h	c
1	1	2	1	2	1
2	2	1	2	3	2
3	2	3	3	3	1
4	3	3	4	2	3
5	2	2	5	1	2
6	3	2	6	2	2
7	3	1	7	3	3
8	1	3	8	1	1
9	1	1	9	1	3

	Block 3			Block 4	
Run	h	c	Run	h	c
1	2	2	1	2	3
2	1	1	2	1	2
3	2	3	3	2	1
4	1	2	4	1	1
5	3	1	5	3	3
6	2	1	6	2	2
7	1	3	7	3	1
8	3	2	8	1	3
9	3	3	9	2	1

DATA ANALYSIS

The experimental data collected, listed in Appendix A, are severely reduced from the plan above. None of the blocks are balanced. Block 1 is the only block with no duplicate runs.

The data for csk, h, and hd are the differences of the specification nominal from the actual readings. The length (len) of the rivet is taken from the part number rather than from actual measurements as previous experimentation shows high correlation between the measured length and the length obtained from the part number.

The mechanics suggested two factors that could (but did not) show a sequence effect. Tape is applied to the surface of the rivet gun which interfaces with the rivet head. As rivets are installed, this tape wears thus diminishing a cushioning effect. The other factor is the firmness the structure into which the rivets are installed. As the installation proceeds from left to right (see Figure 4), the structure becomes progressively less firm.

Multiple regression was used to analyze the data. The factors which contribute to the variability of the rivet head protrusion are, countersink, head diameter, and the h parameter, as shown in Figures 2 and 3 above. Figure 5 shows a plct of the residuals versus the model output. No clear patterns are revealed. The first row of the data was removed from the analysis after viewing standardized residuals showed this row to be an outlier.

Least clear differences are calculated to make block to block comparisons due to the unbalanced data set. LSD_1 is used for comparing blocks 1 and 2, LSD_2 for blocks 1 and 3, LSD_3 for 1 and 4, LSD_4 for 2 and 3, LSD_5 for 2 and 4, and LSD_6 for comparing 3 and 4.

$$t_{0.025}(27) = 2.052$$

$$s = \sqrt{\frac{4 * 16.82^2 + 2 * 20.60^2 + 9 * 37.12^2 + 10 * 20.33^2}{4 + 2 + 9 + 10}}$$

$$= 27.21$$

LSD1 = 2.052 * 27.21 * sqrt(1/5 + 1/3) = 40.78

LSD2 = 2.052 * 27.21 * sqrt(1/5 + 1/9) = 31.14

LSD3 = 2.052 * 27.21 * sqrt(1/5 + 1/10) = 30.58

LSD4 = 2.052 * 27.21 * sqrt(1/3 + 1/9) = 37.22

LSD5 = 2.052 * 27.21 * sqrt(1/3 + 1/10) = 36.76

LSD6 = 2.052 * 27.21 * sqrt(1/9 + 1/10) = 28.83

The Bonferroni Limits are calculated such that BSD1 compares block 1 with block 2, etc, as is done with LSD above. The calculations use $t_{0.025/6}(27) = 2.848$.

BSD = 2.848 * 27.21 * sqrt(1/5 + 1/3) = 56.59

BSD = 2.848 * 27.21 * sqrt(1/5 + 1/9) = 43.22

BSD = 2.848 * 27.21 * sqrt(1/5 + 1/10) = 42.44

BSD = 2.848 * 27.21 * sqrt(1/3 + 1/9) = 51.66

BSD = 2.848 * 27.21 * sqrt(1/3 + 1/10) = 51.01

BSD = 2.848 * 27.21 * sqrt(1/9 + 1/10) = 35.61

The following table shows the results of both of the comparisons by block. Note that if zero is contained in an interval then no clear difference exists between the represented blocks.

	block 2	block 3	block 4
block 1	-2.73	-21.96	-14.90
LSD	(-43.51, 38.05)	(-53.10, 9.18)	(-45.48, 15.68)
BSD	(-59.32, 53.86)	(-65.18, 21.26)	(-57.34, 27.54)
block 2		-19.23	-12.17
LSD		(-56.45, 17.99)	(-48.92, 24.59)
BSD		(-70.89, 32.43)	(-63.18, 38.84)
block 3			7.06
LSD			(-21.77, 35.89)
BSD			(-28.55, 42.67)

It's clear that no differences exist between any of the blocks.

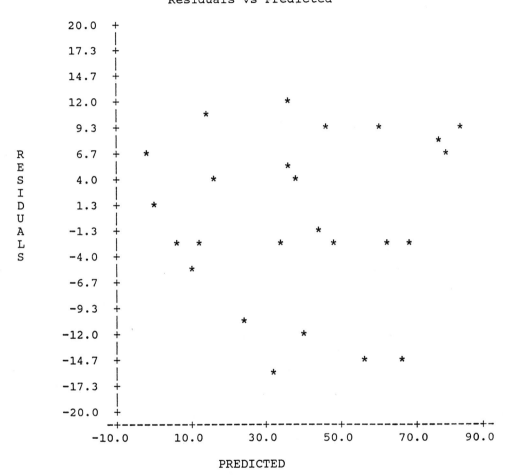

H EXPERIMENT - 26 JAN 89

Residuals vs Predicted

Figure 5. Residuals Plot

7-185

APPENDIX A.   EXPERIMENT DATA

	Prot	csk	h	Len	hd	ce	Run
	35.0	-8.0	0.0	5.0	-1.5	6.0	2
	2.0	11.0	0.0	5.0	-2.5	6.0	3
Block 1	18.0	10.0	1.5	5.0	0.0	6.0	4
	14.0	2.0	0.2	5.0	-4.5	6.0	5
	44.0	0.0	1.4	6.0	-1.0	6.0	6
	47.0	0.0	-0.7	5.0	1.5	5.0	2
Block 2	23.0	10.0	-0.4	5.0	1.5	5.0	7
	6.0	10.0	-0.6	5.0	0.5	5.0	7
	39.0	0.0	-0.5	5.0	-0.5	6.0	1
	82.0	-10.0	-0.7	5.0	-3.0	6.0	2
	1.0	9.0	-0.5	5.0	-1.0	6.0	3
	0.0	10.0	0.1	5.0	-3.0	6.0	3
Block 3	90.0	-10.0	1.0	5.0	-1.5	6.0	5
	63.0	-7.0	0.1	5.0	-2.0	6.0	6
	83.0	-8.0	0.8	7.0	-4.0	6.0	6
	41.0	0.0	0.8	7.0	-4.5	6.0	8
	2.0	11.0	1.0	7.0	-4.5	6.0	9
	25.0	0.0	-0.2	5.0	2.0	5.0	2
	40.0	0.0	-0.7	5.0	1.0	5.0	2
	66.0	-8.0	0.4	5.0	2.5	5.0	3
	40.0	-7.0	-0.8	5.0	-0.5	5.0	4
Block 4	28.0	10.0	1.4	6.0	5.5	5.0	5
	11.0	12.0	1.0	6.0	4.0	5.0	5
	54.0	0.0	1.4	6.0	5.5	5.0	7
	49.0	-11.0	1.3	6.0	4.0	5.0	7
	4.0	11.0	-0.7	5.0	1.5	5.0	8
	58.0	-10.0	-0.1	5.0	2.0	5.0	9

APPENDIX B. RESIDUALS

Row	Observed	Predicted	Residual
2	2.00000	-4.70511	6.70511
3	18.00000	14.00258	3.99742
4	14.00000	30.57593	-16.57593
5	44.00000	46.29594	-2.29594
6	47.00000	34.83098	12.16902
7	23.00000	12.56487	10.43513
8	6.00000	9.14704	-3.14704
9	39.00000	35.31270	3.68730
10	82.00000	74.39116	7.60884
11	1.00000	3.51947	-2.51947
12	.00000	-1.45349	1.45349
13	90.00000	80.53669	9.46331
14	63.00000	65.22723	-2.22723
15	83.00000	76.97873	6.02127
16	41.00000	42.82755	-1.82755
17	2.00000	-4.32410	6.32410
18	25.00000	37.23959	-12.23959
19	40.00000	34.83098	5.16902
20	66.00000	57.13543	8.86457
21	40.00000	54.38330	-14.38330
22	28.00000	31.05339	-3.05339
23	11.00000	21.92986	-10.92986
24	54.00000	44.94712	9.05288
25	49.00000	63.79825	-14.79825
26	4.00000	8.74858	-4.74858
27	58.00000	60.20539	-2.20539

DURBIN WATSON SERIAL CORRELATION = 2.29

CATAPULTING STATISTICS INTO ENGINEERING CURRICULA

S. E. Jones, PhD, University of Alabama

Stephen R. Schmidt, PhD, USAF Academy

Bruce Johnson, MS, HQ USAF/LE-RD

Today's rapidly changing technologies often result in products and processes that are not easily modelled using engineering theory alone. Other tools, such as statistics, can be used to supplement engineering knowlege and thereby speed up product and process development times. The use of a mechanical device such as the catapult (see Figure 1) discussed in this paper has been shown to be a very effective tool for demonstrating the power of blending engineering theory with statistically designed experiments.

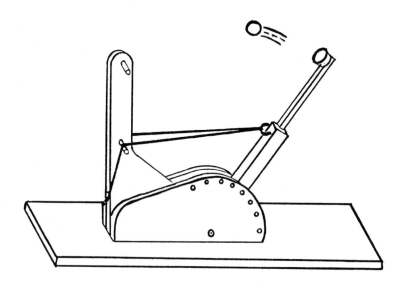

Figure 1. Catapult

Analysis of the Catapult Using only Engineering Theory

The catapult is a device for launching a small projectile (ball) toward a predetermined, downrange impact site R (see Figure 2). The force F (see Figure 3), which rotates the extensible moment arm of the catapult, is provided by a stretched rubber band. The rubber band may be attached to several points on the superstructure of the device and may assume several stretch lengths for a given configuration. The force exerted by the rubber band on the moment arm of the catapult can be a complicated function of the stretch length. However, we assume that the stretched length is a function of the current position of the moment arm, $F = F(\theta)$.

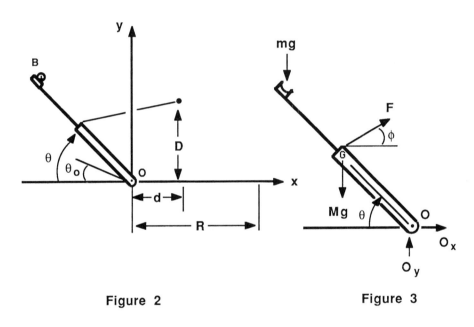

Figure 2   Figure 3

Taking moments about 0, and assuming that there are no applied moments at 0, we find

$$I_0 \ddot{\theta} = r_F F(\theta) \sin\theta \cos\phi - (Mg\, r_G + mg\, r_B)\sin\theta \tag{1}$$

where I_0 is the mass moment of inertia of the compound rod and ball assembly, r_F is the distance from 0 to the rubber band connection, r_G is the distance from 0 to the center of mass of the rod, r_B is the distance from 0 to the center of mass of the ball, M is the mass of the rod, m is the mass of the ball, and g is the local gravitational constant.

Since F = F(θ) and

$$\tan\phi = \frac{D - r_F \sin\theta}{d + r_F \cos\theta}, \qquad (2)$$

equation (1) can be integrated once to give

$$\frac{1}{2} I_0 \dot\theta^2 = r_F \int_{\theta_0}^{\theta} F(\theta)\sin\theta\cos\phi\,d\theta - (Mg\,r_G + mg\,r_B)(\sin\theta - \sin\theta_0) \qquad (3)$$

where we have assumed that the device is released from rest at $\theta = \theta_0$. This equation expresses the current angular rate $\dot\theta$ in terms of the current angular position θ (see Figure 2). When a stop position at $\theta = \theta_1$ is reached (see Figure 3), the angular speed is $\dot\theta_1$ and the ball will be launched with speed

$$v_B = r_B \dot\theta_1 \qquad (4)$$

toward the downrange target at R. We assume that the launch angle is normal to the rod position at $\theta = \theta_1$. From (4), $\dot\theta_1$ can be found, and the expression for it takes the form

$$\frac{1}{2} I_0 \dot\theta_1^2 = r_F \int_{\theta_0}^{\theta_1} F(\theta)\sin\theta\cos\phi\,d\theta - (Mg\,r_G + mg\,r_B)(\sin\theta_1 - \sin\theta_0). \qquad (5)$$

Combining this equation with (4), we can find the initial velocity for the ball when the prescribed stop angle θ_1 is reached.

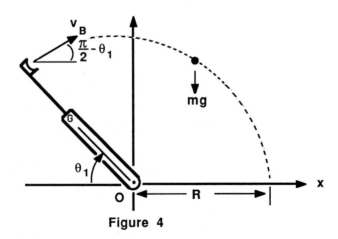

Figure 4

The small ball is launched with initial speed v_B in a direction that is normal to the catapult, as shown in Figure 4. The stop in the mechanism at $\theta = \theta_1$ provides for the initial launch angle, which is $\pi/2 - \theta_1$ radians. The initial launch position has coordinates $(-r_B\cos\theta_1, r_B\sin\theta_1)$ in the rectangular, cartesian coordinate system with origin at 0 (Figure 4).

Neglecting the influence of atmospheric drag, the equations of motion for the ball are

$$\ddot{x} = 0 \quad , \quad \ddot{y} = -g \tag{6}$$

where g is the local gravitational constant. These equations can be easily integrated, and the solution is

$$x = v_B \cos\left(\frac{\pi}{2} - \theta_1\right)t - \frac{1}{2}r_B\cos\theta_1 \tag{7}$$

and

$$y = r_B \sin\theta_1 + v_B \sin\left(\frac{\pi}{2} - \theta_1\right)t - \frac{1}{2}gt^2. \tag{8}$$

At impact, $x = R$ and $y = 0$ at the terminal time $t = \bar{t}$. Inserting these conditions into (7) and (8) and eliminating the terminal time between the resulting equations gives one relation involving v_B and θ_1.

$$r_B\sin\theta_1 + (R + r_B\cos\theta_1)\tan\left(\frac{\pi}{2} - \theta_1\right) - \frac{g}{2\,v_B^2}\frac{(R + r_B\cos\theta_1)^2}{\cos^2\left(\frac{\pi}{2} - \theta_1\right)} = 0. \tag{9}$$

This equation can be combined with (4) and (5) to form a nonlinear system for the unknowns θ_1, $\dot{\theta}_1$, and v_B when the initial launch angle θ_0 is given. We may also view this same system as one in which the unknowns are θ_0, $\dot{\theta}_1$, and v_B when the stop angle θ_1 is prescribed. When the constitutive properties of the rubber band are supplied, this system can be used to determine the launch conditions for a given, downrange site R.

The algebraic structure of the system of equations (4), (5) and (9) is such that they may be easily combined into one equation:

$$\frac{gl_0}{4r_B} \frac{(R + r_B \cos\theta_1)^2}{\cos^2\left(\frac{\pi}{2} - \theta_1\right)\left[r_B \sin\theta_1 + (R + r_B \cos\theta_1) \tan\left(\frac{\pi}{2} - \theta_1\right)\right]}$$

$$= r_F \int_{\theta_0}^{\theta_1} F(\theta) \sin\theta \cos\phi \, d\theta - (Mgr_G + mg\,r_B)(\sin\theta_1 - \sin\theta_0)$$

(10)

This equation is a complicated transcendental involving an integral. The integrand contains a force function $F = F(\theta)$ which is nonlinear for large stretches. The integrand also contains the cosine of the angle ϕ which is itself a complicated function of θ (see equation (2)). There is no reasonable prospect for evaluation of this integral in terms of elementary functions. However, it can be dealt with numerically, and this will be the approach in subsequent work.

One comment must be made at this juncture. For a given value of R and a given force law $F = F(\theta)$, there may be no solution to (10) for any values of θ_0 and θ_1. On the other hand, when the force in the rubber band is sufficiently high, there will be solutions for a wide range of initial and stop angles θ_0 and θ_1.

There are several points of concern in the catapult analysis just presented. First, the force law for the rubber band has not been specified. Each rubber band that is used with the device must be tested and the force as a function of stretch must be determined. Rubber is generally a nonlinear elastic material. So, the force-extension law may be nonlinear.

We will assume that the stretch in the rubber band is uniform. However, for some configurations, the rubber band passes over pegs which can offer enough friction to make the stretch nonuniform. When these pegs are lubricated, this effect can be minimized.

7-192

A second concern is that the moment of inertia I_0 and the center of mass of the extensible moment arm r_G have not been specified. However, both of these quantities are easy to determine experimentally as functions of the rod extension length r_B. The moment of inertia can be computed for any given rod length by treating the compound moment arm as a pendulum and measuring the period of oscillation for various arm lengths. This will compute the moment of inertia, I_0, very conveniently and accurately. Similarly, the position of the mass center as a function of rod length can be determined from an elementary balance experiment.

Finally, the amount of negligible friction which is present in any device of this type is always a point of concern. The analysis does not account for any applied moments due to friction. We assume that the pin surface at 0 can be lubricated to eliminate any frictional moments.

However, an even greater source of difficulty is presented by the interaction between the ball (projectile) and the cup surface. If the ball is pressed into the cup, then friction is induced on the ball surface which may change the direction of the initial velocity vector. The ball will tend to roll out of the cup, as opposed to being launched cleanly at an angle of 90° to the moment arm. We can remedy this situation by either lubricating the cup or by taking care prior to launch not to press the ball into the cup.

In conclusion, a deterministic, engineering analysis for the catapult has been presented. The result of this analysis is a single equation (10) which relates the initial and terminal angles, θ_0 and θ_1, to the applied force $F(\theta)$, the terminal distance R, and the various distances and weights in the system. In order to implement the analysis to select appropriate launch conditions, we must specify all of the input parameters from the problem and write a computer program to extract the information from equation (10). The input parameters consist of all of the distances r_B, r_F, and r_G, the moment of inertia I_0, the rubber band force characteristics, and the range R. Working with students, this project would take the better part of a semester to complete.

Blending Engineering with Statistics

Since it is going to be difficult at best to model the catapult using only an engineering approach, it is advisable to seek an alternative approximation technique to speed up the modeling process. Assume that we know the true functional relationship of distance with each of the variables contained in the catapult. This function could be approximated with a Taylor series model through the use of calculus. For the catapult, as well as today's complex products and processes, the problem is that this functional relationship is typically unknown or difficult to determine. A Taylor series model, however, can still be constructed using empirical data. For the catapult problem, a two level experimental design matrix can be used to generate the appropriate empirical data required to build a first order Taylor series approximation which can include desired 2–way linear interactions. Less than four hours of instruction are normally required to explain a simple approach to accomplish this type of modeling. Air Force Academy faculty and cadets attending an experimental design class recently used this approach for catapult modeling. They used engineering knowledge to brainstorm the four most likely variables that affect distance. These four variables were then included in an eight shot design matrix of highs and lows for each variable. The distance for these eight shots and the design matrix highs and lows were used to construct a first–order Taylor series model. Confirmation tests of the model indicated that ±2–inch accuracy can be achieved over the 200 inch operational range of the device. After initial training, the entire experimental approach was accomplished in two hours.

The obvious conclusions are: (1) The Taylor series modeling approach is fast and accurate; (2) In today's competitive market environment, engineers can potentially increase their ability to reduce product and process development times through the use of statistically designed experiments blended with existing engineering theory; and (3) The use of training devices such as the catapult not only provides hands–on training, but also makes the learning process enjoyable.

OPERATION OF AN ACTIVATED SLUDGE SYSTEM

Authors: Majors James Brickell and Kenneth Knox
         Department of Civil Engineering
         United States Air Force Academy

Purpose: Demonstrate how experimental design can be used in
         environmental engineering. Specifically, show how operation of an
         activated sludge reactor can be better understood and optimized.

THE ACTIVATED SLUDGE SYSTEM

     The activated sludge process is a biological treatment system used
at many wastewater treatment plants. Briefly, biological organisms
(termed sludge) within the reactor are used to aerobically convert
incoming waste (influent) into additional biomass or innocuous carbon
dioxide and water. The activated sludge reactor is followed immediately
by a settling tank, called a secondary clarifier, where the liquid and
solids (sludge) are separated. The sludge from the clarifier is then
either wasted, or recycled back to the reactor to maintain an acceptable
biomass population. Figure 1 shows a schematic of a typical completely-
mixed activated sludge system.

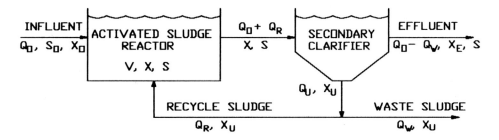

Figure 1. Schematic of activated sludge system

The parameters shown in Figure 1 are defined as follows:

Q_o, Q_u, Q_w, Q_r = flow rate in influent, clarifier underflow,
                  waste, and recycle flows, respectively (m^3/d)

S_o, S = soluble food concentration (measured as BOD_5) in
            the influent and reactor, respectively (mg/L)

X_o, X, X_e, X_u = biomass concentration in influent, reactor, effluent, and clarifier underflow, respectively (mg/L)

V = reactor volume (m^3)

Performance of an activated sludge system is measured using the biochemical oxygen demand, BOD. The BOD is the amount of oxygen required to stabilize the decomposable matter in a water using aerobic biochemical action. In general, a lower BOD is indicative of a higher quality water. A typical municipal wastewater would have a BOD_5 concentration (BOD measured after 5 days at 20°C) of 110 - 400 mg/L. The EPA effluent standards for wastewater, on an average monthly basis, are 30 mg/L BOD_5 (Peavy et al., 1985).

A mass balance analysis of the system described in Figure 1, using the Monod equation to describe the rate of bacterial growth, yields the following analytical relationship:

$$(S_o - S) = \frac{(1 + k_d VX/Q_w X_u)}{(Q_o Y/Q_w X_u)} \qquad \text{Eqn 1}$$

where:

$S_o - S$ = amount of BOD removed by the system

k_d = biological growth rate constant (d^{-1})

Y = fraction of food (S) converted to biomass (X), or

$$Y = \frac{\text{kg X produced per day}}{\text{kg S consumed per day}}$$

EXPERIMENTAL DESIGN

To demonstrate application of experimental design, a simulated activated sludge system was conceived. Key parameters of the system were then varied between expected ranges, and the results noted.

The parameters which affect the treatment efficiency for removing BOD ($S_o - S$) from wastewater can be gleaned from the analytical equation describing the process (equation 1). Metcalf and Eddy, 1979, provide typical ranges for these parameters for an average municipal activated sludge system:

X, reactor biomass concentration	3000 - 6000 mg/L
X_u, clarifier biomass concentration	8000 - 12000 mg/L
k_d, biological growth rate constant	0.040 - 0.075 d^{-1}
Y, fraction of food to biomass	0.40 - 0.80 kg/kg

The system that will be used for this analysis originated from an example problem (10-1) presented by Metcalf and Eddy, 1979, where the reactor volume, V, is 4700 m^3, and the influent flow rate, Q_o, is 21,600 m^3/d. Based on typical wasting rates, the waste sludge flow rate for this system was calculated to be 78.5 - 940.0 m^3/d.

Initially, a series of experiments was designed using the five variables (X, X_u, k_d, Y, and Q_w). A two-level design was used, with the low (-) level at the lower limit of the typical range, and the high (+) level at the higher limit for each variable. A full-factorial design would have required 32 runs (2^5), which may be excessive for an actual plant to perform. Therefore, a quarter-factorial design was used, requiring 8 runs (2^3). Aliasing and the results of each run (based on equation 1) are shown in the design matrix below:

Quarter-Factorial Design Matrix

RUN	X	X_u	Q_w	k_d= XX_u	XQ_w	X_uQ_w	Y= XX_uQ_w	Response (mg/L)
1	+	+	+	+	+	+	+	775
2	+	+	-	+	-	-	-	354
3	+	-	+	-	+	-	-	1001
4	+	-	-	-	-	+	+	102
5	-	+	+	-	-	+	-	1371
6	-	+	-	-	+	-	+	87
7	-	-	+	+	-	-	+	496
8	-	-	-	+	+	+	-	195
Avg(+)	558	647	911	455	515	611	365	$\bar{y}$ = 548
Avg(-)	537	448	184	640	581	485	730	
Δ	21	198	726	-185	- 66	126	-365	
Δ/2	10	99	363	- 93	- 33	63	-183	

Prediction equation (First-order Taylor series approximation):

$$S_o\text{-}S = 548 + 10\bar{X} + 99\bar{X}_u + 363\bar{Q}_w - 93\bar{k}_d - 33\overline{XQ}_w + 63\bar{X}_u\bar{Q}_w - 183\bar{Y}$$

The bars over the variables indicate these are linearly "coded" variables, with the value of -1 at the lower limit, +1 at the higher limit, and 0 at the midrange value. To evaluate the usefulness of the derived Taylor approximation of equation 1, a series of 25 runs was made with the five variables allowed to randomly vary between their typical upper and lower limits. The predicted response was then compared with the "correct" response from equation 1, the results of which are shown in Figure 2. The least squares regression line of the results has a respectable coefficient of determination (R^2) of 0.905.

An inspection of the prediction equation shows that the most important variables for controlling removal of BOD from wastewater seem to be Q_w and Y. A Pareto Diagram, Figure 3, more clearly demonstrates the contribution to BOD removal of each variable. In this figure, X_u and k_d also seem to contribute significantly to the observed responses,

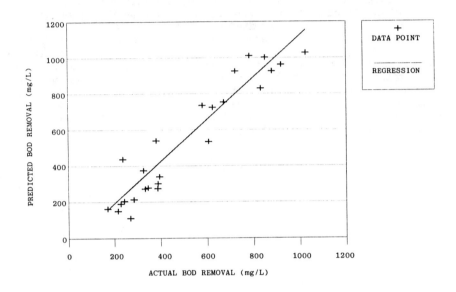

Figure 2. Comparison of predicted versus actual BOD removal rates from the quarter-factorial design model.

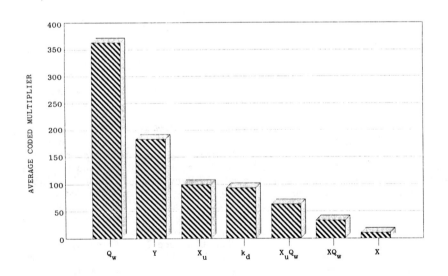

Figure 3. Pareto Diagram for quarter-factorial design.

while X and the tested interactions (XQ_w and X_uQ_w) appear less consequential.

Untested interactions and aliasing could materially impact the prediction. Aliasing cannot be evaluated without further testing. The untested interactions, however, were analyzed using two-factor interaction graphs (Figures 4a-e) between the most significant main effects variables (Q_w, Y, X_u, and k_d). From Figure 4 the following observations can be made:

* Since the interaction graphs are not parallel, significant interaction exists between Q_w and Y (Figure 4a), Q_w and k_d (Figure 4b), and between Y and k_d (Figure 4d).

* Little to no interaction exists between Y and X_u (Figure 4c), and between X_u and k_d (Figure 4e).

Even though it may be impractical using a "real" activated sludge system, a two-level full factorial model was tested on the computer using all five variables (32 runs) in order to more rigorously evaluate the designed experiments approach. Figure 5 shows that the prediction equation does a good job of modeling the actual BOD removal equation, with an R^2 over 0.99. The corresponding Pareto Diagram, Figure 6, shows that with minor exceptions the activated sludge system accurately identified the key variables using the designed experiments approach with only eight runs.

OPERATIONAL EVALUATION OF RESULTS

Based on the results of the quarter-factorial design, to maximize BOD removal from wastewater one should maximize Q_w and X_u, while minimizing k_d and Y. X and the tested interactions are of lesser importance. The interactions between Q_w and Y, Q_w and k_d, and between Y and k_d appear to be significant, but further tests would be required to quantify these relationships.

Operationally, the only variables over which an operator has direct control (by turning a valve, for instance) are X, X_u, and Q_w. k_d and Y are biological rate constants which can only be controlled by relatively extreme measures such as changing the reactor water temperature or changing the type of microorganism used to stabilize the waste products. Therefore a reasonable operational recommendation based on the results of this study would be to maximize the flow rate of waste sludge (Q_w) and the clarifier biomass concentration (X_u) in order to maximize removal of BOD from wastewater. If the biological growth rate constant, k_d, proved to be unstable for some reason, Figure 4b suggests that adjusting Q_w to approximately 275 m^3/d (the intersection of the two lines, accounting for the "coded" variable) would make the activated sludge system robust to k_d. The power of the designed experiments approach is demonstrated here by noting that these conclusions are certainly not obvious from a simple inspection of equation 1.

An actual activated sludge process is extremely complex, however, and other considerations would come into play before undertaking changes to the system. One consideration is cost. For example, increasing Q_w will naturally increase the costs of disposing the waste sludge.

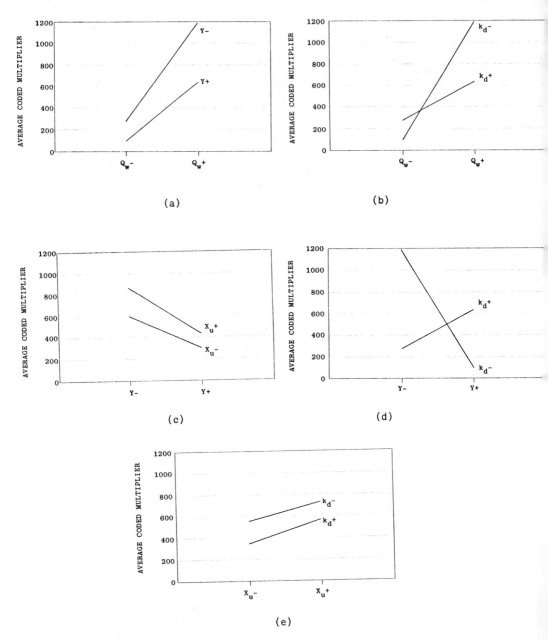

Figure 4. Two-Factor Interaction Graphs.

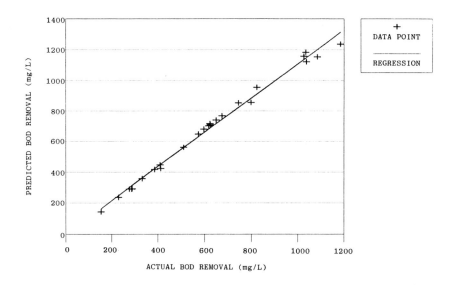

Figure 5.   Comparison of predicted versus actual BOD removal rates from
the full-factorial design model.

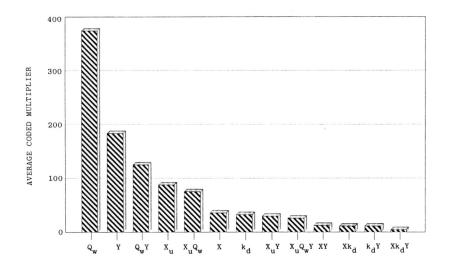

Figure 6.   Pareto Diagram for full-factorial design.

Operational problems may also arise, such as the sludge in the clarifier may form nitrogen gas and float if the residence time in the settler becomes excessive while attempting to increase X_u.

This case study has attempted to demonstrate to power of the designed experiments approach. Using some simple statistical techniques coupled with engineering judgement, complex systems can be economically modeled, and their operation greatly simplified. As a result, adjustments and modifications to the system can be made, and the consequences of these changes predicted with some degree of confidence.

REFERENCES

Metcalf & Eddy, Inc., Wastewater Engineering: Treatment, Disposal, Reuse, 2d ed., McGraw-Hill, New York, 1979.

Peavy, Howard S., Donald R. Rowe, and George Tchobanoglous, Environmental Engineering, McGraw-Hill, New York, 1985.

MINIMIZING THE NUMBER OF LARGE-SCALE COMPUTER SIMULATION RUNS: THE TESTING AND VALIDATION OF THE STATE DEPARTEMENT'S INTERAGENCY WORKING GROUP ACQUIRED IMMUNE DEFICIENCY SYNDROME (AIDS) MODEL

by

Lieutenant Colonel Thomas F. Curry
Captain John J. Tomick
Captain Barbara A. Yost
Department of Mathematical Sciences
United States Air Force Academy

Background

In December 1988, researchers from the Department of Mathematical Sciences were asked to consult on the testing and validation of a simulation model designed to forecast, for most any nation, the effects of the AIDS epidemic. The request was from the State Department Interagency Working Group (IWG) which is chaired by the Deputy Assistant Secretary for the Environment. The IWG is a consortium of several agencies including the Subcommittee for AIDS Models and Methods, the US Agency for International Development, Health and Human Services, the National Institute of Health, the Center for Disease Control, the Department of Energy, and the Department of Defense. The AIDS model was developed during 1989, and this paper contains the experimental design methodology and the results of the initial testing and validation.

The AIDS model is a set of approximately 100 deterministic differential equations developed and programmed by scientists at Los Alamos National Laboratory (LANL) and the Merriam Laboratory at the University of Illinois at Urbana. The model uses risk-based methodology similar to models described in [3]. Data for the model is provided by the US Bureau of Census, Center for International Research (CIR).

The Problem: Uncertain Input Data and Lots of It

In spite of the excellent data collection procedures used by CIR, the AIDS data from Africa and the rest of the world are uncertain. Thus, the question arises as to the value of forecasts that are based on input parameters estimated from such data. The answer depends on the question being asked with the model; the uncertainty in the result is determined by the answer's sensitivity to uncertainties in the input data. It may also depend on uncertainties in the model assumptions and formulation. We discuss the techniques used to analyze the sensitivity of the model to the input data, and how this sensitivity analysis can be used to help inform decisions based on model results. The question of uncertainty due to incorrect model formulation is left to future studies.

The IWG AIDS model requires a large amount of input data. The input data file is over 600 lines long and contains demographic and epidemiological information including population distributions, fertility rates, migration rates, sexual risk behavior categories, sexual contact rates, cocirculating sexually transmitted diseases, condom use, percent of the population that is seropositive (has the HIV virus), risk of infection per sexual contact, rate of progression from infection to AIDS, and blood transfusion and blood screening. Most of these data are enumerated by sex, age, and marital status resulting in over 150 separate input variables that could be analyzed for their effect on the various response surfaces. Using "engineering knowledge," the analysis team selected 97 input variables that were of interest.

Each run of the AIDS model requires approximately 15 minutes on a 386 based microcomputer. If "one-at-a-time" input data perturbation techniques were used, the required 2^{97} runs would require 10^{24} years. Clearly, an efficient design was needed to evaluate the model. In conferences at LANL, we determined that a Plackett-Burman design was well suited for the initial first-order approximation of the gradient of the "local response surfaces."

Experimental Design

To study the relative sensitivity of the results to each of the 97 parameters in the input file, we investigated the response surfaces by varying the input data by small amounts around a baseline set of values for a "generic" country. Using the techniques described in [1,2,4,and 5], we developed a 100 by 100 Hadamard matrix to set up 98 runs. All parameters were varied simultaneously, and the effects of each were ranked (since the parameters each take on their high and low values for half the runs, there is enough information to do t-tests to statistically compare the means of the outputs from the high value and low value runs).

Results

In our study of a recent intermediate version of the model, most inputs were varied about their baseline values by approximately 10% of their range of uncertainty. Using 8 year forecasts, the total population was most sensitive to the initial female population, but all infected populations were most sensitive to the infectivity of men, which was varied only from 0.001 to 0.005 per contact (while its value has been estimated at 0.0001 to 0.05). Infected populations were also very sensitive to the infectivity of women. Thus the model indicates that differences in health and other factors that change transmission probabilities by only small amounts can have a large impact on the spread of the epidemic. Obtaining better estimates of these parameters are key to narrowing confidence bounds on the model results.

In all, six epidemic response surfaces were examined. The five most important input variables driving each response are listed in the following table:

Frequently appearing variables key:
Risk/cntct M = Risk of infection per sexual contact, for males
Risk/cntct F = Risk of infection per sexual contact, for females
Cntct/mth HR F 15-44 = # of sexual contacts per month by high
              risk females age 15-44
Cntct/mth HR M 15-44 = # of sexual contacts per month by high
              risk males age 15-44
STD's female = relative increase of risk of infection due to co-
              circulating sexually transmitted diseases, females

Response variable key:
1 = Total number of people infected
2 = Cumulative AIDS deaths
3 = Number of infected females age 15-44
4 = Percent of population infected
5 = Percent of females age 15-44 infected
6 = Percent of deaths due to AIDS

Frequently appearing variables	Response Variable Ranking						
	1	2	3	4	5	6	Avg
Risk/cntct M	1	1	1	1	1	1	1.0
Risk/cntct F	2	2	2	2	2	2	2.0
Cntct/mth HR F 15-44	3	3	3	3	3	3	3.0
STD's female	6	4	5	6	4	5	5.0
Cntct/mth HR M 15-44	4	7	7	4	8	6	6.0

The Value of the Experimental Design Results

The above table lists only the top five variables controlling the response surfaces. When these results were briefed to the IWG, several health officials stated that this was the first time they had seen any information on what variables were important in driving the AIDS epidemic. They noted that the complete ranking of all 97 input parameters can be used to determine the "best" way to intervene in order to minimize the number of deaths due to AIDS. These results can also be used to guide data collection. Indeed, it is evident that risk of infection per sexual contact must be determined by region before approximate confidence bounds on the forecasts can be calculated.

The experimental design runs also helped us determine model deficiencies. In other testing, by examining the pattern of the parameter settings in failed runs, we determined that when the number of men was increased by 10%, and the number of women was decreased by 10%, the model failed due to there being insufficient numbers of women for the men to marry. This resulted in the programming of a variable marriage rate that can handle depletion of certain age groups in the population. This also suggests that important changes in societal structure and behavior may occur due to segments of a population dying from AIDS.

Lastly, the results of the main effects study have been used to screen the input parameters for analysis of two-way interactions. Specifically, using a "two-to-the-seven-minus-one" experimental design, we examined seven input variables for pairwise interactions. From the marginal mean plots, we determined that (at 96% confidence) there was a statistically significant interaction between cumulative AIDS deaths and the risk of infection per sexual contact for males. This supplies additional support for the need to accurately determine the risk by region.

In summary, we found the use of experimental design techniques to be very valuable. It reduced the time required for model testing from years to weeks, provided us with valuable rankings of main effects, and gave direction to future data collection, and future model analyses.

References

[1] Hadamard, J., "Resolution d'une Question Relative aux Determinants," Bulletin of Scientific Mathematics, Vol. 17, Part 1 (1893).

[2] Hedayat, A., and Wallis, W.D., "Hadamard Matrices and Their Applications," The Annals of Statistics, Vol. 6, No. 6, pp 1184-1238 (1978).

[3] Hyman, J.M., and E.A. Stanley, "Using Mathematical Models to Understand the AIDS Epidemic," Mathematical Biosciences, Vol 90, pp415-473 (1988).

[4] Plackett, R.L., and Burman, J.P., "The Design of Optimum Multifactorial Experiments," Biometrika, Vol. 33, pp 305-325 (1946).

[5] Schmidt, S. R.,and Launsby, R.G., Understanding Industrial Designed Experiments, CQG Ltd Printing, Longmont, CO (1989).

Karl Fischer Moisture Designed Experiment
Rita Whiteley and Robert Lawson
Cell Technology, Inc., Boulder, Colorado

Introduction: The Karl Fischer method of water determination, titrates water with iodine in the presence of sulfur dioxide, methanol, and a suitable base. See the following chemical reaction:
$H_2O + I_2 + SO_2 + CH_3OH + 3RN ------> [RNH]SO_4CH_3 + 2[RNH]I$ where RN = Base.

Use of the Mettler DL18 titrator allows the titration to be followed exactly through the use of a two-pin platinum electrode which has a current source applied to its poles. The voltage measured at the polarized electrode pins is used by the controls as an input signal. When the last traces of water have been titrated, voltage drops to virtually zero, the electrodes are then depolarized by the iodine now present.

The Karl Fischer titration permits determination of available water only. This means water must be brought into solution by suitable means before the titration can be carried out. Sample preparation for freeze-dried ImuVert is very critical. Since a portion of the moisture found in lyophilized ImuVert is tightly bound to the crystalline structure and difficult to make available for titration, an external extraction step must be performed prior to titration.

Objective: Determine how external extraction variables, such as solvents and shake time, effect the measured percent moisture titrated from lyophilized ImuVert.

Factors and Levels: Factors and levels chosen are as follows:

Factors	Levels
Methanol	0, 0.5 ml
Chloroform	0, 0.5 ml
Formamide	0.5, 1.0 ml
Shake Time	30 sec, 3630 seconds

Response: The response looked at was percent moisture with hopes to maximize the measured amount of moisture determined by the Karl Fischer titration.

Experimental Design:

Defining Contrast: I = ABCD
2**K-P Fractional Factorial
Yates Order

Run#	TC	Methanol	Chloroform	Formamide	Shake Time	Response	
						Y_1	Y_2
4	(1)	0	0	0.5 ml	30 seconds	1.2402	1.1005
2	a	0.5 ml	0	0.5 ml	3630 sec.	0.8562	0.9602
3	b	0	0.5 ml	0.5 ml	3630 sec.	0.9883	0.9907
1	ab	0.5 ml	0.5 ml	0.5 ml	30 sec.	1.0212	1.0845
7	c	0	0	1.0 ml	3630 sec.	0.9993	1.0068
8	ac	0.5 ml	0	1.0 ml	30 sec.	0.9625	1.0568
5	bc	0	0.5 ml	1.0 ml	30 sec.	0.8733	1.0253
6	abc	0.5 ml	0.5 ml	1.0 ml	3630 sec.	1.0151	0.9992

The experiment was performed in a randomized order.

Procedure: Prior to sample analysis, the titrator must be calibrated to ensure accurate mechanical operation. Calibration is completed by checking titrant concentration and instrument drift. Both parameters were within statistically acceptable ranges.

A blank vial determination (to determine moisture in the solvents as well as residual moisture attributed to an empty vial) is the next step. This is accomplished by using an empty vial which was processed exactly like lyophilized product; then, testing the blank vial with the same factor levels as will be done for the sample. For example, if run #2 required 0.5 ml formamide, 0.5 ml methanol, no chloroform and a 3630 second shake time--the blank for that run would be done in the same manner.

Sample analysis for the designed experiment is the next stage. For each experimental run, two freeze-dried samples were tested with three to four repeat samplings per vial. Run # was an exception--one sampling/vial was done due to solvent volume. Solvents were added in a specific order as follows: formamide, methanol, and lastly chloroform. Swirling of the sample occurred after the formamide addition to expediate resolution of the freeze-dried product. The experimental design runs that received only the 30 second vigorous vortexing were tested immediately; while, other design runs after vortexing were allowed to shake continuously for one hour on a platform shaker before being tested.

In order for the titrator to determine the percentage moisture, several parameter values need to be entered by the operator. Sample volume to be tested was held constant at 0.4 ml. The blank vial's moisture content was entered based on the blank determination for that given run. Sample weight which is based on the product's dry cake weight, solvent volume, and sample size varied due to the solvent volume used for the external extraction. The following sample weights were used: 2.0 ml solvent--0.0073 g, 1.5 ml solvent--0.0097 g, 1.0 ml solvent--0.0146 g, and 0.5 ml solvent--0.0292 g.

Data/Analysis: Following is an ANOVA table showing the results obtained form the designed experiment.

Index #	Total Observation	Sum of Square	I = ABCD Half Effect	Measures Average	DF	Mean Square	F Ratio
1	2.1057	16.3622	1.011				
2	1.8164	0.0025	0.013	A^1=BCD	1	0.0025	0.608
3	1.9790	0.0163	0.032	B^2=ACD	1	0.0163	3.904
4	2.3407	0.0188	0.034	AB=CD	1	0.0188	4.513
5	1.8986	0.0058	-0.019	C^3=ABD	1	0.0058	1.382
6	2.0143	0.0002	0.004	AC=BD	1	0.0002	0.048
7	2.0061	0.0051	-0.018	BC=AD	1	0.0051	1.219
8	2.0193	0.0355	-0.047	ABC=D^4	1	0.0355	8.518
	Error	0.0333			8	0.0042	
	Total	0.1174			15		

[1] Methanol
[2] Chloroform
[3] Formamide
[4] Shake Time

Using a $F_{T0.05}$ = 5.31, the only significant factor is D (shake time) or the three way interaction ABC (methanol, chloroform, and formamide). However, at the 90 % confidence level, using a $F_{T0.10}$ = 3.46, the B factor (chloroform) and the AB or CD interaction become significant.

See the following graphical display of the single factor effects generated by using low and high points for each factor.

Since the largest single effects (shake time and chloroform) are confounded with the ABC and ACD interactions, another experiment was needed to determine which was causing the significant effect.

Second designed experiment contained the following factors and levels:

Factors	Levels
Chloroform	0.5, 1.0 ml
Shake Time	30, 3630 seconds

Percent moisture was the response examined. Constants in the experiment were: methanol at 0.5 ml and formamide at 0.5 ml.

Experimental design:

2**K Factorial
Yates Order

Run #	TC	Chloroform	Shake Time	Response	
				Y_1	Y_2
2	(1)	0.5 ml	30 seconds	1.2264	1.2141
3	a	1.0 ml	30 sec.	1.1330	1.1467
1	b	0.5 ml	3630 sec.	1.0419	1.1271
4	ab	1.0 ml	3630 sec.	1.1718	1.1533

The experiment was run in a random order.

The data can be summarized in the following ANOVA table and graphs.

I = AB

Index #	Total Observation	Sum of Square	Half Effect	Measures Average	DF	Mean Square	F Ratio
1	2.1690	10.6129	1.152				
2	2.4405	0.0126	0.040	A (CHCL$_3$)	1	0.0126	12.651
3	2.2797	0.0000	-0.001	B (Shake)	1	0.0000	0.003
4	2.3251	0.0064	-0.028	AB	1	0.0064	6.440
	Error	0.0040			4	0.0010	
	Total	0.0229			7		

At the 95 % confidence level only chloroform is an important effect; whereas, at the 90 % confidence level the AB interaction becomes significant. In either case shake time alone has no significance, which means that the important effect noted in experiment one was a three factor interaction.

Conclusions: The most significant effect on the external extraction process preformed on ImuVert is the three solvent interaction. Based on the first experiments single effect plots, 0.5 ml of formamide, 0.5 ml of methanol, and 1.0 ml of chloroform are used for the external extraction.

Since shaking the product versus just vortexing the product was shown to have no effect in the designed experiment, for future testing the 30 second vortex and 60 minute shake will be used. This method did produce smaller coefficients of variation within sample testing per vial.

Attempts were made to remove the methanol to simplify the external extraction since it did not contribute significantly to the external extraction. However, when methanol was removed two phases formed and measurements were harder to make. Therefore, all three external extraction solvent components were kept in the system.

Further optimization of this external extraction process could be done, including trying even less formamide. Also, the two factor interaction AB or CD which is significant at the 90 % level may be important and a deconfounding experiment should be completed.

Figure 1
Single Effect Plots

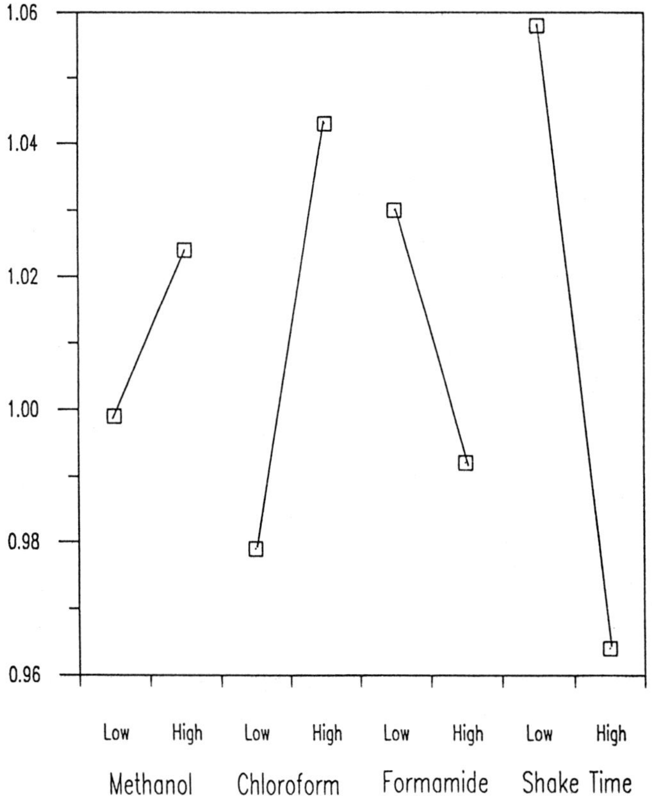

Figure 2
AB or CD Interaction Plot

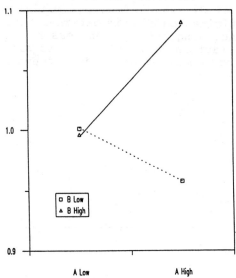

Figure 3
Second Experiment
Single Effect Plots

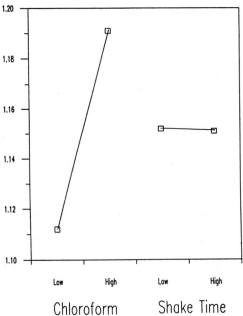

Figure 4
Second Experiment
AB Interaction

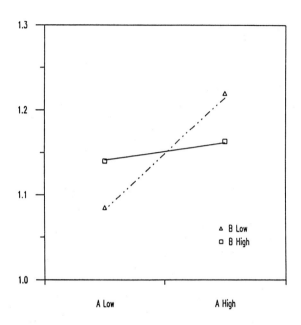

CHAPTER 8: STATISTICAL TABLES

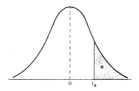

Critical Values of the t Distribution

ν	\multicolumn{5}{c}{α}
	0.10	0.05	0.025	0.01	0.005
1	3.078	6.314	12.706	31.821	63.657
2	1.886	2.920	4.303	6.965	9.925
3	1.638	2.353	3.182	4.541	5.841
4	1.533	2.132	2.776	3.747	4.604
5	1.476	2.015	2.571	3.365	4.032
6	1.440	1.943	2.447	3.143	3.707
7	1.415	1.895	2.365	2.998	3.499
8	1.397	1.860	2.306	2.896	3.355
9	1.383	1.833	2.262	2.821	3.250
10	1.372	1.812	2.228	2.764	3.169
11	1.363	1.796	2.201	2.718	3.106
12	1.356	1.782	2.179	2.681	3.055
13	1.350	1.771	2.160	2.650	3.012
14	1.345	1.761	2.145	2.624	2.977
15	1.341	1.753	2.131	2.602	2.947
16	1.337	1.746	2.120	2.583	2.921
17	1.333	1.740	2.110	2.567	2.898
18	1.330	1.734	2.101	2.552	2.878
19	1.328	1.729	2.093	2.539	2.861
20	1.325	1.725	2.086	2.528	2.845
21	1.323	1.721	2.080	2.518	2.831
22	1.321	1.717	2.074	2.508	2.819
23	1.319	1.714	2.069	2.500	2.807
24	1.318	1.711	2.064	2.492	2.797
25	1.316	1.708	2.060	2.485	2.787
26	1.315	1.706	2.056	2.479	2.779
27	1.314	1.703	2.052	2.473	2.771
28	1.313	1.701	2.048	2.467	2.763
29	1.311	1.699	2.045	2.462	2.756
inf.	1.282	1.645	1.960	2.326	2.576

Copied with permission from *Probability and Statistics for Engineers and Scientists*, Walpole, R.E. Myers, R.H., MacMillian, 1985

Critical Values of the F Distribution

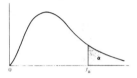

Critical Values of the F Distribution

$$f_{0.05}(\nu_1, \nu_2)$$

| ν_2 | \multicolumn{9}{c}{ν_1} | | | | | | | | |
	1	2	3	4	5	6	7	8	9
1	161.4	199.5	215.7	224.6	230.2	234.0	236.8	238.9	240.5
2	18.51	19.00	19.16	19.25	19.30	19.33	19.35	19.37	19.38
3	10.13	9.55	9.28	9.12	9.01	8.94	8.89	8.85	8.81
4	7.71	6.94	6.59	6.39	6.26	6.16	6.09	6.04	6.00
5	6.61	5.79	5.41	5.19	5.05	4.95	4.88	4.82	4.77
6	5.99	5.14	4.76	4.53	4.39	4.28	4.21	4.15	4.10
7	5.59	4.74	4.35	4.12	3.97	3.87	3.79	3.73	3.68
8	5.32	4.46	4.07	3.84	3.69	3.58	3.50	3.44	3.39
9	5.12	4.26	3.86	3.63	3.48	3.37	3.29	3.23	3.18
10	4.96	4.10	3.71	3.48	3.33	3.22	3.14	3.07	3.02
11	4.84	3.98	3.59	3.36	3.20	3.09	3.01	2.95	2.90
12	4.75	3.89	3.49	3.26	3.11	3.00	2.91	2.85	2.80
13	4.67	3.81	3.41	3.18	3.03	2.92	2.83	2.77	2.71
14	4.60	3.74	3.34	3.11	2.96	2.85	2.76	2.70	2.65
15	4.54	3.68	3.29	3.06	2.90	2.79	2.71	2.64	2.59
16	4.49	3.63	3.24	3.01	2.85	2.74	2.66	2.59	2.54
17	4.45	3.59	3.20	2.96	2.81	2.70	2.61	2.55	2.49
18	4.41	3.55	3.16	2.93	2.77	2.66	2.58	2.51	2.46
19	4.38	3.52	3.13	2.90	2.74	2.63	2.54	2.48	2.42
20	4.35	3.49	3.10	2.87	2.71	2.60	2.51	2.45	2.39
21	4.32	3.47	3.07	2.84	2.68	2.57	2.49	2.42	2.37
22	4.30	3.44	3.05	2.82	2.66	2.55	2.46	2.40	2.34
23	4.28	3.42	3.03	2.80	2.64	2.53	2.44	2.37	2.32
24	4.26	3.40	3.01	2.78	2.62	2.51	2.42	2.36	2.30
25	4.24	3.39	2.99	2.76	2.60	2.49	2.40	2.34	2.28
26	4.23	3.37	2.98	2.74	2.59	2.47	2.39	2.32	2.27
27	4.21	3.35	2.96	2.73	2.57	2.46	2.37	2.31	2.25
28	4.20	3.34	2.95	2.71	2.56	2.45	2.36	2.29	2.24
29	4.18	3.33	2.93	2.70	2.55	2.43	2.35	2.28	2.22
30	4.17	3.32	2.92	2.69	2.53	2.42	2.33	2.27	2.21
40	4.08	3.23	2.84	2.61	2.45	2.34	2.25	2.18	2.12
60	4.00	3.15	2.76	2.53	2.37	2.25	2.17	2.10	2.04
120	3.92	3.07	2.68	2.45	2.29	2.17	2.09	2.02	1.96
∞	3.84	3.00	2.60	2.37	2.21	2.10	2.01	1.94	1.88

Copied with permission from *Probability and Statistics for Engineers and Scientists*, Walpole, R.E. Myers, R.H., MacMillian, 1985

8-2

(continued) Critical Values of the F Distribution

$$f_{0.05}(v_1, v_2)$$

v_2	v_1									
	10	12	15	20	24	30	40	60	120	∞
1	241.9	243.9	245.9	248.0	249.1	250.1	251.1	252.2	253.3	254.3
2	19.40	19.41	19.43	19.45	19.45	19.46	19.47	19.48	19.49	19.50
3	8.79	8.74	8.70	8.66	8.64	8.62	8.59	8.57	8.55	8.53
4	5.96	5.91	5.86	5.80	5.77	5.75	5.72	5.69	5.66	5.63
5	4.74	4.68	4.62	4.56	4.53	4.50	4.46	4.43	4.40	4.36
6	4.06	4.00	3.94	3.87	3.84	3.81	3.77	3.74	3.70	3.67
7	3.64	3.57	3.51	3.44	3.41	3.38	3.34	3.30	3.27	3.23
8	3.35	3.28	3.22	3.15	3.12	3.08	3.04	3.01	2.97	2.93
9	3.14	3.07	3.01	2.94	2.90	2.86	2.83	2.79	2.75	2.71
10	2.98	2.91	2.85	2.77	2.74	2.70	2.66	2.62	2.58	2.54
11	2.85	2.79	2.72	2.65	2.61	2.57	2.53	2.49	2.45	2.40
12	2.75	2.69	2.62	2.54	2.51	2.47	2.43	2.38	2.34	2.30
13	2.67	2.60	2.53	2.46	2.42	2.38	2.34	2.30	2.25	2.21
14	2.60	2.53	2.46	2.39	2.35	2.31	2.27	2.22	2.18	2.13
15	2.54	2.48	2.40	2.33	2.29	2.25	2.20	2.16	2.11	2.07
16	2.49	2.42	2.35	2.28	2.24	2.19	2.15	2.11	2.06	2.01
17	2.45	2.38	2.31	2.23	2.19	2.15	2.10	2.06	2.01	1.96
18	2.41	2.34	2.27	2.19	2.15	2.11	2.06	2.02	1.97	1.92
19	2.38	2.31	2.23	2.16	2.11	2.07	2.03	1.98	1.93	1.88
20	2.35	2.28	2.20	2.12	2.08	2.04	1.99	1.95	1.90	1.84
21	2.32	2.25	2.18	2.10	2.05	2.01	1.96	1.92	1.87	1.81
22	2.30	2.23	2.15	2.07	2.03	1.98	1.94	1.89	1.84	1.78
23	2.27	2.20	2.13	2.05	2.01	1.96	1.91	1.86	1.81	1.76
24	2.25	2.18	2.11	2.03	1.98	1.94	1.89	1.84	1.79	1.73
25	2.24	2.16	2.09	2.01	1.96	1.92	1.87	1.82	1.77	1.71
26	2.22	2.15	2.07	1.99	1.95	1.90	1.85	1.80	1.75	1.69
27	2.20	2.13	2.06	1.97	1.93	1.88	1.84	1.79	1.73	1.67
28	2.19	2.12	2.04	1.96	1.91	1.87	1.82	1.77	1.71	1.65
29	2.18	2.10	2.03	1.94	1.90	1.85	1.81	1.75	1.70	1.64
30	2.16	2.09	2.01	1.93	1.89	1.84	1.79	1.74	1.68	1.62
40	2.08	2.00	1.92	1.84	1.79	1.74	1.69	1.64	1.58	1.51
60	1.99	1.92	1.84	1.75	1.70	1.65	1.59	1.53	1.47	1.39
120	1.91	1.83	1.75	1.66	1.61	1.55	1.50	1.43	1.35	1.25
∞	1.83	1.75	1.67	1.57	1.52	1.46	1.39	1.32	1.22	1.00

Copied with permission from *Probability and Statistics for Engineers and Scientists*, Walpole, R.E. Myers, R.H., MacMillian, 1985

8-3

(continued) Critical Values of the F Distribution

$$f_{0.01}(\nu_1, \nu_2)$$

ν_2	ν_1								
	1	2	3	4	5	6	7	8	9
1	4052	4999.5	5403	5625	5764	5859	5928	5981	6022
2	98.50	99.00	99.17	99.25	99.30	99.33	99.36	99.37	99.39
3	34.12	30.82	29.46	28.71	28.24	27.91	27.67	27.49	27.35
4	21.20	18.00	16.69	15.98	15.52	15.21	14.98	14.80	14.66
5	16.26	13.27	12.06	11.39	10.97	10.67	10.46	10.29	10.16
6	13.75	10.92	9.78	9.15	8.75	8.47	8.26	8.10	7.98
7	12.25	9.55	8.45	7.85	7.46	7.19	6.99	6.84	6.72
8	11.26	8.65	7.59	7.01	6.63	6.37	6.18	6.03	5.91
9	10.56	8.02	6.99	6.42	6.06	5.80	5.61	5.47	5.35
10	10.04	7.56	6.55	5.99	5.64	5.39	5.20	5.06	4.94
11	9.65	7.21	6.22	5.67	5.32	5.07	4.89	4.74	4.63
12	9.33	6.93	5.95	5.41	5.06	4.82	4.64	4.50	4.39
13	9.07	6.70	5.74	5.21	4.86	4.62	4.44	4.30	4.19
14	8.86	6.51	5.56	5.04	4.69	4.46	4.28	4.14	4.03
15	8.68	6.36	5.42	4.89	4.56	4.32	4.14	4.00	3.89
16	8.53	6.23	5.29	4.77	4.44	4.20	4.03	3.89	3.78
17	8.40	6.11	5.18	4.67	4.34	4.10	3.93	3.79	3.68
18	8.29	6.01	5.09	4.58	4.25	4.01	3.84	3.71	3.60
19	8.18	5.93	5.01	4.50	4.17	3.94	3.77	3.63	3.52
20	8.10	5.85	4.94	4.43	4.10	3.87	3.70	3.56	3.46
21	8.02	5.78	4.87	4.37	4.04	3.81	3.64	3.51	3.40
22	7.95	5.72	4.82	4.31	3.99	3.76	3.59	3.45	3.35
23	7.88	5.66	4.76	4.26	3.94	3.71	3.54	3.41	3.30
24	7.82	5.61	4.72	4.22	3.90	3.67	3.50	3.36	3.26
25	7.77	5.57	4.68	4.18	3.85	3.63	3.46	3.32	3.22
26	7.72	5.53	4.64	4.14	3.82	3.59	3.42	3.29	3.18
27	7.68	5.49	4.60	4.11	3.78	3.56	3.39	3.26	3.15
28	7.64	5.45	4.57	4.07	3.75	3.53	3.36	3.23	3.12
29	7.60	5.42	4.54	4.04	3.73	3.50	3.33	3.20	3.09
30	7.56	5.39	4.51	4.02	3.70	3.47	3.30	3.17	3.07
40	7.31	5.18	4.31	3.83	3.51	3.29	3.12	2.99	2.89
60	7.08	4.98	4.13	3.65	3.34	3.12	2.95	2.82	2.72
120	6.85	4.79	3.95	3.48	3.17	2.96	2.79	2.66	2.56
∞	6.63	4.61	3.78	3.32	3.02	2.80	2.64	2.51	2.41

Copied with permission from *Probability and Statistics for Engineers and Scientists*, Walpole, R.E. Myers, R.H., MacMillian, 1985

(continued) Critical Values of the F Distribution

$$f_{0.01}(\nu_1, \nu_2)$$

ν_2	\multicolumn{10}{c}{ν_1}									
	10	12	15	20	24	30	40	60	120	∞
1	6056	6106	6157	6209	6235	6261	6287	6313	6339	6366
2	99.40	99.42	99.43	99.45	99.46	99.47	99.47	99.48	99.49	99.50
3	27.23	27.05	26.87	26.69	26.60	26.50	26.41	26.32	26.22	26.13
4	14.55	14.37	14.20	14.02	13.93	13.84	13.75	13.65	13.56	13.46
5	10.05	9.89	9.72	9.55	9.47	9.38	9.29	9.20	9.11	9.02
6	7.87	7.72	7.56	7.40	7.31	7.23	7.14	7.06	6.97	6.88
7	6.62	6.47	6.31	6.16	6.07	5.99	5.91	5.82	5.74	5.65
8	5.81	5.67	5.52	5.36	5.28	5.20	5.12	5.03	4.95	4.86
9	5.26	5.11	4.96	4.81	4.73	4.65	4.57	4.48	4.40	4.31
10	4.85	4.71	4.56	4.41	4.33	4.25	4.17	4.08	4.00	3.91
11	4.54	4.40	4.25	4.10	4.02	3.94	3.86	3.78	3.69	3.60
12	4.30	4.16	4.01	3.86	3.78	3.70	3.62	3.54	3.45	3.36
13	4.10	3.96	3.82	3.66	3.59	3.51	3.43	3.34	3.25	3.17
14	3.94	3.80	3.66	3.51	3.43	3.35	3.27	3.18	3.09	3.00
15	3.80	3.67	3.52	3.37	3.29	3.21	3.13	3.05	2.96	2.87
16	3.69	3.55	3.41	3.26	3.18	3.10	3.02	2.93	2.84	2.75
17	3.59	3.46	3.31	3.16	3.08	3.00	2.92	2.83	2.75	2.65
18	3.51	3.37	3.23	3.08	3.00	2.92	2.84	2.75	2.66	2.57
19	3.43	3.30	3.15	3.00	2.92	2.84	2.76	2.67	2.58	2.49
20	3.37	3.23	3.09	2.94	2.86	2.78	2.69	2.61	2.52	2.42
21	3.31	3.17	3.03	2.88	2.80	2.72	2.64	2.55	2.46	2.36
22	3.26	3.12	2.98	2.83	2.75	2.67	2.58	2.50	2.40	2.31
23	3.21	3.07	2.93	2.78	2.70	2.62	2.54	2.45	2.35	2.26
24	3.17	3.03	2.89	2.74	2.66	2.58	2.49	2.40	2.31	2.21
25	3.13	2.99	2.85	2.70	2.62	2.54	2.45	2.36	2.27	2.17
26	3.09	2.96	2.81	2.66	2.58	2.50	2.42	2.33	2.23	2.13
27	3.06	2.93	2.78	2.63	2.55	2.47	2.38	2.29	2.20	2.10
28	3.03	2.90	2.75	2.60	2.52	2.44	2.35	2.26	2.17	2.06
29	3.00	2.87	2.73	2.57	2.49	2.41	2.33	2.23	2.14	2.03
30	2.98	2.84	2.70	2.55	2.47	2.39	2.30	2.21	2.11	2.01
40	2.80	2.66	2.52	2.37	2.29	2.20	2.11	2.02	1.92	1.80
60	2.63	2.50	2.35	2.20	2.12	2.03	1.94	1.84	1.73	1.60
120	2.47	2.34	2.19	2.03	1.95	1.86	1.76	1.66	1.53	1.38
∞	2.32	2.18	2.04	1.88	1.79	1.70	1.59	1.47	1.32	1.00

Copied with permission from *Probability and Statistics for Engineers and Scientists*, Walpole, R.E. Myers, R.H., MacMillian, 1985

Studentized Ranges q_T

$\alpha = 0.05$

ν	\multicolumn{9}{c}{p}								
	2	3	4	5	6	7	8	9	10
1	17.97	17.97	17.97	17.97	17.97	17.97	17.97	17.97	17.97
2	6.085	6.085	6.085	6.085	6.085	6.085	6.085	6.085	6.085
3	4.501	4.516	4.516	4.516	4.516	4.516	4.516	4.516	4.516
4	3.927	4.013	4.033	4.033	4.033	4.033	4.033	4.033	4.033
5	3.635	3.749	3.797	3.814	3.814	3.814	3.814	3.814	3.814
6	3.461	3.587	3.649	3.680	3.694	3.697	3.697	3.697	3.697
7	3.344	3.477	3.548	3.588	3.611	3.622	3.626	3.626	3.626
8	3.261	3.399	3.475	3.521	3.549	3.566	3.575	3.579	3.579
9	3.199	3.339	3.420	3.470	3.502	3.523	3.536	3.544	3.547
10	3.151	3.293	3.376	3.430	3.465	3.489	3.505	3.516	3.522
11	3.113	3.256	3.342	3.397	3.435	3.462	3.480	3.493	3.501
12	3.082	3.225	3.313	3.370	3.410	3.439	3.459	3.474	3.484
13	3.055	3.200	3.289	3.348	3.389	3.419	3.442	3.458	3.470
14	3.033	3.178	3.268	3.329	3.372	3.403	3.426	3.444	3.457
15	3.014	3.160	3.250	3.312	3.356	3.389	3.413	3.432	3.446
16	2.998	3.144	3.235	3.298	3.343	3.376	3.402	3.422	3.437
17	2.984	3.130	3.222	3.285	3.331	3.366	3.392	3.412	3.429
18	2.971	3.118	3.210	3.274	3.321	3.356	3.383	3.405	3.421
19	2.960	3.107	3.199	3.264	3.311	3.347	3.375	3.397	3.415
20	2.950	3.097	3.190	3.255	3.303	3.339	3.368	3.391	3.409
24	2.919	3.066	3.160	3.226	3.276	3.315	3.345	3.370	3.390
30	2.888	3.035	3.131	3.199	3.250	3.290	3.322	3.349	3.371
40	2.858	3.006	3.102	3.171	3.224	3.266	3.300	3.328	3.352
60	2.829	2.976	3.073	3.143	3.198	3.241	3.277	3.307	3.333
120	2.800	2.947	3.045	3.116	3.172	3.217	3.254	3.287	3.314
∞	2.772	2.918	3.017	3.089	3.146	3.193	3.232	3.265	3.294

Copied with permission from *Probability and Statistics for Engineers and Scientists*, Walpole, R.E. Myers,

Studentized Ranges q_T

$\alpha = 0.01$

| ν | \multicolumn{9}{c}{p} |
	2	3	4	5	6	7	8	9	10
1	90.03	90.03	90.03	90.03	90.03	90.03	90.03	90.03	90.03
2	14.04	14.04	14.04	14.04	14.04	14.04	14.04	14.04	14.04
3	8.261	8.321	8.321	8.321	8.321	8.321	8.321	8.321	8.321
4	6.512	6.677	6.740	6.756	6.756	6.756	6.756	6.756	6.756
5	5.702	5.893	5.989	6.040	6.065	6.074	6.074	6.074	6.074
6	5.243	5.439	5.549	5.614	5.655	5.680	5.694	5.701	5.703
7	4.949	5.145	5.260	5.334	5.383	5.416	5.439	5.454	5.464
8	4.746	4.939	5.057	5.135	5.189	5.227	5.256	5.276	5.291
9	4.596	4.787	4.906	4.986	5.043	5.086	5.118	5.142	5.160
10	4.482	4.671	4.790	4.871	4.931	4.975	5.010	5.037	5.058
11	4.392	4.579	4.697	4.780	4.841	4.887	4.924	4.952	4.975
12	4.320	4.504	4.622	4.706	4.767	4.815	4.852	4.883	4.907
13	4.260	4.442	4.560	4.644	4.706	4.755	4.793	4.824	4.850
14	4.210	4.391	4.508	4.591	4.654	4.704	4.743	4.775	4.802
15	4.168	4.347	4.463	4.547	4.610	4.660	4.700	4.733	4.760
16	4.131	4.309	4.425	4.509	4.572	4.622	4.663	4.696	4.724
17	4.099	4.275	4.391	4.475	4.539	4.589	4.630	4.664	4.693
18	4.071	4.246	4.362	4.445	4.509	4.560	4.601	4.635	4.664
19	4.046	4.220	4.335	4.419	4.483	4.534	4.575	4.610	4.639
20	4.024	4.197	4.312	4.395	4.459	4.510	4.552	4.587	4.617
24	3.956	4.126	4.239	4.322	4.386	4.437	4.480	4.516	4.546
30	3.889	4.056	4.168	4.250	4.314	4.366	4.409	4.445	4.477
40	3.825	3.988	4.098	4.180	4.244	4.296	4.339	4.376	4.408
60	3.762	3.922	4.031	4.111	4.174	4.226	4.270	4.307	4.340
120	3.702	3.858	3.965	4.044	4.107	4.158	4.202	4.239	4.272
∞	3.643	3.796	3.900	3.978	4.040	4.091	4.135	4.172	4.205

Copied with permission from *Probability and Statistics for Engineers and Scientists*, Walpole, R.E. Myers, R.H., MacMillian, 1985

Power of the Analysis-of-Variance Test

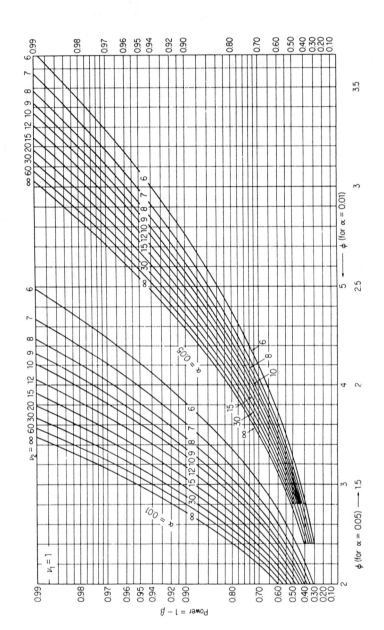

Copied with permission from *Probability and Statistics for Engineers and Scientist*, Walpole, R.E. Myers, R.H. McMillian, 1985

8 - 8

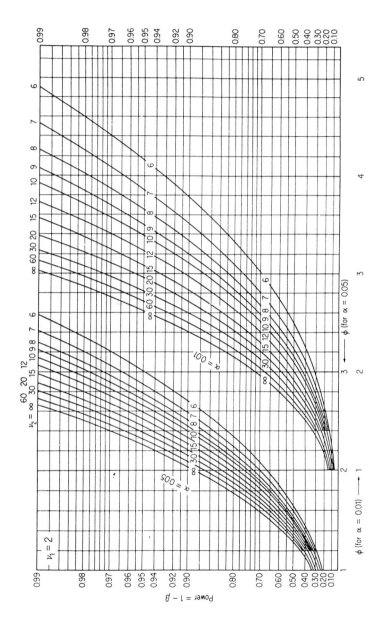

Copied with permission from *Probability and Statistics for Engineers and Scientist*, Walpole, R.E. Myers, R.H., McMillian, 1985

(continued) Power of the Analysis-of-Variance Test

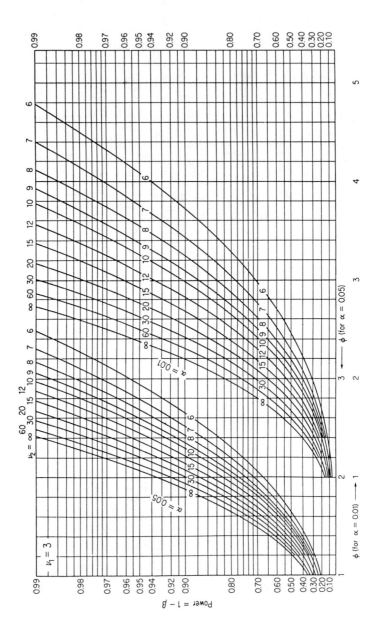

Copied with permission from *Probability and Statistics for Engineers and Scientist*, Walpole, R.E. Myers, R H McMillian, 1985

(continued) Power of the Analysis-of-Variance Test

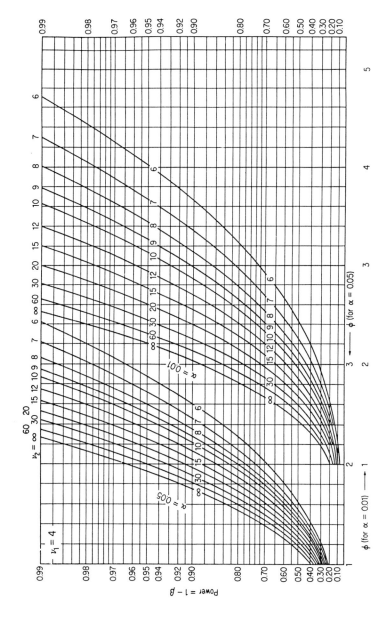

Copied with permission from *Probability and Statistics for Engineers and Scientist*, Walpole, R.E. Myers, R.H. McMillian, 1985

8 - 11

CUMULATIVE PROBABILITIES OF THE
STANDARD NORMAL DISTRIBUTION

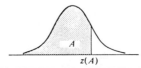

$z(A)$

z	.00	.01	.02	.03	.04	.05	.06	.07	.08	.09
.0	.5000	.5040	.5080	.5120	.5160	.5199	.5239	.5279	.5319	.5359
.1	.5398	.5438	.5478	.5517	.5557	.5596	.5636	.5675	.5714	.5753
.2	.5793	.5832	.5871	.5910	.5948	.5987	.6026	.6064	.6103	.6141
.3	.6179	.6217	.6255	.6293	.6331	.6368	.6406	.6443	.6480	.6517
.4	.6554	.6591	.6628	.6664	.6700	.6736	.6772	.6808	.6844	.6879
.5	.6915	.6950	.6985	.7019	.7054	.7088	.7123	.7157	.7190	.7224
.6	.7257	.7291	.7324	.7357	.7389	.7422	.7454	.7486	.7517	.7549
.7	.7580	.7611	.7642	.7673	.7704	.7734	.7764	.7794	.7823	.7852
.8	.7881	.7910	.7939	.7967	.7995	.8023	.8051	.8078	.8106	.8133
.9	.8159	.8186	.8212	.8238	.8264	.8289	.8315	.8340	.8365	.8389
1.0	.8413	.8438	.8461	.8485	.8508	.8531	.8554	.8577	.8599	.8621
1.1	.8643	.8665	.8686	.8708	.8729	.8749	.8770	.8790	.8810	.8830
1.2	.8849	.8869	.8888	.8907	.8925	.8944	.8962	.8980	.8997	.9015
1.3	.9032	.9049	.9066	.9082	.9099	.9115	.9131	.9147	.9162	.9177
1.4	.9192	.9207	.9222	.9236	.9251	.9265	.9279	.9292	.9306	.9319
1.5	.9332	.9345	.9357	.9370	.9382	.9394	.9406	.9418	.9429	.9441
1.6	.9452	.9463	.9474	.9484	.9495	.9505	.9515	.9525	.9535	.9545
1.7	.9554	.9564	.9573	.9582	.9591	.9599	.9608	.9616	.9625	.9633
1.8	.9641	.9649	.9656	.9664	.9671	.9678	.9686	.9693	.9699	.9706
1.9	.9713	.9719	.9726	.9732	.9738	.9744	.9750	.9756	.9761	.9767
2.0	.9772	.9778	.9783	.9788	.9793	.9798	.9803	.9808	.9812	.9817
2.1	.9821	.9826	.9830	.9834	.9838	.9842	.9846	.9850	.9854	.9857
2.2	.9861	.9864	.9868	.9871	.9875	.9878	.9881	.9884	.9887	.9890
2.3	.9893	.9896	.9898	.9901	.9904	.9906	.9909	.9911	.9913	.9916
2.4	.9918	.9920	.9922	.9925	.9927	.9929	.9931	.9932	.9934	.9936
2.5	.9938	.9940	.9941	.9943	.9945	.9946	.9948	.9949	.9951	.9952
2.6	.9953	.9955	.9956	.9957	.9959	.9960	.9961	.9962	.9963	.9964
2.7	.9965	.9966	.9967	.9968	.9969	.9970	.9971	.9972	.9973	.9974
2.8	.9974	.9975	.9976	.9977	.9977	.9978	.9979	.9979	.9980	.9981
2.9	.9981	.9982	.9982	.9983	.9984	.9984	.9985	.9985	.9986	.9986
3.0	.9987	.9987	.9987	.9988	.9988	.9989	.9989	.9989	.9990	.9990
3.1	.9990	.9991	.9991	.9991	.9992	.9992	.9992	.9992	.9993	.9993
3.2	.9993	.9993	.9994	.9994	.9994	.9994	.9994	.9995	.9995	.9995
3.3	.9995	.9995	.9995	.9996	.9996	.9996	.9996	.9996	.9996	.9997
3.4	.9997	.9997	.9997	.9997	.9997	.9997	.9997	.9997	.9997	.9998

Selected Percentiles

Cumulative probability A:	.90	.95	.975	.98	.99	.995	.999
$z(A)$:	1.282	1.645	1.960	2.054	2.326	2.576	3.090

TABLE OF RANDOM DIGITS

Line	(1)–(5)	(6)–(10)	(11)–(15)	(16)–(20)	(21)–(25)	(26)–(30)	(31)–(35)
101	13284	16834	74151	92027	24670	36665	00770
102	21224	00370	30420	03883	94648	89428	41583
103	99052	47887	81085	64933	66279	80432	65793
104	00199	50993	98603	38452	87890	94624	69721
105	60578	06483	28733	37867	07936	98710	98539
106	91240	18312	17441	01929	18163	69201	31211
107	97458	14229	12063	59611	32249	90466	33216
108	35249	38646	34475	72417	60514	69257	12489
109	38980	46600	11759	11900	46743	27860	77940
110	10750	52745	38749	87365	58959	53731	89295
111	36247	27850	73958	20673	37800	63835	71051
112	70994	66986	99744	72438	01174	42159	11392
113	99638	94702	11463	18148	81386	80431	90628
114	72055	15774	43857	99805	10419	76939	25993
115	24038	65541	85788	55835	38835	59399	13790
116	74976	14631	35908	28221	39470	91548	12854
117	35553	71628	70189	26436	63407	91178	90348
118	35676	12797	51434	82976	42010	26344	92920
119	74815	67523	72985	23183	02446	63594	98924
120	45246	88048	65173	50989	91060	89894	36036
121	76509	47069	86378	41797	11910	49672	88575
122	19689	90332	04315	21358	97248	11188	39062
123	42751	35318	97513	61537	54955	08159	00337
124	11946	22681	45045	13964	57517	59419	58045
125	96518	48688	20996	11090	48396	57177	83867
126	35726	58643	76869	84622	39098	36083	72505
127	39737	42750	48968	70536	84864	64952	38404
128	97025	66492	56177	04049	80312	48028	26408
129	62814	08075	09788	56350	76787	51591	54509
130	25578	22950	15227	83291	41737	59599	96191
131	68763	69576	88991	49662	46704	63362	56625
132	17900	00813	64361	60725	88974	61005	99709
133	71944	60227	63551	71109	05624	43836	58254
134	54684	93691	85132	64399	29182	44324	14491
135	25946	27623	11258	65204	52832	50880	22273
136	01353	39318	44961	44972	91766	90262	56073
137	99083	88191	27662	99113	57174	35571	99884
138	52021	45406	37945	75234	24327	86978	22644
139	78755	47744	43776	83098	03225	14281	83637
140	25282	69106	59180	16257	22810	43609	12224
141	11959	94202	02743	86847	79725	51811	12998
142	11644	13792	98190	01424	30078	28197	55583
143	06307	97912	68110	59812	95448	43244	31262
144	76285	75714	89585	99296	52640	46518	55486
145	55322	07598	39600	60866	63007	20007	66819
146	78017	90928	90220	92503	83375	26986	74399
147	44768	43342	20696	26331	43140	69744	82928
148	25100	19336	14605	86603	51680	97678	24261
149	83612	46623	62876	85197	07824	91392	58317
150	41347	81666	82961	60413	71020	83658	02415

Source: *Table of 105,000 Random Decimal Digits*, Interstate Commerce Commission, Bureau of Transport Economics and Statistics, 1949.

CHAPTER 9: TurboTag

9.1 Introduction

TurboTag is a personal computer (PC) based computer program designed to provide end-to-end design of experiments (DOE) support. Two level, three level, mixed, and screening examples of Taguchi-type orthongonal designs are included. Several exciting analysis tools are provided, including graphical and analytical techniques. The user interface is easily mastered through the use of a spreadsheet based-layered menu system.

TurboTag requires an IBM (tm IBM corp.) PC, XT, AT, or PS2 compatible with at least 256k bytes of RAM, at least one disk drive (5.25, 3.5, and/or a hard disk drive), and CGA, EGA, or VGA graphics. TurboTag will work on a HGA system if a CGA emulator is used. (* TurboTag will be displayed in multi-color on EGA and VGA systems only). The best way to tell if TurboTag will work on your system is to try it!

9.2 TurboTag: Step-by-Step

This section provides a walkthrough of an example L8 design, which can be run on the provided demonstration disk. Note that any text surrounded by | | should be typed exactly by the user (eg. |hello| means: type hello on the keyboard). <CR> means to press the Enter or Return key. <ESC> refers to the Esc (escape) key, </> means the slash key, and <F9> means the F9 key, etc.

9.2.1 Startup

To start TurboTag, simply place your diskette in drive A and type |A:| <CR>, then |turbotag| <CR>. Notice that there are no installation programs to run, and the entire program fits in one small file, leaving plenty of room for your other favorite DOE programs (TurboSim?), and loads of saved designs. Also, if you would like to run TurboTag from your hard disk, you may copy it there and run it the same way.

```
Enter design type (4,8,9,12,16,18,27): 8_
```

When the initial screen pops-up, just wait a couple of seconds and you will be asked if you want to load a design. Since this is your first time, type |N|. Your next choice will be the type of design. For this example, type |8| <CR> (if you are using the demonstration version, this is the only choice you are allowed to make!) Note that 8 refers to the L8 fractional factorial, or orthogonal array.

9.2.2 Initialization

Next you need to tell TurboTag how many main effects you desire. The screen should appear as at right:

For our example type |4| <CR>.

```
15
16
17

>How many main effects [3-7] : 4
```

Since an L8 has three effects automatically assigned, and you have selected four effects, you need to determine which column to alias the fourth effect with. For this example, type |7| <CR> for the ABC column. (For more information on column labels, see chapter 3) Type |2| <CR> for the number of replicates. This means you will have to enter the results of two experimental runs for each trial in the design matrix. Now you will be prompted for the low and high levels for the four main effects in your design. To simplify this example design, type <CR> for each prompt - you could set these to actual factor settings, so your design matches your real world model exactly! If you successfully completed that step, you should have pressed eight <CR>s, and are now being asked if you wish to enter a target. Type |Y| (without a <CR>!) Now, type |50| <CR> for the target value. This value will be what we are shooting for from our design, and will be displayed in the design worksheet, and on various graphs. Note that the target value may be modified later on if you change your mind.

9.2.3 Worksheet and Entering Responses

Now you are ready to complete your design and do some analysis. If you look at the screen, you will see that there is a small box in the upper left corner of the worksheet, below the 8th column, which is labelled 'Y1'. This box indicates your current position in the worksheet, and is the location where any number you input will be placed.

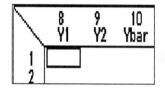

To input your responses just move the box, using the arrow keys, to the spot where you want a number to go, type in the number (you will see it being entered in the lower left corner of the screen following the '>' prompt) and when you are finished, type <CR> to place it in the box. Once you get the hang of this, you can also put the number in the box by pressing any of the arrow keys. This saves time by also moving you to the next place you need to enter a number. When entering a number or label, you can correct your last character by pressing <=> (the backspace key, NOT the left arrow key). For our example, you will need to fill in the eight rows below the 'Y1' and 'Y2' columns. You may notice that if you try to enter numbers where they aren't supposed to be, TurboTag won't let you, so you don't have to worry about putting a number in the wrong spot! Try now to enter the numbers as you see them at right. If you make a mistake on any of the numbers, you can always move the box back to that number and change it.

	8 Y1	9 Y2
1	49.86	56.76
2	51.21	51.41
3	37.23	42.45
4	45.28	44.97
5	42.96	53.67
6	49.05	48.89
7	44.16	53.29
8	53.21	52.68

9.2.4 ReCalc

It's finally time to do some analysis. When you started entering your responses, you may have noticed a star in the upper left corner of the screen. That means that some aspect of the design has been changed and the analysis values need to be updated. To update the values, press <F9> (the F9 function key). The screen will then look like:

Notice that you now have the 'Ybar' (mean or average), 'SD' (standard deviation), and 'S/N' (signal-to-noise, nominal is better) columns filled in. Also, the grand mean and target values are displayed. If you decide to change any of your response values, all you have to do is press <F9> and all the calculated values will be

	8 Y1	9 Y2	10 Ybar	11 Yprd	12 SD	13 S/N
1	49.86	56.76	53.31		4.879	20.77
2	51.21	51.41	51.31		0.141	51.19
3	37.23	42.45	39.84		3.691	20.66
4	45.28	44.97	45.12		0.219	46.27
5	42.96	53.67	48.31		7.573	16.10
6	49.05	48.89	48.97		0.113	52.73
7	44.16	53.29	48.72		6.456	17.56
8	53.21	52.68	52.94		0.375	43.00
9			Ybar 48.57	Grand		
10	YPred					

updated to reflect your changes. On the lower left portion of the screen there are some labels: 'YPred' (the predicted value from the prediction equation), 'Ybar' L & H (the marginal means), and 'LnS' L & H (the marginal natural logarithm of the standard deviation). The values for these labels (as well as the marginal S/N, and the design matrix) live further to the left on your worksheet, and can be seen by moving the box to the left with the left arrow key. Right now there isn't too much to see, but we'll take care of that in the next section.

9.2.5 Menus

So far, you have only been moving the box or entering numbers. To do any more analysis, you need to use the menu system. If you have the prompt line as shown below. (if you don't, just press <CR> until

```
Input responses (Yi), or enter " / " for menus
```

```
Worksheet   Graph      Print      Analyze     Files      Display     Quit
```

you do!), then press </> (the slash key), and the first menu will appear:

The label 'Worksheet' should have a box surrounding it. This means that it is currently selected. To select other menu options, just move the box with the left and right arrow keys. To engage a selected menu option, press <CR>. You may also engage an option by pressing the first letter of the option (eg: |A| for 'Analyze'). The various menu options are discussed in some detail in section 9.3. To return to the worksheet data entry and movement mode (that's just where you were before you pressed </>), press <Esc>. The <Esc> key will also return you to the next higher menu level (some menu options have more menus 'inside' them). Frequently, the <Esc> key will also allow you to abort an option before it really happens.

9.2.6 Analysis

The analysis tools that TurboTag provides are simple yet powerful. There are three main ways to 'look' at your experiment results:

a. Marginal Graphs - analyze the marginal results. You can observe the change in response level (Ybar mean), variation (Ln(S)), or a combination (S/N). To view these marginal graphs, move to the menu mode (if you aren't already there), and engage the 'Graph' option. you should see the options below:

Now engage the '3 Graphs' option, then press <CR>. Next engage the 'Draw' option. You should see these graphs:

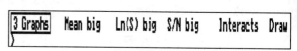

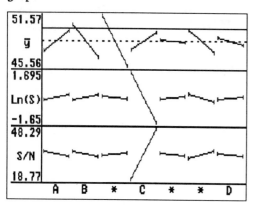

The top graph is the marginal mean graph, and it tells you which effects have the most impact on the response mean. In this case it seems to be an interaction. Note that the dashed line is the response mean, while the solid line is the target value which we input (100). The center graph is marginal Ln(S) graph, and it gives an indication of which effects ('C' this time!) have the largest impact on variation. We would normally like to set the 'knobs' so that variation is minimized, so for example factor 'A' ought to be set at its low setting. The bottom graph is a combination of the other two graphs and is popular with some analysts.

This graph would indicate that perhaps factor 'C' has the biggest overall impact on both the mean (signal) and the variation (noise). Beware of using the S/N alone, since some effects can be masked by the nature of the calculations involved. All three of these graphs may be viewed separately by selecting the 'big' graphing options.

b. Interaction Plots - You may wish to search for possible interactions between main effects, since the main effects are assumed to be independent. For example, let's take a look at the AD and BD interactions (you may examine the other interactions on your own!) First, engage the 'Interacts' option, and press <CR>. Then engage 'Draw'. You will be prompted for the factor 1 column number: type |1| <CR> (that is the 'A' factor column). For the second factor column, type |7| <CR> (that is the 'D' factor column). You Should see the graph:

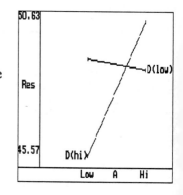

Since the lines are crossed, there is a strong interaction between the two factors. Now, engage 'Draw' again. This time for the first factor column, type |2| <CR> (for factor 'B'), and for the second factor column type |7| <CR> again (the 'D' factor column). You should see the graph :

In this case, the lines are nearly parallel and only a mild interaction is indicated.

c. ANOVA (Analysis of Variance) - ANOVA enables you to determine precisely which factors contribute the most to some attribute of the response you're interested in. TurboTag allows you to perform ANOVA for two level designs only (no problem though, since our example is a two level design!) Return to the main menu by pressing <Esc> to exit the 'Graph' menu. Next engage the 'Analyze' option. First we will run an ANOVA on the response means. Engage the 'Ybar Anova' option. The right portion of your display should look like:

F_e	$\frac{\Delta}{2}$	Ybar Pareto	User Label	Des Fact	Col #
0.03699	-0.1987		*	-AC	5
0.20249	-0.4650		D	ABC	7
0.97432	1.02000		C	C	4
1.28470	1.17125		A	A	1
1.72259	-1.3562		*	-BC	6
3.41193	-1.9087		B	B	2
8.45650	-3.0050		*	-AB	3

Use ← & → to select the significant factors for the pred. eq, then 'Enter'

In this case it looks like all of the main effects are significant, but for the sake of knowledge let's move the noise line to select only the three most significant factors. To do this press the right arrow until your screen looks like this:

Ybar Pareto	User Label
	*
	D
	C
	A
	*
	B
	*

Now press <CR>. Now press <F9> to recalc the worksheet, and press <Esc> twice to return to the worksheet move/ data entry mode. Now use your arrow keys to move the box so that it is in column 13, row 1, as shown below:

Notice that the 'Yprd' (Y predicted) column is now filled, as well as the entry below 'YPred', which is the prediction equation output for the default equation values for factors 'A','B' and 'C'.

	8 Y1	9 Y2	10 Ybar	11 Yprd	12 SD	13 S/N
1	49.86	56.76	53.31	52.65	4.879	20.77
2	51.21	51.41	51.31	51.97	0.141	51.19
3	37.23	42.45	39.84	40.11	3.691	20.66
4	45.28	44.97	45.12	44.86	0.219	46.27
5	42.96	53.67	48.31	48.98	7.573	16.10
6	49.05	48.89	48.97	48.31	0.113	52.73
7	44.16	53.29	48.72	48.46	6.456	17.56
8	53.21	52.68	52.94	53.21	0.375	43.00
9			Ybar 48.57 Grand			
10	YPred					
11	48.57					

Now you can move the box to column 2, row 11, as shown below:

	1 A	2 B	3 *	4 C	5 *	6 *
1	-1	-1		-1		
2	-1	-1		1		
3	-1	1		-1		
4	-1	1		1		
5	1	-1		-1		
6	1	-1		1		
7	1	1		-1		
8	1	1		1		
9						
10			Pred. Eqn			
11	0	0 *****		0 *****	*****	

Under the Pred. Eqn label are column entries with either stars or a 0. The 0 entries are the three factors you selected as significant, 'A','B' and 'C'. You may now go in and change these to whatever values you would like so that you get as close to the target as possible. For now, try 1.0 for 'A', -.25 for 'B', and 0.78 for 'C', and then press <F9>. Move your box to the right so that 'YPred' is displayed, as shown below:

Not bad, huh? Hopefully you can see how powerful this is!

The 'Ln(s)Anova' works pretty much the same way, except no prediction equation is generated. This ANOVA shows you which effects are most significant for reducing the variability in the response.

	3 *	4 C	5 *	6 *	7 D	8 Y1
1		-1			-1	49.86
2		1			1	51.21
3		-1			1	37.23
4		1			-1	45.28
5		-1			1	42.96
6		1			-1	49.05
7		-1			-1	44.16
8		1			1	53.21
9						
10	Pred.	Eqn				YPred
11	*****	0.78	*****	*****	*****	50.00

9.2.7 Conclusion

I hope that the tutorial was helpful - now you should be able to get right in there and do some designing of your own! One more thing you should be aware of, you can load and save your designs (this can save a lot of typing!) Just use the 'Files' option and follow the instructions.

9.3 Menu Option Details

9.3.1 Menu Options - Worksheet

The 'Worksheet' option contains the options shown below:

The first three deal with changing the format of the columns in your worksheet. You can change the width of a column, change the way numbers are displayed (ie: the number of decimal places), and change the column label to your own custom label. Each of these column options requires you to enter the column number you wish to change.

The S/N Mode option allows you to change the current S/N mode to: smaller is better, nominal is better, or larger is better.

9.3.2 Menu Options - Graph

The 'graph' option allows you to select the current graphing mode, and then to actually draw the graph, using the 'Draw' option. The first option, '3 Graphs' draws a combination of the next three options - simultaneously. The next three options draw the marginal mean ('Mean big'), marginal Ln(S) ('Ln(S) big'), and marginal S/N ('S/N big') separately. The 'Interacts' options enables you to look for possible interactions between two main effects. Note taht this option is only allowed for two level designs.

9.3.3 Menu Options - Print

The 'Print' option prints out the worksheet on your printer. Note that this option will NOT print the graphs out. To print graphs on an IBM PC compatible printer, the graphics.com program (comes with your operating system) must be executed before executing TurboTag, then when you wish to print out the current screen (including both the worksheet and the graphs), press <Shift><PrtSc> together.

9.3.4 Menu Options - Analyze

The 'Analyze' option contains the Anova and the 'Set Target' options. The 'Set Target' option enables you to change the current setting of the target value. The 'Ybar Anova' option displays the F (significance) values, delta/2, and Pareto chart for the Ybar values. Additionally, you may move a 'noise floor' line with the left and right arrow keys to determine which values to include in the prediction equation. The 'Ln(s)Anova' is similar, however no prediction equation is generated.

9.3.5 Menu Options - Files

The 'Files' option provides the capability to load and save designs. Note that if you save a design with a prediction equation, the prediction equation will have to be regenerated when the design is loaded. Also, graphs are not saved in the design file.

9.3.6 Menu Options - Display

The 'Display' option allows you to cause the worksheet to occupy the entire screen. Note that in this mode, graphs can't be displayed.

9.3.7 Menu Options - Quit

The 'Quit' option returns you to the operating system. If you select this option, you will be asked if you are sure - since you will lose any work which hasn't been saved.

9.4 TurboSim - a simulation designer and trainer.

To execute TurboSim, just type |turbosim| <CR>. After the intro screen, make the |L| selection to load a simulation file. Then type |simex| <CR> to load our example simulation. Then you can either make production runs or experimental trials. After making experimental runs, you may load your responses into TurboTag to analyze the simulated results. Good Luck! (* Note, the version of TurboSim supplied with this package will only execute previously designed simulations. To design your own simulations, the complete package is required.)

9.5 Steps for Running and Analyzing SIMEX

(The **bold** portion is what you have to enter; **<CR>** means carriage return)

1. A: **TURBOSIM**
2. **L** for LOAD
3. filename: **SIMEX**
4. **3** for view scenario
    ADD: Target = .5, specs [.4, .6]
5. **<CR>** after reading
6. **2** for offline trials
7. **N** for No Direct File
8. Use L8 for factor settings
    A: Thinner (10,20)
    B: Fluid Press (15, 30)
    C: Temperature (68, 78)
    D: Vendor (1,2)
9. Complete all 8 runs with 3 replications each
10. Copy all information in notebook, p. 4 – 26
11. **5** for Quit
12. A: **TURBOTAG**
13. **N** for No Saved Design
14. **8** for L8 design
15. **4** for 4 mains
16. **7** for factor D column
17. **3** for # of replicates
18. Enter –1 for lows and +1 for highs
19. **N** for No Target
20. Enter data using arrow keys
21. **/** for menus
22. **A** for analyze
23. **<CR>** for YBAR ANOVA
24. Copy results in notebook
25. Select noise line with arrow key **<CR>**
26. Complete ANOVA table in notebook
27. Build prediction equation for $\hat{y}$
28. Move analysis window to LNS ANOVA
    <CR>
29. Determine settings to minimize variance
30. Determine remaining factor settings to achieve $\hat{y} = 0.5$
31. **ESC** to get to menus
32. **Q** for Quit
33. **Y** for OK to quit
34. A: **TURBOSIM**
35. **L** for Load
36. filename: **SIMEX**
37. **2** for offline experimental trials
38. **N** for No Direct File
39. Type in optimal settings
40. **4** for # of confirmation runs
41. **N** for no more runs
42. Copy confirmation values
43. **<CR>** **1** for production runs
44. **200** for # of production runs
45. Type in optimal settings
46. Copy Results

GLOSSARY

2-LEVEL DESIGN – an experiment where all factors are set at one of two levels, denoted as low and high (–1, +1 or 1, 2).

3–LEVEL DESIGN – an experiment where all factors are set at one of three levels, denoted as low, medium and high (–1, 0, 1 or 1, 2, 3).

ALIASING – when two factors are set at the same levels throughout the experiment; i.e. the columns are 100% correlated.

ANOVA – Analysis of Variance. A statistical tool based on F ratios that measures if a factor contributes significantly to the variance of a response. Also determines the amount of variance due to pure error.

BALANCED DESIGN – A 2–level experimental design is balanced if each factor is run the same number of times at the high and low levels.

BLOCKING VARIABLE – A variable (factor) which cannot be randomized. The experiment is usually run in blocks for each level of the blocking variable and randomization is performed within blocks.

BOX–BEHNKEN DESIGN – a 3–level design used for quantitative factors and designed to estimate all main, quadratic, and 2–way interaction effects.

BRAINSTORMING – a group activity which generates a list of possible factors and levels, and the method by which the results may be evaluated.

CAUSALITY – the assertion that changes to an input factor will directly result in a specified change in an output.

CENTER POINTS – experimental runs with all factor levels set half-way between the low and high settings.

CENTRAL COMPOSITE DESIGN – a 3–level design that starts with a 2–level fractional factorial and some center points. If needed, axial points can be tested to complete quadratic terms. Used typically for quantitative factors and designed to estimate all main effects plus desired quadratics and 2–way interactions.

CONFIDENCE INTERVAL – a range of values based on a sample mean and standard deviation that has a given probability of containing the true population parameter.

CONTROLLABLE FACTORS – factors the experimenter has control of during all phases; i.e., experimental, production and operational phases.

D–OPTIMAL DESIGN – an experimental design in which the minimum number of runs is based on degrees of freedom needed to analyze the desired effects. Not necessarily orthogonal or balanced, it does minimize the correlation between factors.

DEFINING RELATIONSHIP – a statement of one or more factor word equalities used to determine the aliasing structure in a fractional factorial design.

DEFINING WORDS – factor word equalities in a defining relationship.

DEGREES OF FREEDOM (df) – the number of independent observations minus the number of parameters estimated.

DESIGN ARRAY – an array representing the experimental settings. Usually contains values ranging from −1 to 1 but could be wider if using CCD. The rows represent the runs and the columns represent the factors.

EIGENVALUES – with eigenvectors. Given a matrix multiplied by an eigenvalue, it will produce the same result as the related eigenvector multiplied by the same matrix. Used to describe stationary points on a surface.

EXPERIMENTAL CONDITION – a specific combination of factors and levels that is studied in an experiment. Also referred to as experimental trials or runs.

EXPERIMENTAL DESIGN – purposeful changes to the inputs (factors) to a process in order to observe corresponding changes in the outputs (response).

F RATIO (F) – a ratio of two independent estimates of experimental error. If there isn't an effect, the ratio should be close to 1.

F TABLE – provides a means for determining the significance of a factor to a specified level of confidence, by comparing calculated F ratios to those from the F distribution (pp 8–2 thru 8–5). If the F ratio is greater than the table value, there is a significant effect.

FACTOR – an input to a process which can be manipulated during experimentation.

FISHBONE DIAGRAM – a wire diagram which a group can use to organize its thoughts during a brainstorming session. The backbone of the fish represents the response being measured. The "ribs" represent the types of factors that affect the response.

FOLDOVER DESIGN – a way to obtain a resolution IV design based on two designs of R_{III}. Used when the confirmation runs from a resolution III design differ substantially from their prediction and the experimenter desires to de-alias the two-way interactions from the main effects.

FRACTIONAL FACTORIALS – instead of using a full factorial, some subset of it can be used if the experimenter can assume some interactions will not occur and he/she assigns a factor to that interaction column of the design.

FULL FACTORIAL – all possible combinations of the factors and levels. Given n factors all with two levels, there will be 2^n runs. If the factors have three levels, then 3^n runs would be needed.

GAUSSIAN DISTRIBUTION – the engineering name for the Normal Distribution.

GRADIENT – the slope at a point on a surface.

INNER ARRAY – used in parameter design to identify the combinations of controllable factors to be studied in a designed experiment. Also called a "design array".

INTERACTION – occurs when the combination of two or more factors generates a result that is different from the result produced by the individual factors.

INTERACTION EFFECT – the influence of two or more interacting factors on the results when they are changed from one level to another.

LATIN SQUARE – a classical method used to generate orthogonal designs.

LEVEL – a setting or value of a factor.

LINEAR GRAPH – a tool used by Taguchi to identify sets of interacting columns in orthogonal arrays.

LOSS FUNCTION – a technique for quantifying loss due to product deviations from target values.

MAIN EFFECT – the influence a single factor has on the response when it is changed from one level to another. Also called the "mean response or "average response".

MEAN SQUARE – average squared deviation from the sample average used as an estimate of experimental error.

MEAN SQUARE ERROR – a weighted average of the sum of the variances for each run.

MULTICOLLINEARITY – the existence of strong correlations between input factors.

MULTIPLE REGRESSION – a Taylor series model using several independent variables to predict one dependent variable.

NOISE – unexplained variability in the response.

NORMAL DISTRIBUTION – the "bell-shaped" curve distribution used to calculate probabilities of events that tend to occur around a mean value and trail off with decreasing likelihood.

NORMAL PROBABILITY PLOT – a way to identify significant effects when ANOVA is not possible due to unreplicated data.

NUISANCE VARIABLES – factors not included in the design which, if not held constant or controlled through randomization, will distort the results.

OFF–LINE QUALITY CONTROL – "Interrogating the Process." Methods which focus on problem prevention in product/process design. Achieved through experimental design techniques.

ON–LINE QUALITY CONTROL – "Listening to the Process." Methods which focus on problem corrections in production and process control. Achieved through statistical process control techniques.

ONE–FACTOR–AT–A–TIME EXPERIMENTS – a highly inefficient method where one factor is changed while all the others are kept constant. Once an optimal is found for that factor, it is held at that value and another factor is manipulated. This method assumes no interactions between variables and cannot be used to generate a Taylor Series model for prediction of unexperimental conditions.

OPTIMUM CONDITION – The combination of factors and levels that produces the most desirable results.

ORTHOGONAL DESIGN – a design where the correlation between factors is zero.

OUTER ARRAY – used in parameter design to identify the combinations of uncontrollable, or noise, factors to be studied in a designed experiment. Also called "noise array".

PARAMETER DESIGN – a means to study the effects of noise on controllable factors in a designed experiment. Used for finding a combination of controllable factors which produce results robust to the uncontrollable factors.

PARETO DIAGRAM – a graph which shows the amount of influence each factor has on the response in order of decreasing influence.

PERCENT CONTRIBUTION – identifies the amount of influence each factor has on the results.

PLACKETT–BURMAN DESIGNS – orthogonal, balanced designs that have a multiple of 4 runs. Used for main effects only; i.e., no interactions.

POOLING – in ANOVA, combining factors which have little influence on the results into the error term.

PREDICTION INTERVAL – like a confidence interval, but it estimates where an individual value from a regression model could be, not the average value.

QUALITY – according to Taguchi, "the minimum loss imparted by the product to society from the time the product is shipped." Also referred to as meeting the needs of the customer in terms of performance, maintainability, producibility, etc.

QUALITY CHARACTERISTIC – the response variable (output) from a designed experiment.

R&M 2000 VARIABILITY REDUCTION PROCESS – an Air Force program to increase the quality of weapon systems through statistical methods like design of experiments.

RANDOMIZATION – a system of using random numbers. Used to spread evenly the effects of factors not included in the experiment.

REPLICATION – the number of times a specific combination of factor levels is run during an experiment.

RESOLUTION – a measure of how saturated a fractional factorial design is. The lower the number (the length of the shortest defining word), the more saturated the design.

RESPONSE SURFACE METHODOLOGY – a process of locating an optimal value in a higher order model. Utilizes regression, contour plots an/or method of steepest ascent/descent.

ROBUST DESIGN – a design which determines the specific levels of controllable factors which produce a response insensitive to noise factors.

ROTATABILITY – design characteristic implying that estimated parameter variability is not a function of direction from the center of a design matrix.

RUN – an experimental condition determined by a row of settings in the design matrix. Also referred to as experimental condition or trial.

SATURATED DESIGN – a design where all orthogonal columns have factors assigned to them. Primarily used for screening designs.

SCREENING – testing a set of factors for main effects in hopes of reducing their numbers.

SIGNAL–TO–NOISE (S/N) – a comparison of the influence of controllable factors (signals) to the influence of noise factors. The higher the S/N value, the better.

SIGNIFICANCE LEVEL – the probability of making a wrong conclusion about the importance of a factor, based on statistics.

SIMPLE LINEAR REGRESSION – a model where one independent variable is used to predict one dependent variable.

SPECIFICATION (SPEC) LIMITS – the bounds of acceptable values for a given product or process.

STATIONARY POINT – the corollary in single variable calculus would be a critical point. A stationary point is a point where the gradient is zero; i.e., no slope at the point. The eigenvalues are used to identify a maximum, a minimum, a saddle point or a ridge.

SUM OF SQUARES – the total of the squared differences from a set value, usually the mean.

SYSTEM DESIGN – the selection of materials, parts, products, factors, equipment and process parameters.

T TEST – a statistical test to measure if a significant difference exists between two sample mean values.

TAGUCHI METHODS – experimental design thoughts and processes as developed by Taguchi. Based on the philosophy of the loss function, he uses Signal–to–Noise ratios as the primary analysis tool. Orthogonal design matrices are tabled and originate from fractional factorials, Blackett–Burman and Latin Square designs.

TOLERANCE DESIGN – the specification of appropriate tolerances, product parameters and process factors.

TRADITIONAL METHODS – statistical experimental design thoughts and processes as originally developed by Fisher and others as early as the 1920's. Uses ANOVA as the primary analysis tool, along with orthogonal designs such as fractional factorials, Latin Squares, Plackett–Burman, Box–Behnken, Central Composite and D–Optimal.

TUKEY TEST – a statistical test to measure the difference between several mean values and tell the user which ones are statistically different from the rest.

UNCONTROLLABLE FACTORS – factors that are difficult, undesirable or impossible to change. Also called "noise factors". The uncontrollable factors included in a design matrix must be controllable in the experimental phase.

INDEX

A

B

C